Integrated Air Quality Management
ASIAN CASE STUDIES

Edited by
Nguyen Thi Kim Oanh

CRC Press
Taylor & Francis Group
Boca Raton London New York

CRC Press is an imprint of the
Taylor & Francis Group, an **informa** business

CRC Press
Taylor & Francis Group
6000 Broken Sound Parkway NW, Suite 300
Boca Raton, FL 33487-2742

First issued in paperback 2017

Version Date: 20120525

ISBN 13: 978-1-4398-6225-4 (hbk)
ISBN 13: 978-1-138-07184-1 (pbk)

Library of Congress Cataloging-in-Publication Data

Integrated air quality management : Asian case studies / [edited by] Nguyen Thi Kim Oanh.
p. cm.
Includes bibliographical references and index.
ISBN 978-1-4398-6225-4 (hardback)
1. Air quality management--Asia--Case studies. I. Oanh, Nguyen Thi Kim.

TD883.7.A78I58 2012
363.739'2095--dc23 2011045488

Visit the Taylor & Francis Web site at
http://www.taylorandfrancis.com

and the CRC Press Web site at
http://www.crcpress.com

Contents

Preface

The ultimate goals of air quality management are to ensure that air pollution concentrations do not exceed the defined target levels and that human health and the environment are protected. Air quality management should be based on the integrated approach that combines the use of sound science, the analysis of mitigation options, and the involvement of all key organizations and stakeholders in the public decision-making process. Rapid urbanization and industrialization with the concomitant soaring increase in energy consumption in Asian developing countries over the last few decades have put huge pressure on the environment. The steady growth in vehicle number, the heavy reliance on coal as a cheap energy resource, the use of dirty fuels for residential combustion, and intensive open burning are some of the major factors leading to the progressive deterioration of air quality in developing Asia. Ambient air pollutants are often observed at critical levels, especially in urban areas, which present a significant threat to human health and the environment. There is a growing public awareness of the potential effects of exposure to air pollution in everyday life. Better understanding of the consequent effects of air pollution on health, environment, and eventually the economy has prompted policy makers and institutions to take action to improve air quality across Asia. Consequently, many Asian developing countries have established and implemented some forms of air quality management systems. Despite these efforts, the development of infrastructure, environmental technology, and management practices still lags behind the rate of emission increase; hence, considerable challenges to improving air quality remain.

This book presents the methodological development for integrated air quality management, which employs both technical and policy tools to achieve the air quality goals. These tools are further illustrated by experiences drawn from Asian regional studies. Technical information and policy recommendations presented in this book include the outcomes of several multiyear regional air quality research programs coordinated by the Asian Institute of Technology. The book is structured to combine both fundamental information and advanced knowledge dealing with subjects of integrated air quality management and is intended for an audience of a wide range of backgrounds, including undergraduate and graduate-level students, academic researchers, and policy makers.

The book contains 16 chapters subdivided into an introductory segment and four major technical sections. Chapter 1 presents an overview of major issues of air quality management practices in developing Asia and potential approaches to reduce pollution, including the opportunities for integration of the air quality improvement and climate-forcing mitigation strategies.

Section I (Chapters 2–4) covers the methodology and results of particulate matter monitoring using traditional filter-based and satellite monitoring techniques. The receptor modeling tool for particulate matter source apportionment is discussed in depth and illustrated with applications to the composition data from six Asian cities and other places. Section II presents the applications of the three-dimensional (3D) dispersion modeling tool for urban and regional air quality management focusing on surface ozone, fine particulate matter, and acid deposition (Chapters 5 and 6). The synoptic climatological approach to air pollution potential forecasting with case studies is highlighted in Chapter 7. Several innovative control technologies for gaseous air pollutants are included in Section III (Chapters 8–12). Finally, Section IV (Chapters 13–16) illustrates integrated air quality management through case studies for target source categories, including agricultural residue field burning, vehicle emission, brick kilns, and industrial emission of volatile organic compounds. While drawn from specific examples within Asia, the recommendations from these case studies are also applicable to other regions where similar situations exist.

Nguyen Thi Kim Oanh

Acknowledgments

I am deeply indebted to the contributors who offered their professional expertise and time to make this book a truly informative resource for students and professionals in the field of air quality management. My gratitude goes also to the publishing team of CRC Press and Taylor & Francis, Asia Pacific, for their support, thorough language editing, and production of the book and to the reviewers at the initial stage of the book preparation for their valuable comments. I would like to extend my most sincere thanks to the students in the Air Quality Group at the Asian Institute of Technology and at partner universities, whose names may not appear in the contributors' list but were vital in the realization of this book for their contributions in one way or another. Their dedication to expand the current knowledge of the field and commitment to quality research were, to me, an endless source of inspiration during the completion of this book and throughout my professional career. I am especially grateful to my late mother and my entire family for their continuous encouragement and support.

About the Editor

Nguyen Thi Kim Oanh is a professor in the environmental engineering and management program, School of Environment, Resources, and Development (SERD), Asian Institute of Technology (AIT), Thailand. She obtained her Dipl Eng (Diplom Ingenieur) in meteorology (1978) from the Odessa Hydrometeorology Institute, Ukraine; master's (1991) and doctor of engineering (1994) in environmental technology and management from AIT. Over her career, Prof. Kim Oanh has built up a strong network with experts worldwide to conduct high-quality teaching and research projects successfully. Her research experiences include air pollution monitoring, modeling, emission inventory, and emission control technologies. Her current professional interests also cover the short-lived climate forcers and assessment of climate cobenefit of air pollution control measures. In the capacity of principal investigator (PI) or co-PI, she has conducted a number of air pollution research projects. Of these, 11 involve Asian regional networks, such as the Asian Regional Air Pollution Research Network (AIRPET) project that involved six Asian countries over the period of 2001–2010. Prof. Kim Oanh has authored or coauthored more than 20 book chapters/monographs, over 60 peer-reviewed papers published in technical journals with high-impact factors, and a large number of papers in peer-reviewed international conference and workshop proceedings. She has organized and taught many regional training courses on air quality management and participated in technical capacity building for hundreds of professionals in the field. Prof. Kim Oanh has supervised more than 100 master's and doctoral AIT graduates, who continue to work toward improving air quality in Asia and beyond. Her students have won many "best paper awards" from the U.S. West Coast Section of the Air and Waste Management Association (AWMA) over the years. Prof. Kim Oanh is the recipient of several awards, including SERD/AIT teaching awards, the BAQ (Better Air Quality Conferences) awards of Extra Mile and Air Hero for being the "most active professor in air quality."

Contributors

Sanjeev Arjunagi
Department of Civil Engineering
Indian Institute of Technology Madras
Chennai, India

Hoang Xuan Co
Faculty of Environmental Sciences
Hanoi University of Science
Hanoi, Vietnam

Amornpon Changsuphan
Environmental Engineering and Management
School of Environment, Resources and Development
Asian Institute of Technology
Pathumthani, Thailand

Melliza Templonuevo Cruz
Manila Observatory
Ateneo de Manila Campus
Quezon City, Philippines

Nghiem Trung Dung
Institute for Environmental Science and Technology
Hanoi University of Technology
Hanoi, Vietnam

Jill A. Engel-Cox
Battelle Memorial Institute
Columbus, Ohio, USA
and
Battelle Science and Technology Malaysia
Kuala Lumpur, Malaysia

Zhengping Hao
Research Center for Eco-Environmental Sciences
Chinese Academy of Sciences
Beijing, China

Amy K. Huff
Battelle Memorial Institute
Arlington, Virginia, USA

Tran Nguyen Quang Huy
Environmental Engineering and Management
School of Environment, Resources and Development
Asian Institute of Technology
Pathumthani, Thailand

Asif Iqbal
Environmental Engineering and Management
School of Environment, Resources and Development
Asian Institute of Technology
Pathumthani, Thailand

Thongchai Kanabkaew
Environmental Engineering and Management
School of Environment, Resources and Development
Asian Institute of Technology
Pathumthani, Thailand

Sukanya Kiatphuangchai
Environmental Engineering and Management
School of Environment, Resources and Development
Asian Institute of Technology
Pathumthani, Thailand

Puji Lestari
Faculty of Civil and Environmental Engineering
Institute of Technology Bandung
Bandung, Indonesia

Frank Murray
School of Environmental Science
Murdoch University
Murdoch, Australia

Nguyen Thi Kim Oanh
Environmental Engineering and Management
School of Environment, Resources and Development
Asian Institute of Technology
Pathumthani, Thailand

Didin Agustian Permadi
Environmental Engineering and Management, School of Environment, Resources and Development
Asian Institute of Technology
Pathumthani, Thailand

Ligy Phillip
Department of Civil Engineering
Indian Institute of Technology Madras
Chennai, India

Nguyen Hong Phuc
Environmental Engineering and Management
School of Environment, Resources and Development
Asian Institute of Technology
Pathumthani, Thailand

Nguyen Lu Phuong
Environmental Engineering and Management
School of Environment, Resources and Development
Asian Institute of Technology
Pathumthani, Thailand

Prapat Pongkiatkul
Environmental Engineering and Management
School of Environment, Resources and Development
Asian Institute of Technology
Pathumthani, Thailand
and
Department of Environmental Engineering
Faculty of Engineering
King Mongkut's University of Technology Thonburi
Bangkok, Thailand

Tyagarajan Swaminathan
Chemical Engineering Department
Indian Institute of Technology Madras
Chennai, India

Hailin Wang
Research Center for Eco-Environmental Sciences
Chinese Academy of Sciences
Beijing, China

Xiaoke Wang
Research Center for Eco-Environmental Sciences
Chinese Academy of Sciences
Beijing, China

Baoning Zhang
Environmental Engineering and Management
School of Environment, Resources and Development
Asian Institute of Technology
Pathumthani, Thailand
and
Santec Consulting Ltd.
Calgary, Alberta, Canada

Wei Zhang
Research Center for Eco-Environmental Sciences
Chinese Academy of Sciences
Beijing, China

Wei Zhao
Institute of Process Engineering
Chinese Academy of Sciences
Beijing, China

Guoshun Zhuang
Center for Atmospheric Chemistry Study
Department of Environmental Science and Engineering
Fudan University
Shanghai, China

Yahui Zhuang
Research Center for Eco-Environmental Sciences
Chinese Academy of Sciences
Beijing, China

1

Air Quality Status and Management Practices in Asian Developing Countries

Nguyen Thi Kim Oanh, Didin Agustian Permadi, Nguyen Hong Phuc, and Yahui Zhuang

CONTENTS

1.1 Introduction

Economic growth with industrialization, urbanization, and motorization and consequent changes in land use/cover have been observed in the recent decades in Asia. This economic development brings a better life to people in the region but also puts huge pressure on the environment. The management capacity for the environment in general and air quality in particular has not been developed at the same pace. Dirty fuels are still widely used in combustion devices with low energy efficiency that release large quantities of air pollution. As a result, several cities in developing Asia are known to be among the world's most polluted. Air pollutant levels in many large urban areas in Asia have been reported to exceed the WHO (World Health Organization) guidelines and the respective National Ambient Air Quality Standards (NAAQS). In particular, some of the highest levels of ambient particulate matter (PM) are found in big Asian cities (Hopke et al., 2008; Kim Oanh et al., 2006). For example, the annual average of $PM_{2.5}$ (particles with the aerodynamic diameter below 2.5 μm) at some locations in downtown in Beijing is above 100 μg/m^3 (Kim Oanh et al., 2006), which is well above the WHO guideline value of 10 μg/m^3 (WHO, 2005). Evidence confirmed high levels of surface ozone observed in most Asian large cities (Lee et al., 2002; Permadi and Kim Oanh, 2008; Zhang and Kim Oanh, 2002, and references therein), which is linked to intensive precursor emissions coupled with favorable meteorology conditions, such as high temperature and strong sunshine, which are available in most parts of tropical Asia all year.

Air pollution in urban areas of Asian developing countries presents a significant threat to human health and the environment. This in turn affects the regional economy through increased mortality, morbidity, damage to crops and properties, and loss of tourism. People living in large cities are exposed to high ambient concentrations of air pollutants. Nearly two-thirds of the worldwide estimated 800,000 deaths and 4.6 million lost years of healthy life caused by exposure to urban/outdoor air pollution in the year 2000 occurred in the developing countries of Asia (WHO, 2002b). Indoor air pollution in developing Asia is estimated to cause even more adverse health effects than outdoor air pollution (WHO, 2002b, 2005). Cooking and space heating using low-grade solid fuels are the foremost sources of indoor air pollution in Asian developing countries. Common lack of flues/chimneys to vent combustion smoke leads to high levels of air pollutants in homes; levels are reported in some instances to be a dozen to a hundred times higher than the ambient levels. Millions of people living in the rural areas of Asian developing countries, primarily housebound women and children, are at a higher risk of suffering from the adverse effects of indoor air pollution.

On the regional scale, the meteorological conditions in Asia, with prevalent monsoons, enhance the long-range transboundary transport of pollutants. Intensive emissions, such as smoke from forest fires and acidifying substances (sulfur oxides [SO_x], nitrogen oxides [NO_x]) from large combustion facilities in upwind locations can introduce adverse effects in distant downwind areas. The most notable examples of air pollution long-range transport in Asia include the Southeast Asian (SEA) haze, acid rain, dust storms, and atmospheric brown clouds (ABCs). These transboundary phenomena may cause multiple effects on regional air temperature, precipitation, agriculture, air quality, and human health and currently attract increasing attention from the research community, the public, and policy makers.

This chapter provides an overview of the current situation of both outdoor and indoor air pollution and air quality management (AQM) practices in Asian developing countries, illustrated with regional examples of common and hazardous air pollutants. Major issues together with the opportunities for integration of air quality improvement and climate strategies are highlighted.

1.2 General Description of Selected Asian Cities

The cities described here are those featured in monitoring and modeling applications presented in this book. They include the Bangkok metropolitan region (BMR) of Thailand; Bandung and Jakarta (Indonesia); Beijing (China); Chennai (India); Manila (Philippines); Hanoi and Ho Chi Minh City (HCMC; Vietnam); and Dhaka (Bangladesh). These cities are different in terms of population, topography, meteorology, and other natural conditions. The population in the cities ranges from megasize, with a population of 10 million or above (Beijing, Manila), to a smaller size (Bandung, Hanoi). The demographic conditions together with the diverse climatic conditions (Table 1.1) have significant implications on emission sources and air pollution problems in these cities.

The climate of the cities is dominated by monsoons, with two distinct seasons, dry and wet, but they cover different months depending on the location of the city (Table 1.1). For the cities located in the Northern Hemisphere, the wet season is generally within the summer months (May to October), and the dry season is normally in winter months (November to April). The opposite is true for Bandung and Jakarta, Indonesia, which are located near the

TABLE 1.1
Demographical Information of Selected Asian Cities

City	Topography	Population (year), millions	Area, km^2	Dry Season Average or range				Wet Season Average or range			
				Period	RH, %	T, °C	Precipitation, mm	Period	RH, %	T, °C	Precipitation, mm
Bangkok MR	Flat	10 (2010)	7,762	Oct–May	63–78	27–30	550	Jun–Sept	73–79	28–30	1,150
Bandung	Hilly plateau	3.5 (2005)	168	May–Oct	50–70	23	150	Nov–Apr	63–83	23	1,700
Beijing	Flat (38%), mountain (62%)	20 (2010)	16,800	Sept–May	40	13	170	Jun–Aug	86	25	350
Dhaka MR	Flat	13 (2009)	1,528	Nov–Apr	55–72	19–30	180	May–Oct	71–79	25–33	1,750
Chennai capital	Desolate, mostly flat	6.96 (2006)	200	Jan–May	50–85	26–42	150	Jun–Dec	60–95	25–30	1,080
Jakarta	Flat, flanked by the sea and mountains	9.58 (2010)	662	May–Oct	70–97	25–30	350	Nov–Apr	65–97	26–29	1,450
Metro Manila	Flat, flanked by sea and mountains	11.5 (2007)	636	Nov–May	66–85	22–28	394	June–Oct	72–87	22–27	1,741
Hanoi (urban)	Flat	3.4 (2007)	921	Nov–Mar	81–87	19	187	Apr– Oct	82–86	27	1,480
HCMC	Flat	7.2 (2009)	2,095	Dec–Apr	70–76	26–30	316	May–Nov	73–83	27–29	1,664

equatorial zone in the Southern Hemisphere. Here, the wet season is from November to April, while the dry season is from May to October. Cities such as Bangkok, Jakarta, and HCMC are located along or near the coastline and are thus strongly influenced by the sea-land breezes.

The windroses for December (represent winter) and July (represent summer) in different cities are presented in Figure 1.1. The cities in the Northern Hemisphere (i.e., Beijing, Hanoi, Bangkok, HCMC) have the dominant northerly (NW-NE) winds during winter and the dominant southwesterly (SW) winds during summer. Locally, the monsoon patterns are modified due to the topography and location (e.g., relative to sea or mountain ranges). In Bangkok, N-NE winds are more frequently observed during October–December. The southerly winds dominate during February to May but are also frequently observed in January and June. The prevalence of southerly winds throughout a year suggests a strong influence of sea breeze in BMR, especially when the synoptic scale winds are not pronounced. In general, high percentages of calm conditions are observed in the BMR throughout the year (Zhang and Kim Oanh, 2002). Hanoi has a modified wind pattern during summer that has more southeasterly (SE) winds, although NE winds still clearly dominate in winter. HCMC has a clear dominance of SW winds during summer, while in winter the wind directions vary widely, with some prevalence of northerly winds. In fact, the extension of ridges from the Siberian high pressure that bring in the NE monsoon (cold spells) in Hanoi during winter can rarely induce any significant effects to the southern part of Vietnam because of the presence of a mountain range across the middle part of the country. Jakarta has very high frequencies of calm conditions in both seasons, 59% in July and 73% in December. The most prevalent winds are NE in July (13%) and SW in December (9%). The city is bounded, to its south, by the V-shape Salak mountain range, which decouples local winds from the synoptic-scale wind (Permadi and Kim Oanh, 2008). Bandung windroses (not shown in Figure 1.1) show strong east-west wind directions, westerly during the wet season and easterly during the dry season (Lestari and Mauliadi, 2009). In Dhaka, S-SE winds clearly prevail during summer, while NW-N winds are observed during winter despite the dominance of the large-scale westerly pathways of air masses along the Himalaya range before arriving at the city in winter (Iqbal and Kim Oanh, 2011).

1.3 Current Status of Air Pollution in Asian Developing Countries

1.3.1 Urban Air Quality

Air pollution observed in urban areas may be of a different mixture of pollutants than that found in the countryside or that found indoors. In the developing countries, urban air quality has worsened and approached dangerous levels in several large cities, a situation driven by fast population

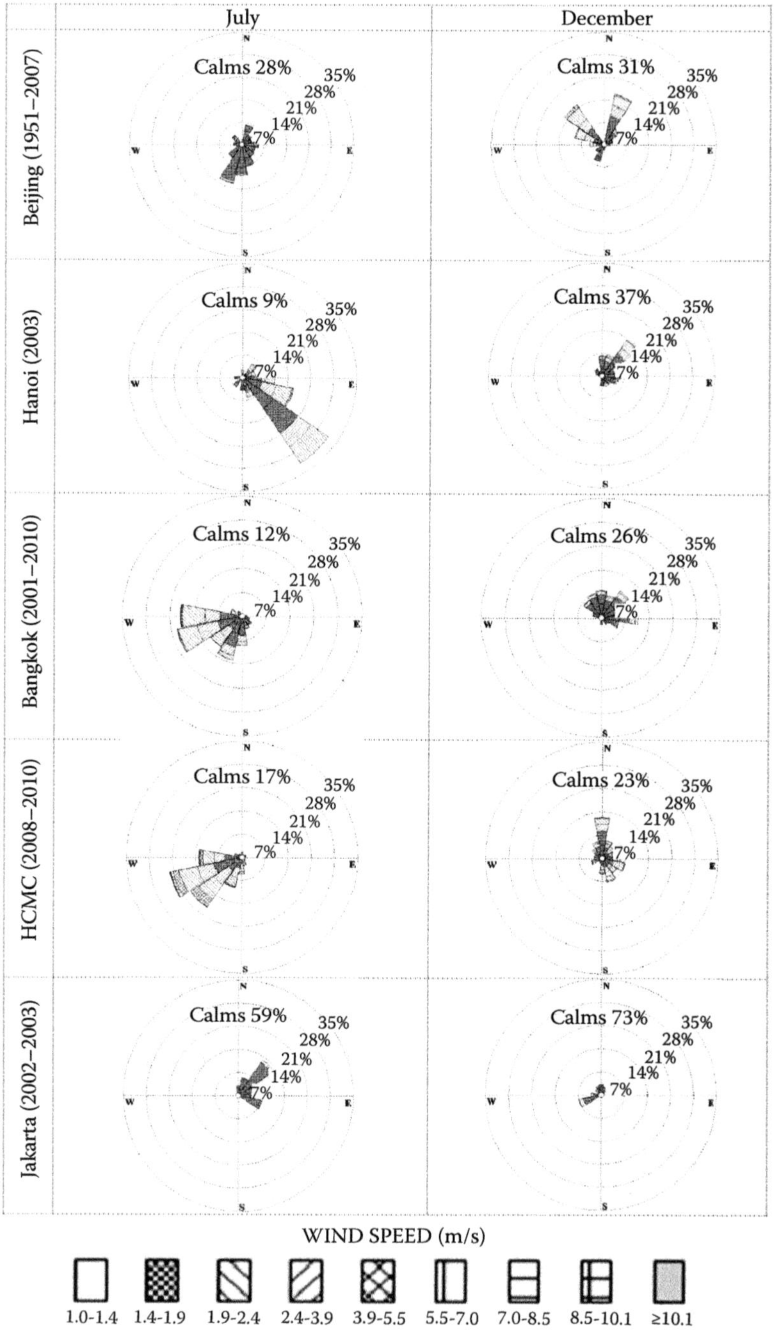

FIGURE 1.1
Windroses in selected cities, arranged in north-to-south order.

growth, industrialization, urbanization, and increased vehicle use. Available measurement data show that the most important air pollutants in developing Asian cities include PM, commonly measured as TSP (total suspended PM) and PM_{10} (particles with an aerodynamic diameter less than 10 μm) and ozone. Fragmented data available for fine particles ($PM_{2.5}$) also show very high levels. Levels of other commonly measured pollutants (CO, SO_2, NO_2, Pb) vary significantly among cities. Other toxic air pollutants such as volatile organic compounds (VOCs; e.g., benzene); semi-VOCs (e.g., polycyclic aromatic hydrocarbons [PAHs], pesticides) are yet to be adequately characterized and reported. This section of the chapter gives an overview of the levels of common pollutants in cities of developing Asia.

1.3.1.1 Particulate Matter

Particulate air pollution is of special concern as it can cause multiple effects on human health and the environment. Particles measured as PM_{10} have long been implicated in causing adverse health effects and increased mortality (Dockery and Pope, 1994), whereas those as $PM_{2.5}$ impose even higher risk (Pope et al., 2009; Schwartz and Neas, 2000). In urban areas of developing countries, 2–5% of total deaths are estimated to be caused by exposure to high PM levels alone (Air Pollution in Mega Cities of Asia [APMA], 2002). Atmospheric particles also interact directly or indirectly with Earth's radiation energy balance and can subsequently affect the global climate (Intergovernmental Panel on Climate Change [IPCC], 2007) and effectively reduce visibility (Watson, 2002).

The available data show that PM is the most notable air pollution problem in Asian cities. The national monitoring networks available in developing Asia focused more on TSP in the past, but recently PM_{10} is also monitored routinely in many countries. There are, however, no systematic $PM_{2.5}$ records, and the fragmented data available have only been generated through various coordinated international projects. This situation is expected to improve in the near future as some Asian countries now have set $PM_{2.5}$ standards, as discussed further in this chapter.

1.3.1.1.1 TSP and PM_{10}

Recorded levels of TSP and PM_{10} in cities of developing Asia often exceeded the respective NAAQS, while PM_{10} normally exceeded the WHO air quality guidelines by a factor of three or more. The TSP levels in megacities in Asia were reported to be the highest in the world (Molina and Molina, 2004). In Indian cities, during the period 1991–2003, the monthly average TSP (January, a dry month) reached the highest levels of above 1,000 μg/m^3, and that of PM_{10} exceeded 400 μg/m^3 at some sites in Delhi and Kolkata (Gupta and Kumar, 2006).

Although a certain decreasing trend in PM levels has been recorded (Figure 1.2), the annual PM_{10} levels in Beijing were still well above even the most lenient NAAQS applicable for industrial areas in 2011 (150 μg/m^3) and

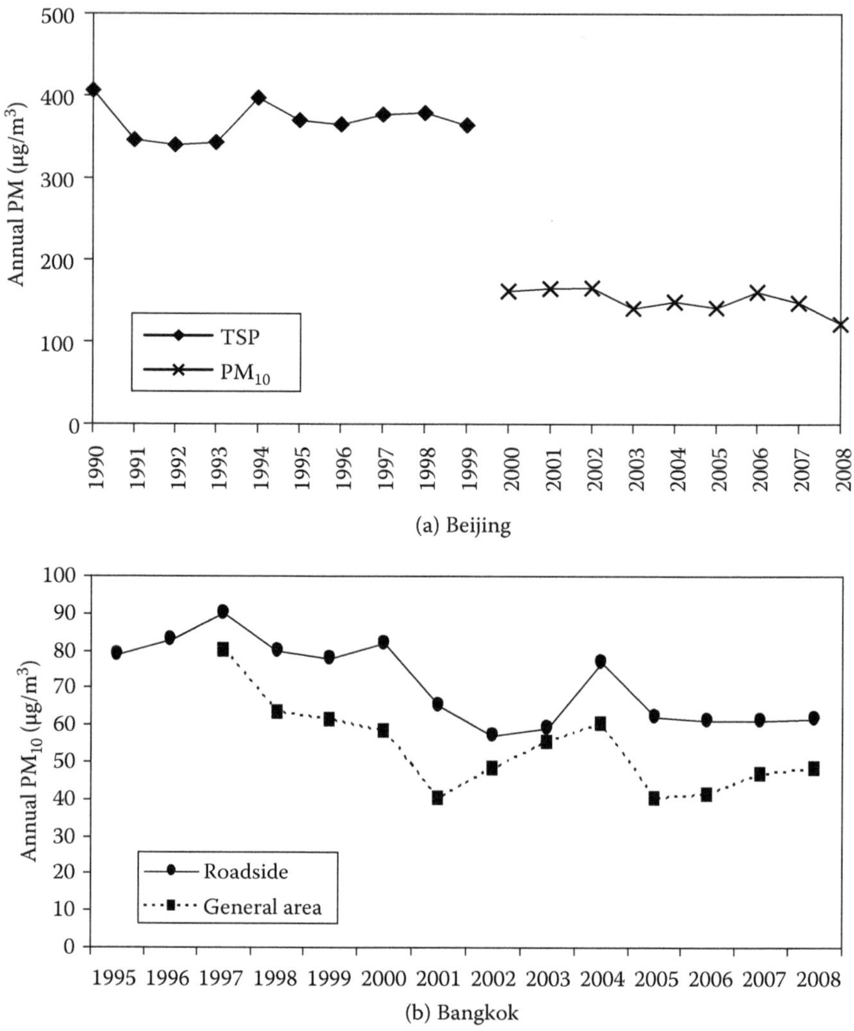

FIGURE 1.2
Annual levels of TSP and PM10 in Beijing and Bangkok. (Bangkok raw data courtesy of PCD (2010); Beijing data from the National Environmental Protection Agency (NEPA,1991–1997); Global Environment Monitoring System (GEMS) (1990–1992); State Environmental Protection Administration (SEPA,1998–2000); Beijing Environmental Protection Bureau (BJEPB, 2000–2009).

the most lenient grade II of China NAAQS released in 2012 (70 μg/m³). In Bangkok, during 1995–2008 the annual roadside PM_{10} showed a reducing trend from the maximum of 90 μg/m³ in 1997 and was stabilized at around 60 μg/m³ during recent years. Lower PM_{10} levels were observed at general ambient areas but were still often above the Thailand annual NAAQS of 50 μg/m³ (Figure 1.2). The frequency of PM exceeding the 24-h NAAQS was also high at the roadside, that is, 4–15% for PM_{10} and 3–6% for TSP during

2006–2008 (Pollution Control Department [PCD], 2003–2008). PM stood as a major problem also in other cities in Thailand, especially in the industrial areas and places affected by biomass open burning. For example, haze episodes observed in Chiang Mai and other northern provinces of Thailand, with the highest 24-h PM_{10} of above 390 μg/m³ recorded in March 2007, were attributed mainly to increased open burning intensity in combination with stagnant air conditions (Kim Oanh and Leelasakultum, 2011). In Delhi, India the annual levels of suspended PM (TSP) during the period 1995–2008 remained fluctuating around 400 μg/m³. The annual levels of respirable suspended PM (RSPM or PM_{10}) were fluctuating around 150 μg/m³ during 1999–2005 but increased afterward and reached above 200 μg/m³ in 2008 (SOE-Delhi, 2010), i.e. well above the current India NAAQS of 60 μg/m³.

A coordinated multiyear monitoring program in six Asian cities conducted by the AIRPET (Asian Regional Air Pollution Research Network) project (Kim Oanh et al., 2006) also reported high PM_{10} levels that varied significantly among the cities and within a city at different characteristic sites. The levels measured in the dry season were well above those in the wet season. The seasonal averages did exceed even the 24-h WHO guideline value of 50 μg/m³ in most of the sites (Figure 1.3).

1.3.1.1.2 $PM_{2.5}$

The measurement data in six Asian cities showed remarkably high levels of $PM_{2.5}$ where the seasonal average values did even exceed the current 24-h WHO guideline of 25 μg/m³ (Figure 1.4). Within each season, the PM levels measured at a site fluctuated significantly, producing a large standard variation (vertical bars show ± 1 SD) owing to the changing emission and meteorology conditions. For example, high fluctuations of PM levels in Beijing may be associated with dust storms, which shows also the regional transport influence (Kim Oanh et al., 2006).

Hopke et al. (2008) presented $PM_{2.5}$ monitoring results, averaged over 4 years (2002–2005), which ranged from below 10 μg/m³ to 48 μg/m³ among the 22 sites in 15 countries in the Asia Pacific region. The sites in New Zealand and Australia recorded the lowest levels, with a period average $PM_{2.5}$ of below 10 μg/m³ and the highest 24-h levels lower than 50 μg/m³. In other locations, the highest 24-h levels were generally above 50 μg/m³, and some sites in China, India, Indonesia, and Sri Lanka even recorded the maximum levels of above 100 μg/m³. The highest 24-h level was measured in an urban site in Bangladesh (293 μg/m³).

The available information shows that $PM_{2.5}$ constitutes a large fraction of PM_{10}. Kim Oanh et al. (2006) reported the $PM_{2.5}/PM_{10}$ ratio generally 0.6 or above for six Asian cities, which was quite in the same range as those reported by Hopke et al. (2008), that is, above 0.7 at some monitoring sites in China and Indonesia. The high fractions of $PM_{2.5}$ in PM_{10} suggest high contributions of fine particles from combustion sources and secondary particles that are formed in the atmosphere. The low ratios of $PM_{2.5}/PM_{10}$ in some locations, that

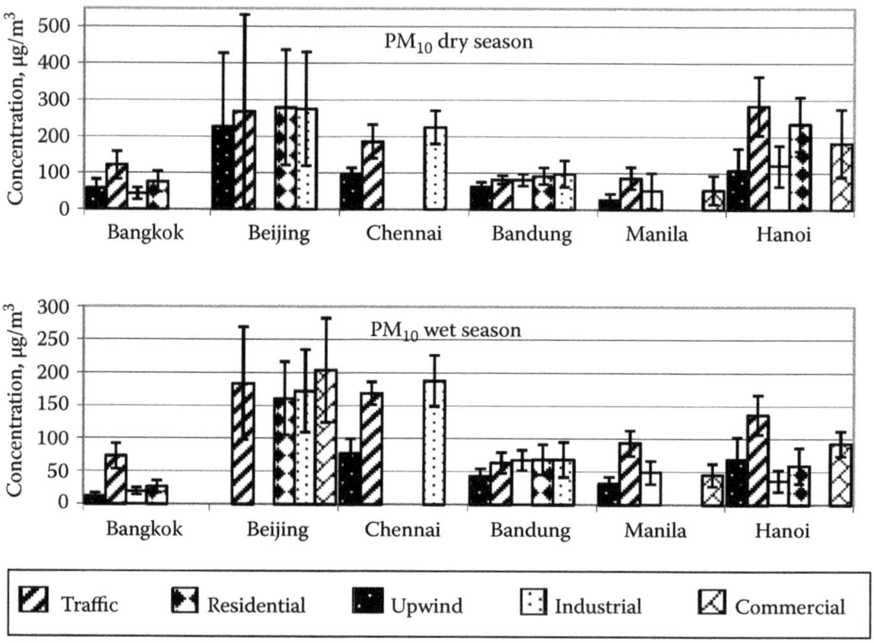

FIGURE 1.3
Seasonal average levels of PM_{10} (2001–2004) in Asian cities. (From Kim Oanh N. T., Upadhyay N., Zhuang Y. H., Hao Z. P., Murthy D. V. S., Lestari P., Villarine J. T., Chengchua K., Co H. X., Dung N. T., and Lindgren E. S. 2006. Particulate air pollution in six Asian cities: Spatial and temporal distributions, and associated sources, *Atmospheric Environment*, 40: 3367–3380. Courtesy of AE, with permission.)

is, 0.2–0.3 at some sites in Pakistan, Sri Lanka, and India (Hopke et al., 2008) or about 0.3 in Chennai (Kim Oanh et al., 2006), confirmed the high contribution of the coarse particles, which were mostly of a noncombustion nature.

1.3.1.2 Ozone Air Quality

1.3.1.2.1 Ground-Level Ozone and Effects

Ozone in the atmosphere can exist in two layers. The stratospheric ozone presents in the upper layer (stratosphere), with the maximum concentrations found at around 25 km, which absorbs the harmful ultraviolet (UV) solar radiation and protects life on Earth. It is the "good ozone," as opposed to the "bad ozone" that exists near Earth's surface (tropospheric ozone). The surface ozone or ground-level ozone is a harmful air pollutant that causes adverse effects to human health, materials, and the environment. In particular, ozone is a well-known phytotoxicant that adversely affects vegetation and hence subsequently reduces the growth and yield of agricultural crops and forests (Emberson et al., 2009; Novak et al., 2003). The tropospheric ozone is also a strong greenhouse

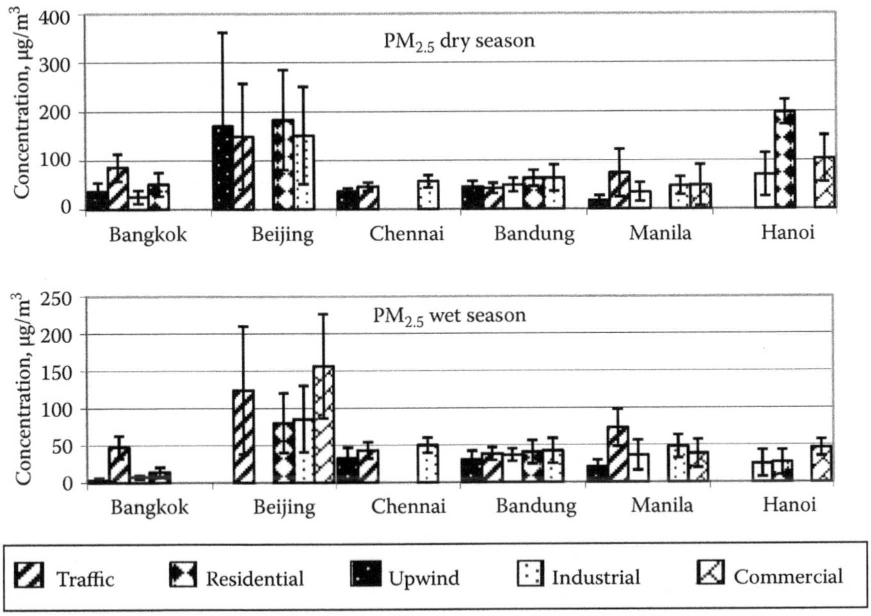

FIGURE 1.4
Seasonal average levels of $PM_{2.5}$ (2001–2004) in Asian cities. (From Kim Oanh N. T., Upadhyay N., Zhuang Y. H., Hao Z. P., Murthy D. V. S., Lestari P., Villarine J. T., Chengchua K., Co H. X., Dung N. T., and Lindgren E. S. 2006. Particulate air pollution in six Asian cities: Spatial and temporal distributions, and associated sources, *Atmospheric Environment*, 40: 3367–3380. Courtesy of AE, with permission.)

gas (GHG) with a strong positive radiative forcing (IPCC, 2007). Due to its widespread effects, ozone is considered one of the most important global air pollutants (WHO, 2006).

Ozone is a secondary air pollutant and is formed in the photochemical reactions of precursors in the presence of sunlight (Sillman, 1999). The main precursors of ozone are the VOCs (including methane), nitrogen oxides ($NO + NO_2 = NO_x$), and carbon monoxide (CO). These precursors are emitted from both natural and human-made sources. Under suitable meteorological conditions ozone, can accumulate to hazardous levels, such as those observed in photochemical smog episodes. It is reported that the background levels of ozone in the Northern Hemisphere increased more than two times compared to the preindustrial period and now has reached levels of 35–40 ppb (parts per billion). The peak ozone levels are now observed as above the WHO guideline (50 ppb, 8 h) in many places in the world (WHO, 2006).

The formation/destruction and accumulation of ozone in urban polluted areas are complex phenomena that involve physical and chemical processes in the lower troposphere and exchange of air with the stratosphere. The influx of ozone from the stratosphere is known to contribute to elevated tropospheric ozone levels, specifically in early spring at middle and high latitudes (Acid Deposition Oxidant Research Center [ADORC], 2006). The relationship between precursor pollutants and ozone differs from one place to another owing to the precursors' emission distribution and meteorology. Normally, high-ozone days are characterized by more intense solar radiation, higher temperature, and lighter surface wind, which are favorable for photochemical production and local buildup of ozone (Permadi and Kim Oanh, 2008). Residence time of ozone in the atmosphere depends on its locations in the atmosphere and geographically. In the boundary layer of the atmosphere (generally below 2 km), the residence time of ozone is about a day. In the upper part of the troposphere, the residence time of ozone may reach a few weeks, which enables the long-range transport that makes both ozone and its precursors a large-scale air pollution problem.

1.3.1.2.2 Ozone Pollution in Asian Cities

Asian urban areas now face increasing photochemical smog pollution, and high ozone concentrations are recorded in many places. The hourly ozone levels in large Asian urban areas often reached above 120 ppb. The peak hourly ozone in Asian big cities, such as Beijing, Bangkok, and Jakarta, was reported to reach above 200 ppb (Chan and Yao, 2007; Permadi and Kim Oanh, 2008; Zhang and Kim Oanh, 2002). In smaller cities (Bandung, Hanoi, and HCMC), the peak hourly levels were lower, that is, in the range of 90–150 ppb (An, 2005; Driejana, 2002; Huy, 2008). In BMR (Thailand), the hourly peak ozone remained high during the 15 years, 1996–2010 (Figure 1.5). The highest hourly level (370 ppb) was recorded in 1997, the year with abnormal meteorology

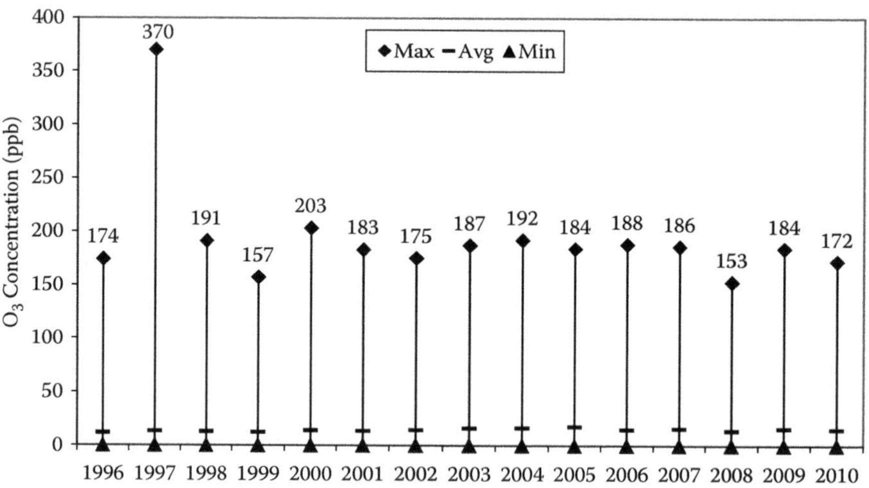

FIGURE 1.5
Statistics of observed hourly O_3 averaged for ambient monitoring stations in BMR, Thailand. (Courtesy of PCD for the raw monitoring data.)

related to a strong El Niño phenomenon and the intensive forest fire in the SEA region (Kim Oanh and Zhang, 2004). The hourly levels were reported to exceed the NAAQS (100 ppb) in 0.3% of hourly measurements in 2008. Ozone is also high in other cities, especially those in the provinces surrounding Bangkok. Considerably high levels of ozone were also observed in remote mountainous areas in Thailand, with monthly levels of 9–55 ppb and annual average levels of 27–28 ppb (Pochanart et al., 2001).

Exposure to high ozone levels observed in Asian populated urban areas may already cause considerable adverse effects. However, the knowledge of the regional ozone pollution level is still limited mainly due to the lack of monitoring data. In most cases, ozone measurements by national monitoring networks are only available in urban areas. The booming of cities in Asia creates mixed land-use patterns with cities surrounded by agricultural fields. This leads to high risks of ozone exposure also for crops in downwind areas of city centers where ozone is potentially high. An integrated monitoring and modeling approach is required to provide the necessary information on spatial and temporal distribution of ozone pollution, which serves as the scientific basis for mitigation of ozone pollution in the region.

1.3.1.2.3 Diurnal Variation of Ozone

Ozone has a strong temporal variation. In urban areas, peaks of the precursor emissions are recorded during the morning and evening traffic rush hours, while the daily hourly maximum ozone is normally observed in the afternoon. The lower levels of ozone precursors (NO_x, VOC, and CO) around noon are caused by (a) the reduction in emissions (e.g., less traffic), (b) the consumption in photochemical reactions producing ozone, and (c) better dilution owing to the elevated boundary layer (higher mixing height). At night, ozone concentrations drop rapidly as photochemical production stops while the deposition and dispersion processes continue. Diurnal variations of surface ozone (Figure 1.6) in Bangkok showed the highest peaks of CH_4, nonmethane hydrocarbons (NMHCs), and CO in the early morning rush hours (around 8:00 a.m.), which are linked to rush hour traffic emissions.

1.3.1.2.4 Seasonal Variation of Ozone

As seen in Figure 1.6, ozone levels in January (dry month) were much higher than August (wet month) even though the peaks of precursors in the morning rush hours between the two months did not differ that much. The differences in the meteorological conditions between the two seasons, such as solar radiation, temperature, and cloud cover, affect the rates of photochemical reactions. Other meteorological parameters such as wind speed and mixing height largely determined the dispersion of O_3 and its precursors in the boundary layer.

Seasonal variations in ambient temperature affect, for example, the biogenic and evaporative VOC emission. Additional precursor emissions from more intensive combustion sources in the dry season (biomass open burning, space heating during cold period) in many parts of Asia are also an

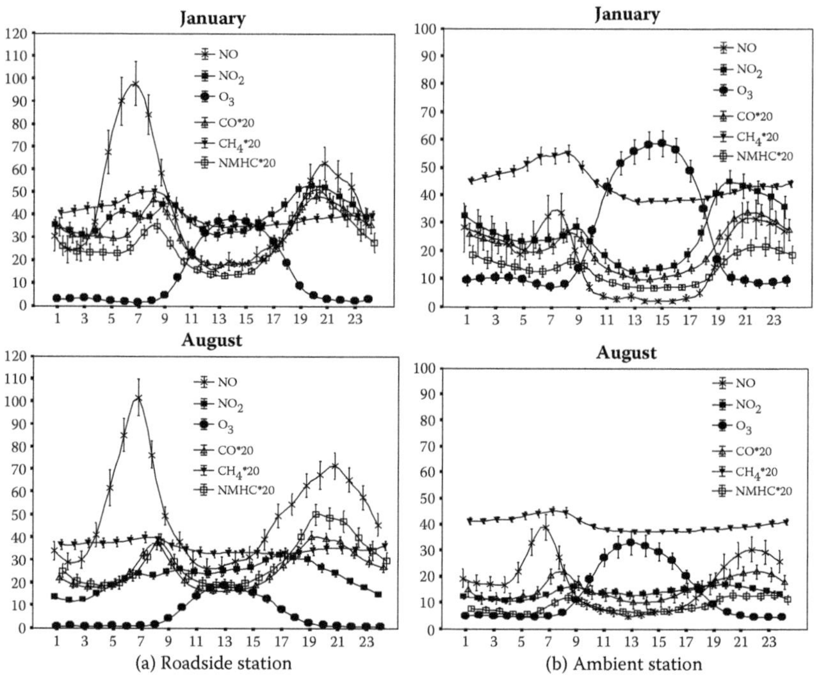

FIGURE 1.6
Diurnal variation of ozone and other pollutants (whiskers show 1 SD) at an ambient station (a) and a roadside station (b) in Bangkok, averaged for January (high-ozone month) and August (low-ozone month) for 1996–2000. (From Zhang B., and Kim Oanh N. T. 2002. Photochemical smog pollution in the Bangkok metropolitan area in relation to O_3 precursor concentrations and meteorological conditions, *Atmospheric Environment,* 36: 4211–4222. Courtesy of AE, reprinted with permission.)

important cause of elevated ozone (and other pollutants) in the dry season. Particularly in SEA, where sunshine is strong all year, the high ozone levels observed in the dry season are linked to more stable atmosphere, additional emission from biomass open burning, and less intensive wet removal process due to lower precipitation in this season. The monthly variation of ozone levels in Thailand (Figure 1.7) showed the highest levels around January–March (dry and hot months) and the lowest around August to September (middle of rainy season). This typical pattern is seen for HCMC and Hanoi, Vietnam. In Jakarta, Indonesia, with its near-equatorial location, the highest ozone levels were observed in October (dry month) and the lowest in January–February (wet month).

1.3.1.2.5 Spatial Distribution of Ozone in Urban Areas

A fresh NO_x emission consists of mainly NO (90% of NO_x) that effectively titrates ozone within a few seconds to minutes (Jenkin and Clemitshaw, 2000). Therefore, in urban areas ozone levels at the roadside (close to traffic emission)

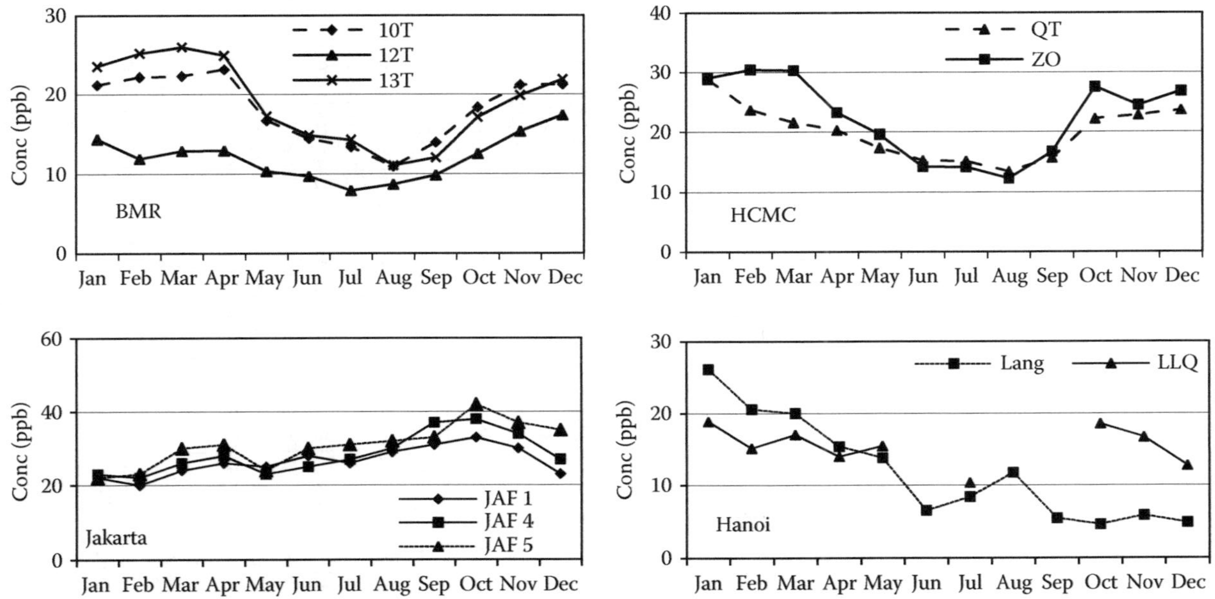

FIGURE 1.7
Monthly average ozone levels based in selected regions in Southeast Asia. BMR: station within Bangkok, Thailand, 2001–2005 (Nghiem, 2008). HCMC: stations in Ho Chi Minh City, Vietnam, 2002–2005 (Huy, 2008). Jakarta: stations in Jakarta, Indonesia, 2002–2004 (Permadi, 2007). Hanoi: stations in Hanoi, Vietnam, 2003 (An, 2005).

are normally low. At some distance away from heavily traveled roads, levels of ozone are significantly higher. Ozone formation involves the so-called NO_x-VOC photolytic cycle, which lasts a few hours; hence, higher levels of ozone are observed downwind of busy city centers. Figure 1.6 also shows that in the vicinity of busy roads (a) during rush hours the levels of primary pollutants (NO, CO, NMHC, CH_4) are significantly higher as compared to the ambient site (b) in both dry (January) and wet (August) months. Ozone levels, however, were significantly lower at the roadside station during both months.

1.3.1.3 Other Pollutants

1.3.1.3.1 Conventional Gaseous Pollutants (SO_2, CO, NO_x)

National monitoring networks deploying automatic stations generally also measure other common gas pollutants such as SO_2, NO_2, and CO. The available data show the levels varying greatly between the urban areas owing to the difference in the energy production and consumption patterns (e.g., high sulfur content in fuel and consumption of more fossil fuels in some cities). Generally, these pollutants are still within the prescribed NAAQS levels. For example, in Thailand, measurements in recent years showed that only 8-h CO occasionally (0.01% of measurements) exceeded the standard (9 ppm) at roadside; hourly NO_2 at roadside also exceeded the NAAQS level (170 ppb) occasionally (one measurement in each year of 2006–2008), while SO_2 and 1-h CO were well below the standards. In other Asian cities, NO_2 also seemed to be more problematic than SO_2 when compared to the NAAQS. In Dhaka, in 2003, for example, annual SO_2 was 7 ± 8 ppb (NAAQS: 30 ppb), while NO_2 was 59 ± 58 ppb, that is, above the NAAQS level of 53 ppb (SOE-BLD, 2005). High frequencies of NO_2 exceeding the hourly and daily standards were observed at roadside stations in Jakarta and to a lesser extent in Bandung, Indonesia, while SO_2 and CO were within the limits in both cities (Driejana, 2007). In Beijing, for example, the annual average of SO_2 generally was reduced from 1990 to 2008 and in recent years was below the NAAQS Grade II, applicable for urban areas of 60 $\mu g/m^3$. The NO_2 levels were also reduced, and over the last decade the levels were below the NAAQS applicable for urban areas in 2011 of 80 $\mu g/m^3$ but were well above the level of China new NAAQS (Grade II: 40 $\mu g/m^3$) released in 2012 (Figure 1.8). A similar trend for SO_2 has also been observed in Delhi, India with the annual levels declining from above 20 $\mu g/m^3$ in 1995 to below 10 $\mu g/m^3$ in 2008 (i.e., well below the Indian NAAQS of 50 $\mu g/m^3$), whereas annual NO_x levels have steadily increased since 1997 reaching above 50 $\mu g/m^3$ in 2008 (SOE-Delhi, 2010).

1.3.1.3.2 Lead

The levels of Pb in the urban areas dropped drastically after the phasing out of leaded gasoline. In Bangkok, for example, before 1991 the Pb levels measured at the roadside were well above the monthly NAAQS level (1.5 $\mu g/m^3$) but were gradually reduced after premium unleaded gasoline was introduced

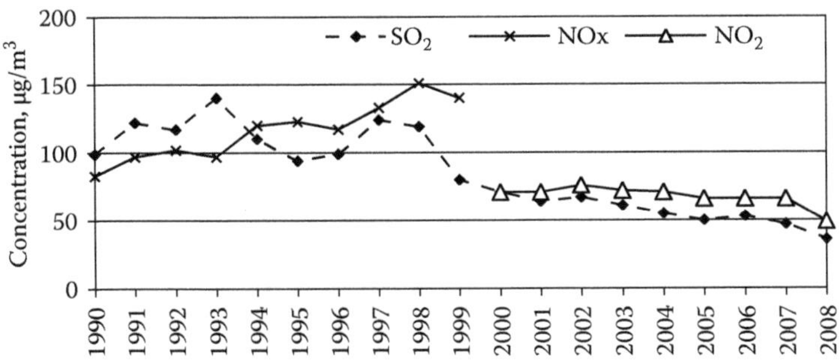

FIGURE 1.8
Annual average SO_2 and NO_x/NO_2 in Beijing. *Note:* Annual China NAAQS (urban area) for SO_2, 60 μg/m^3; for NO_2, 80 μg/m^3. (Based on raw data from NEPA (1991–1997); GEMS (1990–1992); SEPA (1998–2000); BJEPB 2000–2009.)

in May 1991 and were further reduced after regular unleaded gasoline was introduced in September 1992. Since the phasing out of leaded gasoline in January 1996, the levels have been well below the standard (Supat, 1999). In Vietnam, after the phasing out of leaded gasoline in July 2001, Pb in the HCMC air was reduced from a monthly average of 3 μg/m^3 to below 1 μg/m^3 in early 2002. A similar reduction was also reported for Hanoi, with the 24-h Pb levels reduced from around 20 μg/m^3 in 2000 to around 2 μg/m^3 in 2002–2003 (Long, 2006).

1.3.1.3.3 Volatile Organic Compounds

Hazardous pollutants in Asia are rarely monitored and reported. These consist of chemical, physical, and biological agents of different types. They are present in the atmosphere in much smaller concentrations than the conventional pollutants but may appear quite high at some locations. The VOCs are a group of hazardous air pollutants that are of concern in many urban areas of Asian developing countries. The BTEX (benzene, toluene, ethylbenzene, xylenes) compounds are the most monitored among the VOCs. People living in urban areas or nearby industrial estates are likely exposed to higher concentrations of BTEX than other places. Several studies conducted at the Asian Institute of Technology (AIT) showed high levels measured at roadsides in SEA cities. In Hanoi, BTEX levels varied significantly during a day (Truc and Kim Oanh, 2007), reflecting the variations in the traffic density, with high VOC levels observed during morning and afternoon rush hours and lower levels at noon (Figure 1.9). In HCMC, the levels measured at a moderately traveled road also varied significantly during a day, ranging between 6 and 53 μg/m^3 for benzene, 14 and 170 μg/m^3 for toluene, 3 and 24 μg/m^3 for ethylbenzene, and 7 and 80 μg/m^3 for xylenes (Giang, 2008).

In Thailand, VOCs are considered as important issue in both cities and industrial areas. The problem is well perceived by the public; for example, about 40% of environmental complaints in the Maptaput industrial estate,

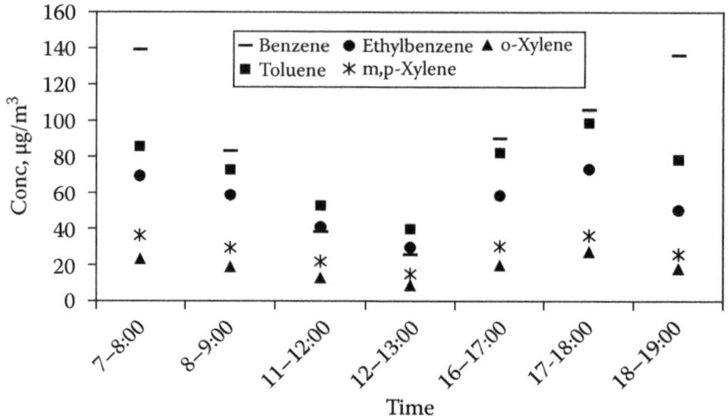

FIGURE 1.9
Roadside hourly BTEX levels of a busy street in Hanoi in November–December, 2004. (From Truc V. T. Q., and Kim Oanh N. T. 2007. Roadside BTEX and other gaseous air pollutants in relation to emission sources, *Atmospheric Environment*, 41: 7685–7697. Courtesy of AE, reprinted with permission.)

received at PCD, were related to VOC pollution (PCD, 2004). Ambient VOCs are now regulated in Thailand, with annual air quality standards established for nine compounds effective since September 14, 2007 (PCD, 2008). Benzene was the most problematic and had an annual average in 2006 that ranged between 1.6 and 5.1 $\mu g/m^3$ (NAAQS: 1.7 $\mu g/m^3$), and the levels observed in Bangkok were higher than in the Maptapud industrial estate. The second VOC of concern is 1,3-butadiene, which also exceeded the annual NAAQS level (0.33 $\mu g/m^3$) in the industrial estate and at roadside stations in Bangkok. 1,2-Dichloroethane also exceeded the NAAQS level (0.40 $\mu g/m^3$) in the industrial estate, while other regulated VOC compounds were well below the standards (PCD, 2006). A study by AIT for BTEX measurements at three sites in the 2008 dry season in the Maptapud industrial estate also reported high levels of benzene. Traffic, petrochemical industry, solvent volatilization, and open biomass burning were among the major identified VOC sources (Wingpud, 2009).

1.3.1.3.4 Semi-VOC

Semi-VOCs are present in the ambient air in both gaseous and PM phases. Examples of these pollutants include PAHs, pesticides, polychlorinated biphenyls (PCBs), and others. Many are highly toxic and genotoxic. In a suburb site in BMR, Thailand, the levels of 16 PAHs (in both gas and PM phases) were reported to be high during the dry period, with intensive rice straw field-burning activities, that is, around 400 ng/m^3, excluding naphthalene, as compared to about 10–40 ng/m^3 during the wet season (Klinmalee, 2008). In China, high PAH levels in the airborne $PM_{2.5}$ and PM_{10}, 299 ± 75 ng/m^3 and 372 ± 81 ng/m^3, respectively, were measured in ambient air of a village during the cooking periods, an increase by more than 70% over the non-cooking periods. Incomplete combustion of fuels, mainly crop residue such as wheat, cotton, corn, and beans (60%) and wood fuel (20%) used in cooking and space heating in the village, was considered the main cause for the elevation of PAHs (Wang et al., 2008).

Limited available information showed high levels of the toxic organochlorine pesticides (OCPs) and PCBs in the ambient air of developing Asia. Iwata et al. (1994) reported high levels of OCPs in Thailand and Vietnam, especially the DDT group, which reached a few nanograms per cubic meter (i.e., two to three orders of magnitude higher than for Oceania and Japan). Several studies at AIT showed that the levels of these hazardous chemicals were still high in ambient air in Thailand despite bans on their use for a number of years. Past intensive uses of these hazardous chemicals for agricultural and household purposes have resulted in their ubiquity in the environment. Levels of OCPs measured in BMR (Pentamwa and Kim Oanh, 2008b) are given in Table 1.2. The total OCPs (17 compounds) of 5–6 ng/m^3 measured in the outdoor air of urban areas of BMR were higher than those in suburban and rural areas. Among the OCPs, the heptachlor group was the most abundant (around 2–3 ng/m^3) at all measurement sites (urban, suburban, and rural). Early studies conducted by AIT (Buesing, 2001; Pentamwa, 2001) also showed high levels of the heptachlor group (about 0.23–0.54 ng/m^3) measured at different sites in BMR; these levels are about 100 times higher than those reported for southern Ontario, Canada, in 1988–1989 (Hoff et al., 1992). Levels of OCPs at background sites in East Asia during 2004–2006, reported by a coordinated monitoring effort (MOE, 2007), were also high and significantly above those obtained for Japan, for example, especially for DDT and hexachlorobenzene (HCB) (Table 1.2).

PCB levels in the ambient air may also be considerable in the region, but the information is very limited. Pentamwa and Kim Oanh (2008b) reported the total PCBs measured in BMR in the dry season of 2006–2007 as 5–7 ng/m^3, with a dominance of lighter PCB compounds (PCB31, PCB28). AIT's earlier studies (Buesing, 2001; Pentamwa, 2001) also reported total PCBs of 2–3.5 ng/m^3 (10 compounds) in different sites of BMR; these levels are well above the level reported for southern Ontario, Canada, in 1988–1989 (Hoff et al., 1992) of about 0.2 ng/m^3.

The persistent organic pollutants (POPs) such as PCB and DDT are toxic and bioaccumulative and hence can affect human health not only through direct exposure but also through their accumulation in terrestrial and aquatic ecosystems. Being persistent, they are present in the environment over considerable time periods, during which they are transferred between the environmental media and participate in long-range transport through the air, water, and migratory species and across international boundaries and deposited far away from their emission sites. High levels of these compounds in the air environment of developing Asia are of concern due to their health impacts not only for the present population but also for future generations.

TABLE 1.2

Levels of OCP in the Ambient Air (Both Gas and PM Phases) in Asian Developing Countries as Compared to Other Countries (pg/m^3)

	Bangkok, Thailand, 2007[a]			SEA, 2005–2007[b]	Japan, 2007[b]	Republic of Korea, 2007[b]	China		India, 2006[e]		North America	
	Urban (Chatuchak)	Suburban (Klongluang)	Rural (Nongsuea)	Remote (5 Countries)	Remote (Hateruma Isl.)	Rural (Goisan)	Rural (Taihu Lake)[c]	37 Cities[d]	Urban (9 Sites)	Rural (6 Sites)	Rural, 1996[f]	40 Sites, 2000–2001[g]
Pesticides	*N* =12	*N* = 4	*N* = 6	*N* = 3/Site	*N* = 6	*N* = 3	*N* = 21	*N* = 4/City	*N* = 1/Site	*N* = 1/Site	*N* = 6	*N* =2/Site
α-HCH	30–160	nd–800	nd–160	—	—	—	21–164	—	22–1,691	12–167	—	—
β-HCH	nd–160	nd–770	nd	—	—	—	—	—	0–149	1–28	—	—
γ-HCH	—	—	—	—	—	—	18–96	—	135–3,562	31–437	32–65	—
δ-HCH	50–150	nd	nd	—	—	—	-	—	8–214	4–42	—	—
ε-HCH	nd–60	nd	nd	—	—	—	-	—	—	—	—	—
HCB	210–410	0–1,610	nd–80	87–330	64–125	81.5	18–109	1,478	—	—	—	50–133
ΣHeptachlor	2,290–2,820	150–4,330	460–1,680	nd–25.5	0.36–4.9	nd–0.4	18–173	39	—	—	0.2–6.5	—
Aldrin	500–570	200–520	120–1,400	nd–0.11	—	—	—	—	—	—	50–201	—
Isodrin	1,140–1,310	nd–520	0–480	—	—	—	—	—	—	—	—	—
Dieldrin	70–380	nd	nd	0.51–36	3.2–6.7	<0.1	—	—	—	—	211–401	—
Endrin	110–120	nd	nd	<0.1–11	0.22–0.67	<0.1	—	—	—	—	—	—
ΣChlordane	300–390	50–2,440	20–210	0.8–94.2	3.4–12.6	2.7	—	550	23–570	12–94	38–86	0.7–95
Σ*DDT*	nd–80	nd	nd–70	2.9–189.8	0.7–10.8	1.8	169–3,724	560	53–2,986	21–1,015	622–1,133	nd–838
α-Endosulfan	0–40	nd	0–170	—	—	—	nd–888	33	3–680	1–992	—	nd–685
β-Endosulfan	0–10	nd	0–140	—	—	—	—	14	19–143	nd–160	—	nd–119
Mirex	—	—	—	0.1–1.8	0.13–0.32	<0.05	—	—	—	—	—	—

Sources: [a]Pentamwa and Kim Oanh (2008b); [b]MOE (2007); [c]Qiu et al. (2004); [d]Liu et al. (2009); [e]Zhang et al. (2008); [f]Alegria et al. (2000); [g]Shen et al. (2005).

Note: *N*, number of measurement sites; —, not analyzed; nd, not detected. ΣHeptachlor includes heptachlor, *trans*-heptachlor-epoxide, *cis*-heptachlor-epoxide, heptachlor-exo-epoxide; ΣChlordane includes *trans*-chlordane, *cis*-chlordane, *trans*-nonachlor, *cis*-nonachlor, and oxychlordane; Σ*DDT* includes *p,p′-DDE*, *p,p′-DDD*, *p,p′-DDT*, *o,p′-DDE*, *o,p′-DDD*, *o,p′-DDT*.

1.3.2 Indoor Air Pollution

Indoor air pollution levels depend on a large number of factors, including the presence of indoor sources and the emission rate, ventilation rate, penetration of outdoor pollutants into the indoor environment, and pollutant sink or removal rate on indoor surfaces. Many sources, related to both combustion and noncombustion activities, emit air pollution indoor. In poorly ventilated homes, indoor fine PM concentrations can exceed the limits, which are set for outdoor air, by a dozen to a hundred times. Exposure is particularly high among women and children in rural areas, who spend most of the time near fireplaces in homes. On a global scale, indoor air pollution is responsible for the death of 1.6 million people every year (WHO, 2005).

Typical indoor air pollutants that originate from indoor sources can be grouped in four categories: (1) combustion pollutants (gaseous pollutants and PM that may potentially be emitted from all types of combustion processes, including tobacco smoke); (2) noncombustion VOCs (evaporation of VOCs from building materials, household products, paints, or contaminated soil); (3) hazardous chemicals/materials applied indoors (pesticides, asbestos); and (4) biological agents (mildew, molds, fungus, or bacteria, house dust mites, etc.). Sources and health effects of the pollutants of the first group are relatively better known and rather well documented in literature worldwide. Levels of other pollutant groups, especially in developing Asia, have not been widely reported, but all are health hazards (WHO, 2005, 2010).

The exposure to air pollution in private homes is not yet regulated by law; therefore, indoor air is often considered beyond the scope of urban AQM and has yet to become a central focus of research, development aid, and policy making. Accordingly, in spite of growing public awareness of the adverse effects from indoor air pollution, most AQM measures still heavily address outdoor air issues, which may be largely different from those of indoor air quality. For example, most ambient air quality standards in the world do not include biological contaminants, which are important in the indoor air context.

1.3.2.1 Combustion Pollutants

1.3.2.1.1 Domestic Combustion

Smoke from fuel burned indoors (for cooking and space heating) contains a range of health-damaging pollutants, such as fine PM, VOCs, CO, and carcinogenic PAHs. WHO has assessed the contribution of risk factors to the burden of disease and ranked indoor air pollution as the 8th most important risk factor, responsible for 2.7% of the global burden of disease (WHO, 2002b). Indoor air pollution from cooking has long been discussed, and the health effects have been widely reported (Kim Oanh et al., 2000; Smith, 1987; WHO, 2002b).

In Asian developing countries, cooking and space heating using dirty fuels are the leading causes of indoor air pollution. Solid fuels such as wood, agricultural waste, cattle dung, or low-quality smoky coal are still commonly used in rural and periurban areas. Simple nonairtight cookstoves with low efficiency, such as tripods or three-stone stoves, are still popular in rural areas; these consume more fuel and generate higher emissions of incomplete combustion products per unit of energy produced. Poorly ventilated kitchens lead to a high buildup of combustion pollutants indoors during cooking periods. In the regions where cold winters exist, space heating with open fires is done with reduced ventilation to keep in the warmth, which leads to even higher indoor pollution levels. In homes of developing countries, PAH levels were reported in the range of 100–10,000 ng/m^3, compared to 20 ng/m^3 commonly measured in busy traffic areas or 20–100 ng/m^3 in areas of cigarette smoking (United Nations Environment Program [UNEP], 1991).

Rural homes in a hot climate are commonly naturally ventilated; hence, combustion smoke can spread from the kitchen, which in many cases is not separated from the main home, to other parts indoors. Sengkhammy and Kim Oanh (2007) conducted a survey of indoor air quality in selected homes in a village of Vientiane, Laos, and showed that most homes (63%) in the village had the kitchen attached to the main house. Charcoal was the major fuel used for cooking (80%), followed by liquefied petroleum gas (LPG; 13%), while wood fuel and electricity had low shares. The highest indoor $PM_{2.5}$ levels were found in the kitchen and living room of a townhouse (home 2) that cooked with both wood fuel and charcoal and had the kitchen attached to the living room (Figure 1.10). Among other surveyed residences, higher levels of $PM_{2.5}$ were found in home 1 (which used charcoal only) and lower levels in other homes (charcoal, LPG, and electricity were used). During cooking, the average 10-min CO concentration was much higher in both the kitchen and the living room of home 2, reaching more than 12–15 ppm. In other homes that cooked with charcoal or electricity, the CO levels were generally below 10 ppm. During noncooking periods, CO levels dropped to below 5 ppm in all homes. High CO and black carbon (BC) measured in homes during cooking periods suggested that incomplete fuel combustion was the main cause.

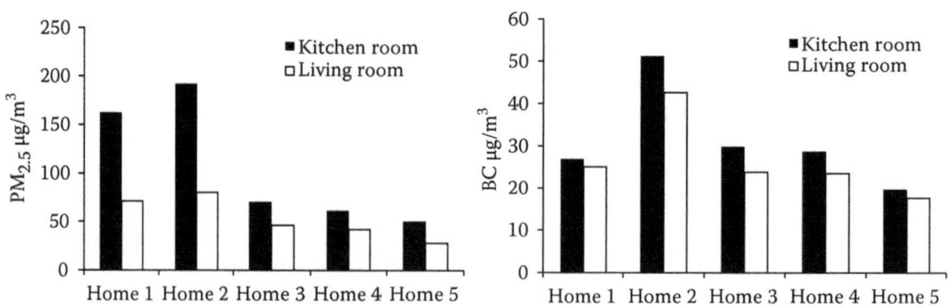

FIGURE 1.10
$PM_{2.5}$ and black carbon levels in kitchen and living rooms, Vientiane, Laos. (From Sengkhammy B., and Kim Oanh N. T. 2007. Indoor air pollution in Vientiane, Laos, in relation to cooking fuel type and house design. In *Proceedings of the International Conference on Air Quality Management in Southeast Asia,* Ho Chi Minh City, Vietnam, November 21–23, 2007.)

The association between lung cancer mortality and smoky coal use for cooking and space heating in poorly vented homes was reported for Xuan Wei County, Yuannan Province, China. The unadjusted rate of annual lung cancer mortality per 100,000 in a commune where 89.7% of the population used smoky coal was 109.3 cases, well above the rate in a reference commune (2.3 cases) where no smoky coal was used (Mumford et al., 1989). The measured air pollution levels in the kitchen during cooking with smoky coal were high for PM_{10} (38–39 mg/m^3), benzo(a)pyrene (BaP: 10.3 μg/m^3), and the mutagenicity associated with PM_{10} (Mumford et al., 1987). A study (Large et al., 2009) demonstrated that the smoky coal associated with lung cancer in Xuan Wei contained an unusually high content of silica, mostly in fine grains, and suggested that the most likely cause of the cancer was the interaction between silica and volatile contents in the coal.

Although the current trend of switching to cleaner fuels (LPG, electricity) for cooking is observed in the urban population in developing Asia, the poorest sections of urban society and the rural population are often still dependent on low-grade fuels such as dirty coal briquettes or noncommercial biofuels. The problems of indoor air pollution due to domestic combustion for cooking and space heating are well known, but the solution to such a complex set of problems is not simple. With low income levels, most people in urban slums and in rural areas cannot afford cleaner fuels even if available. Thus, the poorest still continue to rely on the most polluting fuels, while more wealthy households shift to LPG or electricity. A wide range of interventions or measures are available to reduce indoor air pollution from fuel combustion, which can be classified into three groups: (a) source emission reduction (by improved stoves, use of cleaner and more efficient fuels such as LPG, biogas, electricity, and solar energy); (b) improvement of ventilation to reduce exposure (use of airtight stoves with chimneys, use of a chimney to vent the smoke from the kitchen, provision of a window in the kitchen, etc.); and (c) changes in user behavior (properly dry fuel before use, open windows while cooking, avoid sitting in the smoke plumes, keep the infirm and young children away from smoke). The changes in user behavior in many cases can bring a difference at no cost at the user levels; hence, education and information should be provided to raise the awareness and know-how.

1.3.2.1.2 Tobacco Smoke

About 1.5 billion people in the world are tobacco smokers. In 2000, the low- and middle-income countries were estimated to account for 82% of all smokers (WHO, 2000). Environmental tobacco smoke (ETS) is a complex mixture of toxic pollutants: fine particles, nicotine, nitrosamines, PAHs, CO, NO_x, and many others (U.S. Environmental Protection Agency [USEPA], 1995). Levels of respirable particles indoors with smokers were noted as over four times that of outdoors (Godish, 1991). A number of carcinogenic pollutants, including radon and PAHs, also attach to these fine particles. ETS causes adverse health effects not only to tobacco smokers but also to secondhand or passive smokers (i.e., exposed nonsmokers). A ban on smoking in public places is practiced in many Asian countries now, which, once successfully enforced, can substantially reduce the public exposure to ETS.

1.3.2.2 Pollutants Released from Building Materials

Indoor air problems may occur in modern offices and urban homes even without combustion, especially in airtight buildings with poor ventilation. Such problems have been reported in many developed countries, causing the so-called sick building syndrome. As airtight air-conditioned buildings are becoming more popular in developing Asia, this indoor air pollution issue will be similar to that of developed countries. Major pollutants of this group include chemical vapors, radon from soil, and asbestos from construction materials.

1.3.2.2.1 Chemical Vapor

Many household products contain organic chemicals that release vapors indoors while they are in use or stored. These include various building materials, furnishings, paints, vanishes, solvents, adhesives, office equipment, and others. The released chemical vapors can exist in the gas phase (VOCs), be adsorbed on airborne particles, or adsorbed/absorbed onto indoor materials. The indoor materials with sorbed vapors (carpets, gypsum boards, etc.) would serve as temporary sinks and off-gas the vapors slowly, which results in chronic exposure (Chang et al., 1998). New buildings have higher emission rates than aged ones, but some building materials can release VOCs slowly for years.

Formaldehyde (HCHO), a colorless and pungent-smelling gas, is one of the most important and ubiquitous pollutants indoors. The most significant indoor sources of HCHO are pressed wood products using HCHO-based adhesives, such as particleboard, plywood paneling, and medium-density fiberboard. Formaldehyde can cause both acute and chronic health effects. At concentrations of 0.1–5 ppm, it irritates the mucous membrane of the eyes and the upper respiratory tract. HCHO may trigger attacks in asthma sufferers. Epidemiological studies of exposed workers and mobile home residents indicated that HCHO may cause cancer in humans (Godish, 1997). In the United States, the Occupational Safety and Health Administration (OSHA) regulates the permissible levels of professional exposure to an 8-h average of 0.75 ppm. Reported levels of other VOCs indoors are at a few micrograms per cubic meter individually, but total VOC concentrations may reach above 1 mg/m^3. The extent and nature of the health effects depend on the chemical compounds and levels and length of exposure periods; the worst case is cancers.

Use of low-quality particleboard, solvent-based paints, low-quality carpets, and the like in homes and offices in Asian developing countries would result in high levels of chemical vapors indoors, but the situation is still largely neglected. The scarcity of information on this important indoor pollution type in the region is a barrier to awareness-raising efforts for mitigation. While waiting for the emission standards of household products to be established and properly enforced in developing Asia, the educational measures can help users protect themselves from exposure.

1.3.2.2.2 Radon

Radon is a gas produced in the radioactive decay process of radium, which itself is a product of uranium decay and is found in almost all soils and rocks. Radon gas enters homes through dirt floors, cracks in concrete walls and floors, floor drains, and sumps. Radon is ubiquitous in the ambient air, with average concentrations in the range of 0.20–0.25 pCi/L (picocuries per liter). Indoor levels may be several times or several orders of magnitude higher than ambient air levels. Any home may have a radon problem, but the gas itself poses no hazard. However, radon undergoes radioactive decay, producing a series of short-lived progenies. Most of the radon progenies become attached to particles present in the indoor air. When these particles are inhaled, a fraction of radon progenies is deposited in the lungs. Inside the lung, some radon progenies emit alpha particles that are absorbed in the nearby lung tissues and cause major genetic damage. People exposed to high radon levels have an increased risk of developing lung cancers (WHO, 2009), with even greater risks for smokers. The USEPA guideline for radon is 4 pCi/L for remedial action (USEPA, 2009), while WHO (2009) recommends a new reference level of 2.7 pCi/L (100 Bq/m^3) to minimize the exposure health hazard. It is estimated that radon causes thousands of preventable lung cancer deaths each year. An average radon level of 1.3 pCi/L worldwide could be linked to 13–25% of acute myeloid leukemia cases (Godish, 1991). In the United States, the annual death rate due to radon ranged from 7,000 to 30,000 and averaged 14,000 (USEPA, 1995).

The radon pollution potential varies from one region to another depending on geological conditions. There is yet to be comprehensive research on radon levels in homes in Asian developing countries. A few studies in Thailand, for example, showed that the average indoor radon levels measured in different provinces ranged from 21 to 52 Bq/m^3 (0.6–1.4 pCi/L). One study also reported a high indoor level of 200 Bq/m^3 (5.4 pCi/L) in a southern province (Wanabongsea et al., 2005, and references therein). It is noted that homes with good basement ventilation, such as those standing on columns, which are popular in many tropical countries, can effectively prevent radon entry, hence would have less indoor radon exposure risks.

1.3.2.2.3 Asbestos

Asbestos is the commercial name of six naturally occurring fibrous minerals belonging to two groups: serpentine (chrysotile) and amphibole (actinolite, amosite, anthophyllite, crocidolite, and tremolite). Three types of asbestos have been used commercially: chrysotile, amosite, and crocidolite. Chrysotile (white asbestos) was the most common asbestos and comprised 90% of all world commercial asbestos. The crocidolite (blue asbestos) is commonly identified as the most dangerous mineral, but amosite (brown asbestos) may be equally harmful.

The most dangerous asbestos fibers are too small to be visible. After they are inhaled, they can remain and accumulate in the lungs. Four types of diseases are associated with exposure to asbestos: lung cancer, mesothelioma (a cancer of the chest and abdominal linings), asbestosis (irreversible lung scarring that can be fatal), and nonmalignant pleural disease (formation of fibrous and calcified plaques). Symptoms of these diseases do not show up until many years after the initial exposure. Most people with asbestos-related diseases have had occupational exposures to high concentrations, and smokers have been reported to be at higher risk of cancers.

Several asbestos products have been banned around the world, but a total ban has yet to be implemented. As of 2004, crocidolite was banned in most Asian countries, amosite was banned in some countries, while chrysotile use is still largely allowed (Ladou, 2004). In Asia, the per capita annual use of asbestos ranged from 0 kg in Singapore to 1.9 kg in Thailand. The wealthy industrialized countries showed a steady decrease in asbestos use. For example, in Japan the use was reduced from a peak of 3.1 kg/capita/year in 1974 to 0.6 kg/capita/year in 2004. But, in poorer developing countries the per capita use is increased (Ladou, 2004).

Asbestos, with its excellent fire resistance, was widely used in the past for building construction and is now still commonly found in pipe and furnace insulation materials, asbestos shingles, millboard, textured paints and other coating materials, and floor tiles of old homes (i.e., mostly in the bound form). These asbestos-containing materials (ACMs) are unlikely to release fibers unless disturbed. Improper attempts to remove ACM can release dangerous asbestos fine fibers into the air and result in exposure. In many Asian developing countries, where demolition and renovation of old buildings are intensive, the risk of asbestos exposure is likely to be high. This is particularly relevant when construction workers have little knowledge of the presence of ACM at working sites. Awareness raising and proper enforcement of regulations for building demolition and renovation activities are essential to avoid excessive professional exposure to airborne asbestos fibers.

1.3.2.3 Hazardous Chemicals Used Indoors: Pesticides

Pesticides are the most common hazardous chemicals used indoors. They are used in and around homes to control insects, termites, rodents, fungi, and microbes (antimicrobials/disinfectants). Common insecticides used indoors normally belong to synthetic pyrethroids and organophosphate pesticide groups. They are generally less toxic compared to POP pesticides and other OCPs used in agriculture (WHO, 2002a). It is important to note that pesticides are toxic by nature and should be used properly. Health concerns associated with the exposure to pesticides indoors include both acute toxic effects that occur immediately after applications and potential carcinogenic effects of some pesticides. Improper applications could result in excessive levels in the indoor air. Numbers and concentrations of pesticides detected indoors have been shown to be typically higher than those detected outdoors. Epidemiological studies reported that the exposure to pesticides in their home environment can cause young children to have chronic health effects. Some pesticides used indoors in Asia, for example, are already banned in other parts of the world (Robert et al., 2003).

Pentamwa et al. (2010) conducted a shop shelf survey in BMR, Thailand, and revealed 12 different active ingredients among 22 available commercial products for indoor applications, with pyrethroid-containing pesticides the most common. The organophosphate pesticide group (e.g., dichlorvos, a highly toxic pesticide) was still found, although with a low share (i.e., less than 1%). Most households (93%) had used at least one pesticide product indoors over the past year, mainly to control mosquitoes. The survey found that the aerosol spray was the most common application method and usually partially used pesticide containers were stored in living rooms, where the chemicals were most often applied. These practices certainly contribute to high exposure of people to these toxic chemicals. Pentamwa and Kim Oanh (2008b) monitored the airborne levels of 25 OCPs and 7 pyrethroids (both in gas and PM phases) in BMR homes and reported high levels of several toxic compounds, well above the outdoor levels, especially in homes located in crowded urban areas. The levels of total pyrethroid compounds detected indoors were 0.3 ng/m^3, compared to 0.1 ng/m^3 found in the outdoor (urban air) samples. Indoor air also contained high levels of toxic POP pesticides despite the bans on their use in effect for several years. The indoor OCP (including POP pesticides) levels for urban homes were above 10 ng/m^3, of which the majority were from heptachlor (about 5 ng/m^3), followed by the chlordane group (0.5 ng/m^3) and the endosulfan group (0.1 ng/m^3). These toxic chemicals were also found in the floor surface wipe samples, which amounted to a few micrograms per square meter. Homes in rural and suburban areas were reported to have lower OCP levels (i.e., about 15–20 times less than the urban homes).

1.3.2.4 Biological Agents

Contaminants of biological origin are also called biogenic particles or viable particles and consist of bacteria, molds, fungus, bacteria, viruses, pollen, and others. Numerous sources of indoor biological contaminants have been identified, including people and animals as carriers of the agents (viruses, bacteria); soil and plant debris; and saliva and dander from household pets. The health effects of biogenic agents are mainly related to their infectious reactions, causing influenza (including avian flu), severe acute respiratory syndrome (SARS), measles, tuberculosis, and chicken pox, as well as allergic reactions, such as hypersensitivity pneumonitis, allergic rhinitis, and certain types of asthma. House dust mites, for example, are the source of one of the most powerful biological allergens. These tiny mites (250–300 μm) are translucent and are not visible to unaided eyes. House dust, which is heavily contaminated with the fecal pellets of these mites, is one of the strongest allergenic materials found indoors (Godish, 1991).

The initial colonization of viable particles originates from both the indoor and outdoor environment. In homes, these sources are located at wet or moist places, such as bathrooms, flooded basements, furniture and poorly maintained humidifiers, and air conditioners. Contaminated central air-handling systems can become breeding grounds for a range of biological agents and subsequently distribute them throughout the home. One type of bacteria that has been linked to central air systems is *Legionnella pneumophila,* which is a major cause of legionellosis or Legionnaire's disease, a respiratory illness, worldwide (USEPA, 1995). These bacteria are ubiquitous, can survive in water for a long time, and thrive in cooling towers, including those of large central air-conditioning systems.

In the last few decades, there has been a strong increase in the number of people suffering from allergies in Western developed countries. One of the suspected reasons for this increase is modern airtight building designs, creating an increased concentration of both biogenic and chemical allergens. Such airtight buildings are becoming increasingly popular in many urban areas of Asia. The hot and humid climate of the tropical Asian region could result in high levels of biological agents in homes, but there is yet a comprehensive study to address this phenomenon.

1.3.2.5 Other Growing Issues of Indoor Air Quality in Asia

As Asia becomes further developed, indoor air quality issues related to modern airtight and mechanically ventilated public buildings and vehicle interiors will be of increasing importance. In many Asian countries, to cope with the hot and humid climate, air conditioners that consume a huge amount of energy are used extensively in offices, shopping centers, vehicles, homes, and other places. Minimization of fresh air intake is a measure to save energy; hence, in many cases almost the same cooled air would be circulated indoors through air handling systems over extended periods. Air pollutants, especially those generated from indoor sources, may be trapped inside such mechanically ventilated environments and build up to high levels indoors. Klinmalee et al. (2008) reported high levels of pollutants inside a shopping center, especially on crowded weekends. Indoor $PM_{2.5}$ levels were 83 ± 27 μg/m^3 on weekends, compared to around 50 ± 14 μg/m^3 on weekdays. High CO_2 levels measured inside the shopping center, especially over weekends (1,113 ± 260 ppm), indicated the need for increased fresh air intake. BTEX levels were also high inside the shopping center, most significantly benzene (29 ± 25 μg/m^3) and toluene (273 ± 235 μg/m^3), while ethylbenzene and xylene levels were generally less than 10 μg/m^3. High levels of these toxic pollutants suggest a high exposure and health risk, especially when people spend more time at shopping centers during weekends. The issue related to the vehicle interior air quality is addressed in Chapter 8 of this book.

1.3.3 Transboundary Pollution Problems

Air pollutants emitted at a location may circulate the globe within a few days to weeks depending on meteorological conditions. In many instances, air pollutants are found thousands of kilometers away from their emission sources. The air pollution is sent around the globe, and the impacts of human activities are felt on scales from local to global. Through the transboundary transport phenomena, the air pollutants emitted within a state or country would introduce adverse effects in other states or countries. Regional and transboundary air pollution have been a topic of scientific research for several decades, and the importance has been recognized increasingly (United Nations Economic Commissions for Europe [UNECE], 2004). The pollutants that have been observed and model predicted of having a high potential for long-range transport include fine particles, acidifying substances (SO_2, NO_x), ozone and its precursors, heavy metals (e.g., mercury), and POPs. Transboundary transport should be properly considered in formulating national AQM strategies, especially when a large part of ambient air pollution has a foreign origin.

Transboundary air pollution is an international challenge that requires international cooperation to bring countries together to solve the issue. Cooperation on emission inventories, ecological research, monitoring, and modeling to determine pollutant pathways are some examples of joint efforts to deal with transboundary air pollution. Advanced techniques such as satellite, aircraft, and harmonized ground monitoring networks can be used to trace the pollution plumes.

A number of intensive emission source regions exist in Asia, and the monsoons enhance the long-range transport air pollution. Hybrid Single Particle Lagrangian Integrated Trajectory (HYSPLIT) modeling applied to track the origin of air masses also confirmed, for example, the potential long-range transport of pollution to Bangkok (Pongkiatkul and Kim Oanh, 2007) and Dhaka (Iqbal and Kim Oanh, 2011). Results of three-dimensional (3D) air quality models showed the plumes of ozone pollution extending from large cities in SEA (Nghiem and Kim Oanh, 2008). This section provides a discussion of the major transboundary air pollution issues in the Asian continent: SEA haze, ABCs, Northeastern Asian dust, and acid deposition. It is noted that POPs as well as the trace elements from coal combustion and various manufacturing processes, particularly Hg, also have a high potential for long-range transport, but lack of comprehensive measurements hinder any in-depth analysis.

1.3.3.1 Southeast Asian Haze

Smoke from big forest fire events, which frequently take place during the dry season in SEA, cause the most severe transboundary haze pollution problem in the ASEAN region (Koe et al., 2001). The smoke contains high levels of toxic gases (VOCs, PAHs, CO) and fine particles that would result in a substantial exposure risk to the large population in the region. The fine haze particles also absorb, scatter, and reflect solar radiation and affect cloud formation, which subsequently affects atmospheric visibility and climate forcing. The SEA haze is reocurring, but the most severe episode was observed in 1997–1998 during an El Niño and draught period when nearly 9.7 million hectares of forests were destroyed (Asian Development Bank [ADB], 1999). At the peak of the haze, the ambient PM pollution reached unprecedented levels in the region. For example, in late July 1997, the PM levels observed in Sarawak, Malaysia, were up to 15 times higher than normal. In Thailand, during the haze episode in late September to early October 1997, the 24-h PM_{10} peaked at above 200 μg/m^3. In August 2005, another severe haze was observed when some areas in Malaysia had an air pollution index reading above the hazardous level (300) and on some days even above the emergency level (500). When the smoke hit southern Thailand, the hourly PM_{10} levels peaked sharply and rose higher than 200 μg/m^3 (Pentamwa and Kim Oanh, 2008a). The ASEAN Agreement on the Transboundary Haze Pollution, put in force in 2003, aims to tackle transboundary haze pollution resulting from land and forest fires. The agreement exemplifies ASEAN's resolution to tackle transboundary haze pollution comprehensively (http://www.as eansec.org/15129.htm).

1.3.3.2 Atmospheric Brown Cloud

The ABCs appear as large air masses that are a few kilometers thick and contain pollution at levels similar to those found in polluted air in heavily populated regions (Ramanathan and Crutzen, 2003). The ABC was named the South Asian haze when discovered during the Indian Ocean Experiment (INDOEX) intensive field observation in 1999. Later, it became evident that such clouds exist downwind of any populated region with intensive emissions, hence the term *atmospheric brown cloud* is commonly used. In Asia, this haze extends over South, Southeast, and East Asia. It consists of a mixture of aerosols (BC, organic carbon [OC], sulfates, nitrates, etc.) and is thought mainly to originate from burning of fossil fuels and biomass over entire Asian region. ABCs are transported far from the source regions and hence can introduce multiple effects on regional air quality, air temperature, precipitation, agriculture, and health (UNEP, 2008).

1.3.3.3 Asian Dust Storm

The yellow sand or the dust storm that occurs frequently over arid desert areas of East Asia, mainly over the Gobi Desert and arid areas of China and Mongolia, is reported to deteriorate the PM air quality in China, Hong Kong, Taiwan, Korea, Japan, and Mongolia and even passes over the Pacific to reach the western coast of America (Kim et al., 2003; Ma et al., 2004; Sun et al., 2004, and references therein). This yellow dust is called *Kosa* in Japan, *Huangsha* in China, and *Whangsa* in Korea. It appears as a distinct yellow cloud on satellite images. The highest frequency of occurrence is observed during spring (March–May) when the soil in the arid areas has thawed but is not yet covered by vegetation. A dust storm occurs when strong winds sweep up large quantities of dust particles and suspend them in the air. Large particles settle out near the source, hence causing the greatest impact on the local and neighboring regions. Fine dust particles may be lifted as high as 1–3 km into the atmosphere and remain in the air for a period of 5–10 days, during which they are transported over a large distance. The resulting dust clouds are often so thick that they obscure the sun, reduce visibility, slow traffic, and close airports. As the dust storms move over urban centers, they pick up particles from other anthropogenic sources.

Although the dust storms are associated with natural emissions, a major factor leading to intense dust storms in recent years is believed also to link to inappropriate development activities, which turned the once-fruitful agricultural land into desert due to overuse of the land for farming and grazing. Various cooperative projects and forums have been created to enhance the exchange of information and policy dialogue among environmental experts in Northeast Asian countries. These efforts to address transboundary air pollution, including yellow sand, have yet to produce solid and concrete long-term plans or binding implementation schemes (Jho and Lee, 2009).

1.3.3.4 Acid Deposition

The acidic gases of SO_2 and NO_x are emitted primarily from fossil fuel combustion and are the major anthropogenic precursors of acid deposition. They may remain in the atmosphere for several days; during that time they can travel over large distances, in some cases more than 1,000 kilometers. During their atmospheric residence time, they are transformed into acidic compounds by a complex series of chemical reactions. They may be deposited from the atmosphere as dry or wet particles or as vapors, causing acidifying effects on Earth's surface. This phenomenon is known as *acid deposition,* which incorporates both wet (with precipitation) and dry deposition. The term *acid rain* is frequently used, and it is normally perceived as the wet acid deposition that has pH lower than that of the pure rainwater (pH 5.6). Acid deposition kills aquatic life, trees, crops, and other vegetation; damages buildings and monuments; corrodes metals; reduces soil fertility; and can leach toxic metals into drinking water sources. The impact of acid deposition on an ecosystem depends on its assimilation capacity for the excess acidity, usually caused by NO_3^- and SO_4^{2-}, which in turn depends on the components of the ecosystem (International Institute for Applied Systems Analysis [IIASA], 2004). An ecosystem can tolerate acid deposition as long as it is below a critical deposition load, that is, below the level that would cause chemical changes that lead to long-term harmful effects on essential ecosystem properties.

Satellite images show high SO_2 and NO_x concentrations over several locations in the East Asia region. The intensive SO_2 and NO_x emissions in these upwind locations have a high potential to cause acid deposition in the downwind regions. Dry deposition, for example, may be important during the dry season in SEA when the region is under the influence of the NE monsoon. Acid deposition in Asia is monitored through EANET (Acid Deposition Monitoring Network in East Asia; http://www.eanet.cc/) and the Malé Declaration networks (http://www.rrcap.unep.org/male/), which report some high-acidity rainwater, with pH of 5.0 or less, in a number of monitoring sites.

To mitigate acid deposition, the emission sources of the precursor gases should be effectively abated. Addition of alkaline materials, such as lime compounds, soda ash, and the like, into the acidified environmental resources (e.g., a lake) can neutralize the pH, but it takes many decades for all the damaged ecosystems to recover. Acid rain was one of the first transboundary issues of concern, emerging during the 1960–1970s in North America and Europe. Now, most industrialized countries have effectively cut down SO_2 emissions, while NO_x emissions remain a problem in many places, mainly due to the increase in mobile sources. In Asia, energy use has surged, and the use of cheap fuels such as coal and oil with a high sulfur content has increased rapidly. In the stationary sources, current control efforts also largely address SO_x emissions. For mobile sources, the trend of low-sulfur fuels will effectively curtail SO_2 emission, while the three-way catalyst for gasoline cars will reduce NO_x emission. However, the fast growth of vehicle fleets in Asian cities may offset these emission control efforts.

1.4 Emission Sources and Contribution

1.4.1 Major Anthropogenic Emission Sources

Fuel combustion is the principal anthropogenic air pollution source that tends to increase along with population size and economic growth. Fuel type is a useful indicator of potential emissions: coal and biomass as high-emitting solid fuels; gasoline/diesel, kerosene as medium-emitting liquid fuels; and LPG and natural gas (NG) as low-emitting gaseous fuels. The Health Effects Institute (HEI, 2010) analyzed the data compiled by the International Energy Agency (IEA) and showed an increase of primary energy consumption in Asia by 75% during the period from 1990 to 2005 (i.e., corresponding to an annual rate of 3.9%). In China, for example, the energy consumption almost doubled during this period, with a steep increase since 2001. This rapid growth of primary energy consumption, coupled with low-quality fuels and limited application of the emission control measures, has lead to a drastic increase in air pollution emissions.

TABLE 1.3

Anthropogenic Emission Sources of Air Pollutants in Asian Developing Countries (Not Including Forest Fires)

	Pollutant Emitted						
Source Types	**SO_2**	**NO_x**	**CO**	**PM**	**BC**	**VOC**	**Heavy Metals**
Stationary							
Coal/heavy oil-based power plant	3	3	1	1–2	1–2	1	1
Coal/heavy oil burning in small industry and homes	2	2	1–2	2	1–3	1–2	1–2
Industrial manufacturing process[a]	1–3	1–3	1–2	1–3	1–2	1–3	1–3
Solid/hazardous waste incinerators (depending on technology)	1–2	1–2	1–2	1–2	1–2	1–2	1–2
Biomass fuel burning in small industry and homes	1	1	3	2–3	1–2	2–3	
Diesel-powered backup generators	1–2	1–2	1	1–2	1–2	1–2	1–2
Traffics and Related							
Diesel-powered vehicles	1–2	2–3	1	2–3	2–3	1–3	1–2
Gasoline car (without catalyst)	1	1–2	3	1		2–3	3[b]
Gasoline car (with catalyst)	1	1	1	1		1	
Four-stroke motorcycle			3	1		2	3[b]
Two-stroke motorcycle			3	2		3	3[b]
Gasoline station, solvents						2	
Others (Not Including Forest Fires)							
Agroresidue and grassland burning	1	1	2–3	2–3	1–2	2–3	
Solid waste open burning	1	1	2–3	2–3	2–3	2–3	1–2

Note: The relative importance as air pollution sources is indicated as follows: 1, less important; 2, important; 3, very important.

[a] The importance degree varies with the manufacturing processes (cement, petrochemical, etc.).

[b] If leaded gasoline is used. Assuming that cars with a catalyst use unleaded gasoline (a prerequisite), then lead emission is not significant.

Major anthropogenic combustion sources of air pollution in Asia include mobile sources (on-road and off-road); stationary sources (power plants, industries, incinerators, residential); and open burning (forest fires, grassland, agricultural residue, solid waste). A summary of air pollution sources in Asian developing countries and their relative importance is presented in Table 1.3.

1.4.1.1 Mobile Sources

1.4.1.1.1 Urban Road Traffic

The steady growth in number of vehicles in Asia leads to the increased emission of toxic air pollutants, including fine particles, CO, NO_x, VOCs, and semi-VOCs (PAHs). Exposure to traffic emission in cities is of greater concern as the pollutants are released at the breathing level in populated areas, especially when emitted in street canyons with poor dispersion conditions. Uncontrolled vehicle exhausts, particularly those of diesel-powered and two-stroke gasoline engines, are the largest sources of air pollution in many Asian urban areas. The motorbike population in Asia is the largest in the world since they are the cheapest means of individual motorized transportation and are most affordable for the expanding working class. Many Asian cities are also too crowded to accommodate a drastic expansion of car fleets. In Hanoi, Vietnam, for example, motorcycles (four stroke) comprise over 90% of the on-road auto fleet (Truc and Kim Oanh, 2007), which increased 16% per year during the 2000–2008 period (Hanoi People Committee [HPC], 2008). In several Asian cities, two-stroke motorcycles that have high emission factors of VOCs, CO, and fine particles are still popular and hence may contribute a significant emission share. If leaded gasoline is still used, the organic lead emitted with the unburned fuel from two-stroke motorcycles is even more toxic than the inorganic lead formed in the combustion (Weaver and Chan, 1994).

Leaded gasoline has been successfully phased out in most Asian developing countries. However, high VOC emission from incomplete combustion of reformulated unleaded gasoline in old engines that are not equipped with a proper control device (catalytic converter) could be another issue. In fact, vehicle fleets in Asian developing countries often consist of a wide range of year models and technologies. Emission control measures are not normally enforced for old vehicles, and inadequate inspection allows dirty vehicles to remain in use. In particular, the use of secondhand, high-mileage, and poorly maintained vehicles results in large emissions per unit of fuel consumed. In Bangkok, for example, old diesel-powered bus "superemitters" have an average $PM_{2.5}$ emission of above 8 g/kg as compared to that of the "normal" fleet of around 2 g/kg. These superemitters still constitute 15% of the city diesel fleet (Subramanian et al., 2009). Elemental carbon (EC) is the dominant fraction of PM emitted from these old vehicles (Kim Oanh et al., 2010), as shown in Figure 1.11. Frequent congestion, seen as a high-priority problem in many urban areas in Asia (e.g., China, India, Bangladesh, Indonesia, Philippines, Thailand, and Vietnam) adds another dimension to the urban air pollution in terms of both emission and commuter exposure.

1.4.1.1.2 Other Mobile Sources

At present, most of the efforts to control mobile source emission are still targeted at on-road vehicles. Emission from off-road vehicles may be significant

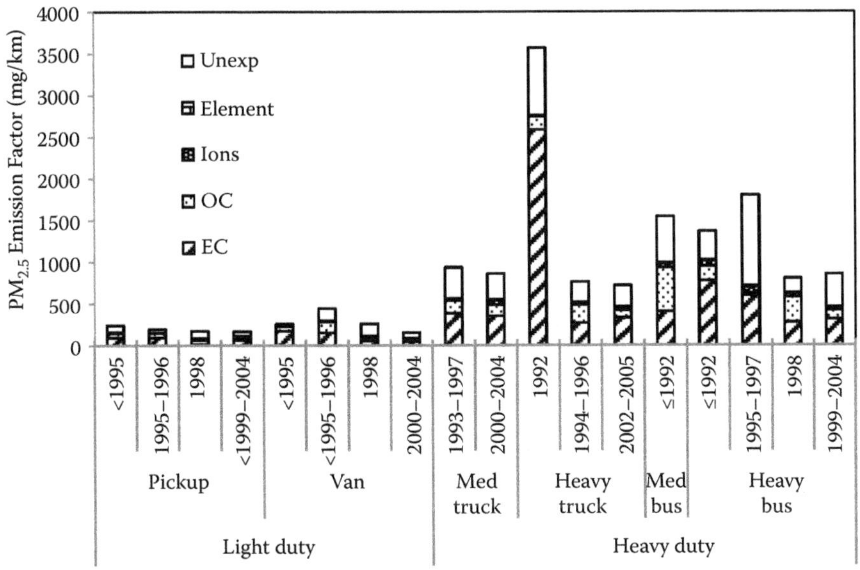

FIGURE 1.11
Emission factors and composition of $PM_{2.5}$ from different diesel vehicle groups in Bangkok. (From Kim Oanh N. T., Thiansathit W., Bond T. C., Subramanian R., Ekbordin W., and Ittipol P. 2010. Compositional characterization of $PM_{2.5}$ emitted from in-use diesel vehicles in Asia, *Atmospheric Environment*, 44(1): 15–22. Courtesy of AE, reprinted with permission.)

but remains largely unregulated. Thus, a range of polluting diesel-powered construction and agricultural machinery, and boats in some cities, continue to pollute the air. Ship emissions in ports and in international water are largely unregulated. Low quality of fuel oil used in marine transport may release a large amount of particles, including BC, NO_x, VOCs, and SO_2 (Corbett et al., 2007). Streets et al. (2000) estimated that shipping trade in Asia grew by an average of 5.4% annually between 1988 and 1995, and SO_2 emission grew by 5.9% annually, rising from 545 Gg in 1988 to 817 Gg in 1995. The SEA region, surrounded by sea, would be heavily affected by emissions from ships.

1.4.1.2 Stationary Sources

1.4.1.2.1 Industry/Power Plants

The heavy reliance on coal as a cheap fuel in power plants and in large-scale industrial boilers causes significant air pollution emissions in most Asian countries. The current control efforts are focused more on these large sources; hence, their emission is increasingly regulated. The trend to relocate these sources away from populated areas and to concentrate them in industrial estates can also reduce the exposure risk. Much less attention is being paid to small and medium industries/workshops (in many cases on the family scale), which are often found scattered in populated areas of developing Asia. These small manufacturing facilities are typically based on backward and heavily polluting facilities, but their emission is largely uncontrolled. In many cases, high concentrations of pollutants are observed locally, which lead to high risks of population exposure.

1.4.1.2.2 Domestic

Residential heating and cooking with dirty fuels are important pollution sources in Asian rural and periurban areas. In particular, high emission of products of incomplete combustion (PIC) including fine particles, CO, VOCs, semi-VOCs from residential combustion in Asia implies high exposure and health risks to a large population, as discussed in Section 1.3.2.

1.4.1.2.3 Other Stationary Sources

Other important stationary point sources include municipal solid waste incinerators and hospital waste (bioinfectious hazardous wastes) incinerators. The latter are popular, and many of them are of small scale, installed in local hospitals, and therefore would emit substantially if not properly managed. Often neglected is also the unregulated emission from backup diesel generators operated during power failures. Unstable power supply in many Asian cities prompts operations of numerous generators installed in hotels, supermarkets, institutions, and even private homes.

1.4.1.3 Open Burning

Biomass open burning is a large emission source of air pollution, especially during the dry season. Section 1.3.3 provided the information on SEA haze episodes that are caused by large forest fire events in the region. This section focuses more on other types of open burning of importance in Asia, namely, agricultural residue field burning and solid waste burning.

1.4.1.3.1 Agroresidue Field Burning

In Asia, crop residue field burning can contribute substantially to air pollution at the local, regional, and global scales. This field-burning activity, uncontrolled and with incomplete combustion, emits a large amount of toxic air pollutants (fine particles and inorganic and organic gases) and GHGs (Andreae and Merlet, 2001; Kanabkaew and Kim Oanh, 2010). The field-burning activity is projected to rise in the near future owing to not only the increased cultivation areas but also the improved living standard of farmers. Wealthier farmers do not anymore rely on the agroresidue as a source of energy, and for a faster rotation of crops, they prefer to burn agroresidue in the field after harvesting. The fraction of rice straw, which is by far the largest agricultural residue in Asia, subjected to field burning was estimated to be about 14% in India (Gadde et al., 2009). A survey conducted in the central part of Thailand revealed the field-burning fraction of rice straw was above 90% (Tipayarom and Kim Oanh, 2007).

In China, about 19–24% of crop residue is field burned (Cao et al., 2006; Yang et al., 2006), contributing approximately 100 Gg of BC and 396 Gg of OC to share 6.6% BC and 9.3% OC of the respective national total emission in 2000 (Cao et al., 2006). Emissions from field-burning activities deteriorate local air quality and have been reported to cause adverse health effects (Torigoe et al., 2000; Yang et al., 2006). Booming Asian cities are normally surrounded by agricultural land; hence, the field burning of agricultural wastes results in widespread ambient pollution not only in rural and periurban areas but also nearby urban areas. More details on the emission from rice straw field burning and the management practices are given in Chapter 13 of this book.

1.4.1.3.2 Solid Waste

In developing countries of Asia, municipal solid waste open burning is a major source of toxic air pollutants. Solid waste open burning is commonly done in three locations: in the backyard (at generation sites), at the transfer stations, and at the final disposal sites (landfill fires). The burning at source occurs more intensively in the periurban areas and in small towns due to insufficiency of solid waste collection services. In rural areas, solid waste is not normally burned as it is used for composting or is directly disposed to the environment. Smoldering conditions of the solid waste burning, especially for landfill fires, release substantial amounts of toxic air pollutants; however, these are not yet properly quantified. Lack of emission factors together with difficulty in estimation of the accurate amount of solid waste burned present a major obstacle to the emission quantification (Shrestha et al., 2011). Better solid waste collection and management in the future would minimize this source, but for the time being it still creates noticeable and considerable air pollution. In some countries, solid waste backyard burning is officially banned, but the effectiveness of enforcement is still an issue.

1.4.1.4 Other Sources

A range of other common sources typically present in Asian cities are street cooking, meat grilling, candle burning, and excessive burning of incense sticks and joss paper as a part of religious activities, as well as cremation in some places. In addition, evaporative VOC emission from gasoline stations may be substantial in many cities, especially if the vapor is not recovered during tank unloading and vehicle refueling. Road dust (resuspended from unpaved roads or dusty paved roads) and construction dust contribute substantially to coarse PM levels in Asian cities, as suggested by receptor modeling studies (Kim Oanh et al., 2006).

1.4.2 Sectorwise Emission Shares

National or subnational up-to-date source emission inventory (EI) is rarely available in Asia. To provide an overall view of the emission of different sectors in selected Asian countries, the national emission data have been extracted from the global EI for the year 2006, which is provided by the Center for Global and Regional Environmental Research (CGRER) (http://www.cgrer.uiowa.edu/EMISSION_DATA_new/index_16.html). The emission data for various cities in Table 1.4 have mainly been extracted from reported EI studies produced at AIT for dispersion modeling, which is discussed in Chapter 5 of this book. The city-level EI includes more detailed emissions from some important sources (e.g., the solid waste open burning and gasoline stations), which are not explicitly presented in the national emission data obtained from the Web site.

Relative contributions of various sectors to the national emission differ from one country to another. China, the country with the highest emission listed in Table 1.4, has a higher contribution from power plants and industry compared to other sectors like transportation and residential and biomass open burning (including forest fires and crop residue field burning). The residential sector contributes the largest emission in Bangladesh, Vietnam, Indonesia, India, and, to some extent, the Philippines. Industry has a higher contribution to the emission in Thailand and Philippines, while transportation has a high contribution in Vietnam and Thailand. In most urban metropolitan regions presented in Table 1.4, the largest emission is from traffic, as expected, especially for VOCs, CO, and NO_x. In the Dhaka division, residential and open burning still appear to be major PM emission sources (Iqbal and Kim Oanh, 2011).

1.5 Air Quality Management Practices

1.5.1 Integrated Air Quality Management

The integrated air quality management (AQM) program should be based on the principles of air quality objectives, the use of sound science, the analysis of mitigation options, and the involvement of all key organizations and stakeholders in the public decision-making process. The ultimate goal of AQM is to ensure that ambient air pollution concentrations do not exceed the defined target levels, and that human health and the environment are protected. The program requires a careful balancing of the cost of source control measures against health, environmental, and economic benefits. AQM issues also need to be prioritized within the context of other local influences on public health and the environment, such as the contribution of indoor air pollution and unsafe water and sanitation conditions. The main components of an AQM

TABLE 1.4

Annual Anthropogenic Emissions from Asian Countries in 2006 (Gg/year)

		Bangladesh		Thailand		Vietnam			Indonesia			
Pollutants	**China**	**National**	**Dhaka[a]**	**National**	**BMR[b]**	**National**	**Hanoi[c]**	**HCMC[d]**	**National**	**Jakarta[e]**	**Philippines**	**India**
NO_x	21,646	245	0.33	1,467	329	464	27	73	1,896	128	379	5,405
NMVOC[f]	25,938	888	0.58	3,571	269	1,945	27	223	8,225	141	2,313	12,979
CO	182,632	5,117	5.4	12,416	464	12,760	165	582	26,704	567	4,397	73,364
SO_2	31,100	157	0.41	1,327	240	401	4	165	1,499	17	956	5,669
BC	1,924	56	—	84	—	110	—	—	229	—	39	427
OC	3,940	286	—	394	—	465	—	—	1,247	—	206	1,536
PM_{10}	18,223	438	0.28	475	38	737	—	—	1,838	—	231	4,002
$PM_{2.5}$	13,266	391	—	388	—	657	—	—	1,610	—	199	3,111
National sectorwise emission (Gg/year)	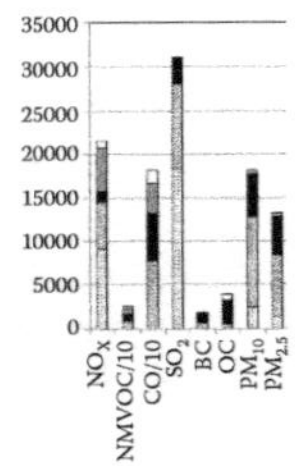	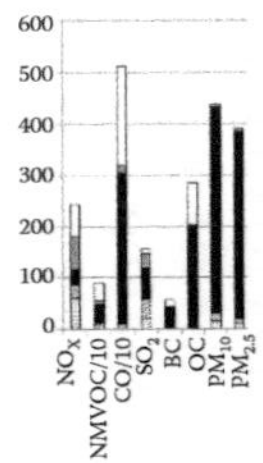		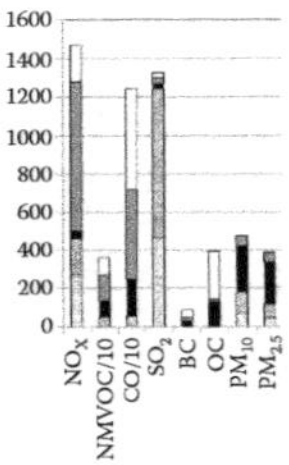		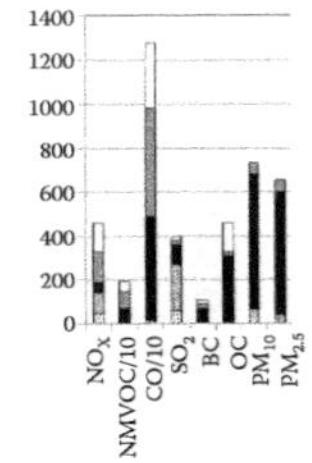			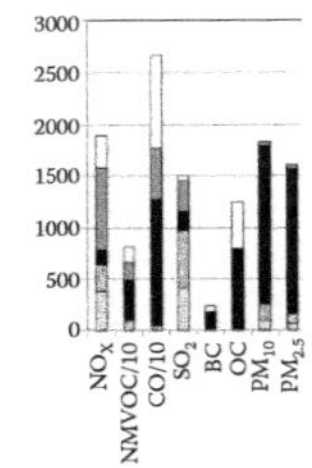		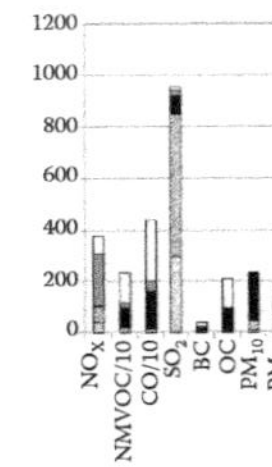	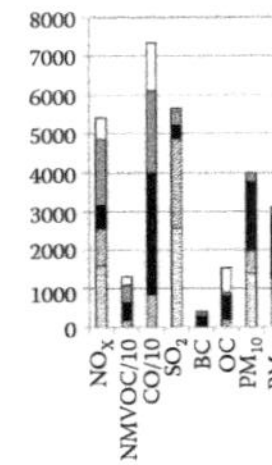
Legend	Power; Industry; Residential; Transportation; Biomass burning											

Source: National emission extracted from http://www.cgrer.uiowa.edu/EMISSION_DATA/index_16.htm; emission is for 2006. The urban emission was extracted from AIT studies.

Note: BMR, Bangkok metropolitan region; HN, Hanoi City; HCMC, Ho Chi Minh City.

[a] Dhaka division for year 2006 (Iqbal and Kim Oanh, 2011).

[b] Bangkok metropolitan region for 2000 (PCD, 2000).

[c] Hanoi metropolitan region for 2003 (An, 2005).

[d] Ho Chi Minh City for 2005 (Huy, 2008).

[e] Jakarta for 2003 (Permadi, 2007).

[f] NMVOC, non-methane volatile organic compounds (VOCs)

framework include the establishment of appropriate air quality goals or targets, development of clean air implementation plans, implementation and enforcement of efficient actions to realize the plans, and establishment of institutional arrangements to achieve the clean air goals. While recognizing that the specific process used may have to be adjusted or modified to fit local circumstances, there are several essential elements that need to be addressed in any integrated AQM process:

- Set the ambient air quality standards or guidelines (goal/target) and the time frame for attainment.
- Conduct monitoring to assess current air quality conditions in relation to various existing source categories.
- Develop and regularly update the emissions inventory.
- Apply air quality modeling tools to simulate the impacts of current and future emission on ambient air quality.
- Identify the emission reductions necessary to meet the desired air quality targets.
- Identify and prioritize the emission control measures in view of anticipated growth patterns, future emissions scenarios, and their cost effectiveness.
- Implement the emission control measures considering the command and control (CAC) and market-based instrument (MBI).
- Regularly revise the air quality goals.

The AQM is a dynamic process with the goal or target revised regularly. This provides a continuous process of air quality improvement. A number of coordinated activities have been taken in Asia to establish and improve AQM. One of the first initiatives was taken through the Urban Air Quality Management Strategy in Asia (URBAIR) project in the 1990s, resulting in the suggestion of action plans for Kathmandu, Jakarta, Manila, and Mumbai (Shah et al., 1997). Some other important coordinated initiatives in this field include Air Pollution in the Megacities of Asia (APMA, 2002) project, initiated in 2000 by UNEP and WHO, and the Clean Air Initiative for Asian Cities (http://cleanairinitiative.org/portal/index.php).

There has been a growing public awareness of the high concentrations of air pollution experienced in everyday life, especially in urban areas. Better understanding of the consequent adverse effects of high air pollution on health, the environment, and the economy has prompted national authorities to increase efforts to improve air quality in various countries and cities in Asia. Accordingly, some forms of the AQM system have been developed. Section 1.5.2 highlights current practices of AQM and related issues in developing Asia.

1.5.2 Current Practices and Issues of AQM in Asia

1.5.2.1 Ambient Air Quality Standards

Ambient air quality standards (AAQSs) are the legal limits of pollutants in the air during a stipulated time period. The AAQSs express the desired objectives or goals of ambient air quality that would protect public health and ecosystems. After setting the standards, a concrete action plan should be developed and implemented for the attainment within a specified time frame. Setting AAQSs is the first step of AQM as the standards define the required actions. Therefore, these legal limit values should be rationally selected, taking into account the natural conditions and current socioeconomic development status of a country. Hence, air quality goals or standards vary between countries. Within a country, the AAQSs also should change with time and be revised regularly. It is important to recognize that the standards are not the absolutely safe levels but just the acceptable air pollution levels under given circumstances. The standards are not necessarily to be the same as the air pollution criteria, which are based on dose-response relationships derived from the latest scientific research. The standards are also not necessarily the same as the WHO guidelines, which are developed to protect health and the environment and are largely based on the criteria taking into account uncertainty factors resulting from the insufficiency of the scientific data (WHO, 2006).

Asian nations have adopted a range of air quality standards (CAI-Asia, 2010). The NAAQS in most countries have been set for six common pollutants (CO, SO_2, NO_2, PM, O_3, and Pb) for both short-term and long-term exposure periods, but the values differ significantly from country to country. China and India have different standards set for different zones, accepting lower levels in the pristine or ecologically sensitive areas and higher levels in other areas. Thailand also has a more lenient 1-h SO_2 standard for an industrial area. Table 1.5 summarizes the NAAQS of some Asian countries for PM, ozone, and Pb. TSP and PM_{10} are currently regulated in most Asian countries, while several countries (Bangladesh, China, India, Japan, Singapore, and Thailand) also have standards for $PM_{2.5}$. Having the $PM_{2.5}$ level regulated is certainly a step forward, but intensive resources should be made available for its monitoring to assess attainment.

The majority of Asian countries have more lenient standards than the WHO guidelines, especially for PM. In addition, NAAQS do not include exposure periods less than 1 h; for example, WHO guidelines for CO are available for short-term exposure starting from 15 min (to 8 h), while most countries have 1-h and 8-h standards; a few countries (China, Vietnam, Indonesia) also set a 24-h CO standard. Similarly, the WHO guidelines for SO_2 are given on short-term exposure (i.e., 10 min and 24 h), while the NAAQS are generally set for 1 h, 24 h, and 1 year. Ozone standards are generally available for both 1-h and 8-h exposure periods, but a few countries still have only the hourly standards.

Several countries also have standards set for other pollutants. Thailand, recognizing the importance of VOC pollution, has since 2007 promulgated annual

TABLE 1.5

Ozone (ppb) and Particulate Matter (μg/m³) Ambient Air Quality Standards in Selected Asian Countries

Pollutants		Ozone (O_3), ppb[a]		TSP, μg/m³			PM_{10}, μg/m³			$PM_{2.5}$, μg/m³		Lead (Pb), μg/m³		
Averaging time		1 hour	8 hour	1 hour	24 hour	Annual	1 hour	24 hour	Annual	24 hour	Annual	24 hour	3 month/ seasonal	Annual
Bangladesh		120	80	—	—	—	—	150	50	65	15	—	—	0.5
China[b]	Grade I	75	47	—	120	80	—	50	40	35	15	—	1.0	0.5
	Grade II	93	75	—	300	200	—	150	70	75	35	—	1.0	0.5
India[c]		90	50	—	—	—	—	100	60	60	40	1	—	0.5
Indonesia		120	—	—	230	—	—	150	—	—	—	2	—	1
Philippines		70	30	—	230	90	—	150	60	—	—	—	1.5	1
Thailand		100	70	—	330	100	—	120	50	50	25	—	1.5[d]	—
Vietnam		90	60 (40)[e]	300	200	140	—	150	50	—	—	1.5	—	0.5
Japan		60[f]	—	—	—	—	200[g]	100[g]	—	35	15	—	—	—
United States[h]		—	75	—	—	—	—	150	—	35	15	—	0.15	—
WHO guidelines		—	50	—	—	—	—	50	20	25	10	—	—	0.5

[a] Approximately converted to parts per billion if NAAQS are given in micrograms per cubic meter.
[b] China: NAAQS released in 2012, Grade I is for natural protection area and other areas which need special protection; Grade II is for residential, commercial, industrial, and rural areas.
[c] India has two levels of standards (ecologically sensitive areas and other areas). The differences are only for SO_2 and NO_2.
[d] Monthly standard.
[e] The 24-h ozone standard is given in parentheses (40 ppb).
[f] Photochemical oxidants (e.g., ozone and peroxyacetyl nitrate) produced by photochemical reactions.
[g] Defined as suspended particulate matter (SPM) with size less than 10 μm.
[h] USEPA-designated 2008 standards.

standards for nine VOC compounds and conducts monitoring for the pollutants regularly in industrial and urban sites (PCD, 2007). China has set standards for B*a*P and for several heavy metals (GB 3095, 2012) while India has daily and annual standards for NH_3 and annual standards for benzene, particulate B*a*P, arsenic, and nickel (http://cpcb.nic.in/). Japan also has NAAQS for several VOCs and dioxins (http://www.env.go.jp/en/).

In most countries, the NAAQS are periodically updated to ensure high stringency and to include more air pollutants. On the regional scale, harmonization of NAAQS of Asian countries is desirable, especially for transboundary pollutants, but is perhaps the most challenging task. Many countries in the region have comprehensive sets of NAAQS documented, but the generation of suitable monitoring data for the attainment assessment remains a major issue because this activity requires large resources. Insufficiency of air quality monitoring data, especially those with necessary quality assurance/quality control (QA/QC), is most problematic for effective AQM in many Asian developing countries.

1.5.2.2 Monitoring Ambient Air Quality

A monitoring system that generates ambient air quality data with relevant QA/QC is an important technical tool of AQM. Together with EI and air quality modeling, monitoring provides and documents the air quality information. Early in the 1980s, many Asian countries still relied on manual monitoring methods that mainly collected air quality data at temporary sites over short sampling periods. Often, the data obtained were fragmented and not harmonized due to diverse measurement methods deployed. Since the beginning of the twenty-first century, air quality monitoring capabilities in Asian developing countries have been improved significantly. Many countries and cities now have complex air quality monitoring networks equipped with automatic analyzers that provide online measurements for a number of air pollutants (Table 1.6). These expensive and sophisticated monitoring networks commonly have been started through international collaborations with developed countries. The networks are expected to generate timely data and information for efficient AQM actions, such as assessing the attainment of the NAAQS so that appropriate measures can be taken in time to reduce emission or exposure. Recently, the online display electronic boards and Web sites are also made available for several urban areas to provide the public access to air quality information. However, the availability of modern monitoring equipment does not always guarantee the generation of data with acceptable quality. In some cases, the monitoring networks do not function properly; hence, either no data or no reliable data (without necessary QA/QC) are produced. The cost of operation and maintenance of monitoring stations, lack of spare parts, poor capacity for calibration and data management, and low commitment are some of the major reasons for the malfunction

TABLE 1.6

Summary of National Air Quality Monitoring Networks in Selected Asian Countries

Country	Number of Automatic Stations	Major Pollutants	Other Pollutants [a]	National Authority	Internet Sources of Information
China	688 (Beijing 27)	SO_2, NO_2, PM_{10}, O_3, CO_2 and meteorology	Dust and sandstorm	City Environmental Monitoring Center (City EMC)	http://www.bjepb.gov.cn
Bangladesh	4 (Dhaka 1)	PM_{10}, $PM_{2.5}$, SO_x, NO_x, CO, and meteorology		Department of Environment	http://www.doe-bd.org
India	342 (Chennai 6)	SO_2, NO_x, TSP, O_3, CO, PM_{10}, and meteorology	PAHs, heavy metals	Central Pollution Control Board (CPCB)	http://www.cpcb.nic.in/Network.php
Indonesia	33 (Jakarta 5, Bandung 5)	NO_2, SO_2, PM_{10}, CO, O_3, and meteorology		Ministry of Environment (MOE)	http://www.menlh.go.id
Thailand	56 (Bangkok 17)	PM_{10}, CO, SO_2, NO_x, and O_3 and meteorology	VOC, $PM_{2.5}$	Thailand Pollution Control Department (PCD)	http://www.aqnis.pcd.go.th/station/allstation.htm http://www.pcd.go.th/indexEng.cfm
Philippines	10 (Manila 1)	PM_{10}, SO_2, CO, NO_2,O_3, and meteorology	$PM_{2.5}$, methane, NMHC, BTX	Department of Environment and Natural Resources (DENR)	http://www.adb.org/Documents/Reports/Urban-Air-Quality-Management/philippines.pdf
Vietnam	20 (Hanoi 7, HCMC 9)	PM_{10}, CO, SO_2, NO_2, NO_x, O_3, NO, and meteorology		Ministry of Natural Resources and Environment (MoNRE)	http://www.urbanemissions.info/model-tools/sim-air/hanoi-vietnam.html

[a] Other pollutants may be monitored at other stations and with different frequencies.

of several networks. In fact, a few of them were terminated following the phasing out of external supports.

A number of dense monitoring networks, in terms of population coverage, are currently available in some countries in developing Asia. However, existing monitoring networks still mainly focus on urban areas at a few "hot spots" where the primary pollutants levels are the highest. Secondary pollutants, such as ozone, which are normally built up at downwind areas away from city centers, are not properly characterized by these networks. Also, data for hazardous air pollutants and $PM_{2.5}$ are still fragmented and mainly made available through short-term research projects. To provide a better overview of air pollution with a large geographical coverage, satellite data can be applied as a low-cost option. The potential use of the available satellite data for air quality monitoring with associated advantages and limitations are discussed in Chapter 4 of this book.

Development of air quality monitoring networks to generate air quality data with QA/QC should be one of the highest priorities of integrated AQM in Asian developing countries. Such data, once generated, should be used meaningfully for AQM activities. The quality of the data should be seen as a priority so that the effective use of the data can be proven and the enormous resources spent to generate such data are justified. It is wasteful to generate poor-quality data that are of no use, which in turn discourages investment to improve monitoring networks.

1.5.2.3 Emission Inventory and Air Quality Modeling

Emission inventory is a quantitative assessment of emission loads from relevant sources or sectors and is the basis for identifying the most important emission sources to prioritize pollution control efforts. Further, an EI with adequate spatial and temporal distributions is required to prepare the input data for air quality dispersion models. The modeling tool is used to estimate the ambient air concentrations and to analyze the anticipated effects of future emission control options. Development of EI is always a challenging and resource-consuming activity of AQM. National specific and operational EIs are not available in most of the Asian developing countries, as highlighted in Section 1.4. In some instances, EIs are made available through research projects, but they are not regularly updated as required to reflect the changes in the emission sources with time. These temporal emission changes are especially significant in developing countries because of booming economic activities.

There are a few regional and global emission databases available at present, such as the CGRER, Emission Database for Global Atmospheric Research (EDGAR), Regional Emission Inventory in Asia (REAS), and Greenhouse Gas and Air Pollution Interactions and Synergies (GAINS), which are downloadable from the respective Web sites, as detailed in the work of Kim Oanh and Permadi (2009). These data are useful and are frequently applied in various regional or global air quality and climate modeling attempts. However, they present the annual estimate, available on coarse spatial resolutions, and are not regularly updated. To be applicable for urban and national AQM purposes, EIs should be prepared using detailed activity data and relevant emission factors for typical sources in given areas. Such EI databases should have adequate QA/QC, be traceable, and be updated.

The receptor modeling approach to the source apportionment studies is quite popularly applied in air quality research in Asia (Hopke et al., 2008; Kim Oanh et al., 2009, and references therein). This is particularly useful for identifying the major source categories that contribute to ambient pollutants, commonly PM and VOCs. Receptor modeling results are relevant to a specific site and over the period that the ambient data have been generated for the model input. The receptor modeling results can supplement but not substitute for the EI database. Neither can they be used to analyze impacts of future emission changes on air quality directly, which instead requires the dispersion modeling tool. However, in most Asian developing countries the capacity for operating complex 3D dispersion air quality models is still mainly confined to universities and research institutions. Many such attempts used the emission input data extracted from the global emission database available online. Further discussion of air quality modeling for various urban and regional domains is given in Chapters 5 and 6 of this book.

1.5.2.4 Source Emission Control

Emission reduction is the key action to improve air quality. Asian developing countries have begun to take measures to reduce emissions from specific sources, most notably motor vehicles, factories, and power plants. Emission standards are set for mobile sources and stationary sources; however, the effectiveness of enforcement varies widely between countries. The capacity of source monitoring for enforcement has generally been improved in the region. Major emission sources in Asian developing countries were highlighted in Section 1.4, which lists a range of issues related to superemitters in traffic fleets, open burning, or small and medium polluting industries. This section only focuses on some current emission control efforts in developing Asia.

1.5.2.4.1 Traffic Emission

To improve air quality, Asian countries have developed road maps that define the progressive implementation of the European vehicle emission standards (EURO) for new vehicles. Accordingly, by 2016 the new light-duty vehicles sold in Asia should at least have met the standards equivalent to EURO2, but the majority of countries impose more advanced technologies, such as EURO3 or above (CAI-Asia, 2011). In Thailand, starting from 1998 new heavy-duty diesel vehicles registered in the country had to meet the EURO1 standard. The EURO2 standard was implemented in 2003, and from 2007 new heavy-duty diesel vehicles had to meet the EURO3 standard (PCD, 2009). In China, nationwide, the new light-duty vehicles had to meet EURO1 in 2000 and gradually advance to EURO4 by 2010. Similarly, the new heavy-duty vehicles had to meet the EURO1 standard starting in 2000 and progressively reach the EURO5 standard in 2012. Large Chinese cities such as Beijing, Shanghai, and Guangzhou promulgate municipal emission standards more stringent than the national standards (Beijing Environmental Protection Bureau [BJEPB], 1993–2009). Vietnam has a plan to adopt EURO4 from 2017 and EURO5 from 2022 for the vehicle fleet (CAI-Asia, 2011). The emission standards are also imposed for in-use vehicles in developing Asia, but the effectiveness of inspection and maintenance programs continues to be an issue.

Fuel switching is currently promoted in developing Asia (e.g., switching to compressed natural gas [CNG] for public transportation). In New Delhi, the entire public transportation fleet (buses, taxis, and auto rickshaws) was successfully converted to CNG-powered in 2003, and by 2005 the city fleet consisted of almost 90,000 CNG vehicles (Reynolds and Kandlikar, 2008). Switching from diesel-powered to CNG would reduce the emission of primary particles including BC, hence bringing an air quality and climate cobenefit. Thailand widely promotes CNG and ethanol fuels (Energy Sector Management Assistance Programme [ESMAP], 2008); consequently, the share of vehicles running on the alternative fuels grows rapidly. By 2010 in BMR, the majority (>95%) of taxis and Tuk-Tuk used CNG/LPG or dual fuel (gasoline and CNG), while for other types of vehicles the shares of CNG were lower but already considerable (i.e., about 40% of buses and 14% of private cars) (Department of Land Transport [DLT], 2009). Other Asian developing countries with abundance of agricultural products also promote biofuels (ethanol and biodiesel) to partially replace fossil fuel in the transport sector with an increasing trend of production and use (Gonsalves, 2006).

Regulations on fuel quality are imposed in most Asian countries. After the successful phase-out of leaded gasoline, the regulation now focuses on sulfur and benzene content. In Thailand, sulfur content in diesel fuel has been reduced steadily; for example, 500 ppm was enacted in July 1998, 350 ppm was enacted in January 2004, and 50 ppm sulfur was proposed for 2012 (ESMAP, 2008; CAI-Asia, 2011). The levels of aromatics, including benzene, in gasoline are also gradually reduced (Warapetcharayut et al., 2007). The road maps for sulfur content reduction have been prepared by Asian countries, but the proposed levels are still considerably high (CAI-Asia, 2011) (i.e., as high as 3,500–7,000 ppm in some countries for the target year 2010, although the majority already had plans to reduce the sulfur content to 500 ppm or below by 2012). This high sulfur content continues to be a barrier to retrofitting air pollution control devices onto existing vehicles or applications of advanced vehicle technologies.

The most notable improvement in air quality has been reported for ambient Pb levels following the phase out of leaded gasoline in Thailand (Supat, 1999), Vietnam (Long, 2006), and Indonesia (Chapter 15 of this book). Replacement of two-stroke engines by four-stroke engines for motorcycles and three-wheelers in Thailand (Warapetcharayut et al., 2007) or the ban of dirty two-stroke three-wheeler "baby taxis" in Dhaka (State of Environment, Bangladesh [SOE-BLD], 2005) are some examples of effective AQM measures. However, further efforts are required in association with integrated traffic management to reduce vehicle driving, shorten trips, and reduce congestion eventually. The rapid growth of vehicle fleets in many Asian countries is likely to offset some emission reduction brought about by the current measures. In addition, most of the control efforts are still for on-road vehicle emissions, while the off-road mobile sources are often overlooked.

1.5.2.4.2 Stationary Sources

Relocation of industries away from populated urban areas and into industrial estates is being imposed in Asian countries (e.g., China, Thailand, and Vietnam). This may be an effective measure to reduce pollution levels and the exposure risk in populated urban areas. However, this merely redistributes emission sources, so would not be as effective at the regional scale. Current technology intrusion progressively introduces cleaner combustion technologies in large industries and power plants in developing Asia. Such technology intrusion does not happen at the same pace in small and medium industries; they continue to pose problems for local and regional AQM.

Use of cleaner fuel for industrial combustion (i.e., switching from coal-based to NG-based boilers) is implemented in China, where about 16,000 coal-fired space-heating boilers and other appliances have been switched to NG or are being eliminated. Flue gas control devices are also installed in the existing industries to reduce emissions at the source. Beijing has implemented a number of initiatives to improve the air quality (BMG, 1998–2008), such as relocating and retrofitting major polluting industries and implementing more stringent vehicle emission standards. A Web survey of 13 thermal power plants in Beijing showed that about 50% are coal-based (pulverized coal). Other plants mainly use NG (NG combined cycle), except for one plant that uses biogas and a production unit of another plant that uses diesel oil. The coal-based power plants are mainly equipped with wet scrubbing systems for desulfurization and ESPs (electrostatic precipitators) or baghouses for PM control. The majority of plants are equipped with NO_x removal devices ($DeNO_x$), including SCR (selective catalytic reduction), SNCR (selective noncatalytic reduction), or low-NO_x burners. One coal-fired power plant uses the electron beam ammonia (EBA) process for simultaneous desulfurization and $DeNO_x$. Among the surveyed facilities, only one coal-fired plant applied CO_2 capture.

In many parts of developing Asia, the available emission control techniques for stationary sources still focus more on easy-to-treat pollutants that can be controlled at a relatively low cost. Thus, deployed control devices are commonly more efficient for removal of coarse PM rather than fine PM and more for SO_2 removal than for NO_x removal. VOC emission control has yet to receive adequate attention. Therefore, problems such as those imposed by fine particles would remain not only because of the increase in the emission of primary PM but also because of the formation of secondary particles in photochemical reactions involving VOC and NO_x emissions. Ozone, a major product of these photochemical reactions, is also expected to rise.

Several other sources, such as open burning of solid waste and agricultural residue, are largely overlooked in AQM in many countries and therefore need more rigorous management measures. The residential sector remains a major contributor in some countries (see Table 1.4), but the emission reduction measures are not yet widely implemented. Cleaner domestic combustion would improve indoor air quality and at the same time reduce the pollution loads to the local and regional environment.

1.5.3 Integrated AQM: Examples of Good Practices

Integrated AQM involves applications of a combination of legal, technical, and economic instruments. The AQM should be comprehensive and should aim to facilitate the setting of air quality priorities and support the development of capability in the relevant institutions. There are a few successful examples of AQM, such as the phasing out of leaded gasoline in most Asian countries or the ban of dirty vehicles in some urban areas. A combination of CAC, MBI, and educational measures are commonly used in these cases to produce almost immediate and measureable improvement of ambient air quality, such as the observed drastic drops in Pb and PM pollution. In particular, the ban of (about 50,000) dirty two-stroke three-wheeler baby taxis in Dhaka, Bangladesh, in January 2003 reduced the biweekly average PM_{10} by 27% (from 200 to 146 μg/m^3) and $PM_{2.5}$ by almost 40%, from 88 to 52 μg/m^3 (SOE-BLD, 2005).

Other management measures, such as progressive enforcement for cleaner vehicles and fuel technologies, would require an extended time period to assess the impacts on the ambient air quality. In some cases, a reduction in primary pollutants may result in an increase in secondary pollutant (ozone, secondary particles) levels; hence, the complexity of the air pollution phenomena should be considered in development of emission reduction strategies. Case studies of integrated AQM strategies for target sources in Asia (rice straw field burning, industrial VOC emission, traffic emission, and brick manufacturing) are presented in Chapters 13–16 of this book.

The integrated AQM actions taken in Beijing in preparation for the 2008 Olympic games improved the air quality drastically (BJEPB, 1993–2009). Principally, the hierarchical AQM strategy in Beijing consists of long-term control measures, contingency control measures, and temporary control measures (BMG, 1998–2008). The long-term control measures include (1) end-of-the-pipe emission control for point and line sources; (2) fuel shift from coal to NG, geothermal, wind, and solar energy; (3) energy conservation campaign; (4) economic infrastructure reform to reduce the proportion of polluting industries, and (5) ecological planning or construction for green coverage to reduce fugitive dust from bare land and material recycling to

TABLE 1.7

Average Daily (24-h) Levels of Major Air Pollutants (μg/m³) in Beijing during the Olympics Periods in 2008

Periods	PM_{10}	SO_2	NO_2	CO
August 8–24 (Olympics)	57	8	23	800
September 6–17 (Paralympics)	71	12	34	1,000
24-h NAAQS, urban areas	150	150	80	—

Source: Beijing Environmental Protection Bureau. 1993–2009. *Beijing Environmental Quality Communique for Various Years*, Beijing Environmental Protection Bureau, Beijing.

reduce emission from waste incineration. The contingency measures are reserved for the case of episode occurrences due to stagnant meteorological conditions. The temporary control measures are extra steps taken for further pollution control, such as those implemented during the period of the Olympics (Olympic Games and the Paralympic Games in 2008), which have been shown to improve urban air quality significantly (Table 1.7). During the period of more than 2 months, nearly one-half of vehicles were kept off-road, and the most polluting plants in the Beijing, Tianjin, and Hebei region were shut down. During the Paralympics, a slight relaxation was allowed that permitted private cars to drive without restriction outside the Sixth Ring Road. Accordingly, during this period some elevation of air pollutant levels was observed (Table 1.7). After the games finished, however, most of the temporarily closed industrial plants and 80% of the vehicles resumed their normal activities (BJEPB, 1993–2009). Despite the improvement of the air quality during the Olympics period, the annual average PM_{10} level for 2008 in the city still exceeded the annual NAAQS.

1.5.4 Integration of Air Quality Improvement and Climate Strategies

The interaction between air pollution and climate has been a topic of growing interest. Some traditional pollutants (e.g., BC particles and tropospheric ozone) are also recognized as strong, short-lived climate forcers (IPCC, 2007); hence, the reduction of their atmospheric levels will also result in reduction of climate forcing. Carbon dioxide, the major GHG of interest, is chiefly emitted from fuel combustion, as are the majority of primary air pollutants. Therefore, emission reductions by improving energy efficiency, energy saving, and use of renewable energy are potential co-control measures that benefit both air quality and climate. Quantification of the cobenefit is still a challenge as it involves a comprehensive assessment of the emission, the resulting atmospheric concentrations, the consequent effects on the climate and air quality, and finally the effects on human health and the ecosystem. Much uncertainty remains, and not all those processes and their links are completely understood.

Aunan et al. (2004) suggested that measures for GHG reductions often have other benefits, especially in air quality, for China. Most of this synergy is a result of energy system structural changes, fuel switching, and energy efficiency improvements, which all influence air pollutant emissions. Air pollution control in Beijing, for example, has triggered a series of actions with the climate and ecological cobenefits from carbon emission reduction (CDM, 2004). The renewable energy sources, such as hydropower, wind, geothermal, and solar energy, are becoming popular now in Beijing. Afforestation efforts in Beijing for fugitive dust mitigation also resulted in carbon sequestration and biodiversity in the city. Similar actions in other parts of Asia are also expected to result in climate cobenefit.

The "Integrated Assessment of Black Carbon and Tropospheric Ozone" by UNEP-WMO (2011) suggests that control of BC particles and tropospheric ozone through rapid implementation of proven emission reduction measures would have immediate and multiple benefits. The short-lived climate forcers (SLCFs: methane, BC, and ozone) remain in the atmosphere for only a relatively short time, so are fundamentally different from the long-lived GHG. Reductions in near-term warming can be achieved by controlling the SLCFs, whereas CO_2 emission reductions, beginning now, are required to reduce the long-term climate forcing. However, the targeted sectors/sources for the emission reductions are different. Major CO_2 reduction strategies mainly target the energy and large industrial sectors and therefore would not necessarily result in significant reductions in emissions of BC or the ozone precursors. In contrast, significant reduction of the SLCFs can be achieved for a range of small sources, such as residential cookstoves, landfills, livestock, agroresidue field burning, and others. Early adoption of specific practical emission reduction measures for methane (an ozone precursor) and BC, with coemittants, would have important "win-win" benefits for near-term climate, human health, agriculture, and the cryosphere. These measures would reduce projected global mean warming by around 0.5°C by 2050, avoid 0.7 to 4.7 million annual premature deaths from outdoor air pollution, and increase annual crop yields by 30 to 135 million tonnes due to ozone reductions in 2030 and beyond (Shindell et al., 2012). The greatest health and crop benefits would be felt in Asia, where most of the reduction measures are expected to take place.

The current awareness of the serious effects of air pollution on health and the economy has already prompted developing countries in Asia to establish and implement various forms of AQM systems. Many Asian countries have also started formulation of climate policies to cope with the climate change and its consequences. However, the AQM programs in the region are still largely concentrated on local emission reduction, and the cobenefit of climate-forcing mitigation is not yet adequately considered. It is important to integrate multiple objectives in the policy planning through the cobenefit approach, which takes advantage of synergies between local air pollution problems and the regional/global consequences. The cobenefit approach provides a win-win opportunity and is expected to be more cost effective than conventional approaches.

1.6 Summary

Asian cities are diverse in demography and geography but are generally under the influence of the monsoon climate. As a consequence, the division into dry and wet seasons is useful for air quality characterization purposes. The most significant ambient air pollutants in Asian developing countries are PM, followed by surface ozone, which often exceed the NAAQS and WHO guidelines in large urban areas. Current efforts have already resulted in stabilization of air pollution levels in some cities, and in a few instances a reducing trend of the coarse PM pollution has been observed. Fragmented data showed high levels of fine PM, ozone, and other toxic air pollutants such as VOCs, semi-VOCs (PAHs, pesticides), in developing Asia, but available databases are not yet sufficient for a comprehensive assessment. Indoor air pollution remains a critical issue in many areas in Asia, especially in poor rural areas, where dirty solid fuels are burned in poorly ventilated kitchens or homes. Regionally, important phenomena of the transboundary transport of air pollution, including the SEA haze, ABC, dust storm, and acid deposition, persistently exist, so collaborative efforts are required for effective management.

Traffic is the major pollution source in urban areas of Asian developing countries, and the fast growth of fleets may offset the gain made through implementation of cleaner vehicles and fuels, especially in highly congested cities. On the national scale, combustion in industry and power plants contributes significantly to air pollution in some countries, while the residential sector appears to be a major contributor in other countries. Open field burning of crop residue and solid waste contributes significantly to ambient air pollution in periurban and rural areas but is often overlooked by current AQM systems.

Thanks to the growing awareness on adverse effects of air pollution to human health, crops, and the environment, many developing Asian countries have already established and implemented some forms of AQM systems. Significant progress with some successful AQM practices has been recorded, most notably the phase-out of leaded gasoline. However, considerable challenges to improving air quality in Asia remain. These include, for example, improvement of monitoring networks and timely use of air quality data and information for AQM, development of EI and operational modeling capacity, and rigorous enforcement of appropriate control measures to bring down the source emission. The foremost task is the development of necessary human resources to conduct the AQM activities.

The integration of air quality and climate policies is possible yet challenging, as it requires comprehensive and coordinated development in both areas. The international development finance and climate change negotiations may provide funding opportunities for Asian developing countries, through the cobenefit approach, to reduce both air pollution and climate forcer emissions.

References

Air Pollution of Mega Cities in Asia (APMA). 2002. *Urban Air Pollution Management and Practice in Major and Mega Cities of Asia*, edited by Haq G., Han W. J., and Kim C., Korea Environment Institute, Seoul, Korea.

Acid Deposition Oxidant Research Center (ADORC). 2006. *Tropospheric Ozone a Growing Threat*, Acid Deposition Oxidant Research Center, Niigata, Japan.

Alegria H. A., Bidleman T. F., and Shaw T. J. 2000. Organochlorine pesticides in ambient air of Belize, Central America, *Environmental Science and Technology*, 34: 1953–1958.

An D. D. 2005. Photochemical smog modeling for air quality management in the Hanoi metropolitan region, Vietnam. Master's thesis, Asian Institute of Technology, Pathumthani, Thailand.

Andreae M. O. and Merlet P. 2001. Emission of trace gases and aerosols from biomass burning, *Global Biogeochemical Cycles*, 15: 955–966.

Association of Southeast Asian Nations (ASEAN). 2003. ASEAN HAZE agreement to enter into force on 25 November 2003. ASEAN Secretariat, Jakarta, Indonesia. http://www.aseansec.org/15129.htm

Asian Development Bank (ADB). 1999. *Planning for Fire Prevention and Drought Management Project*, Asian Development Bank, Jakarta, Indonesia.

Aunan K., Fang J., Vennemo H., Oye K., and Seip H. M. 2004. Co-benefits of climate policy—lessons learned from a study in Shanxi, China, *Energy Policy*, 32(4): 567–581.

Beijing Environmental Protection Bureau (BJEPB). 1993–2009. *Beijing Environmental Quality Communique for Various Years*, Beijing Environmental Protection Bureau, Beijing. Retrieved from http://www.bjepb.gov.cn/

Beijing Municipal Government (BMG). 1998–2008. BMG public notice on air pollution control measures and tasks, Stage 1–Stage 15, Beijing Municipal Government. http://www.bjepb.gov.cn

Buesing A. M. C. 2001. Comparative study of monitoring methods for airborne organochlorine pesticides in tropical region, master's thesis, Asian Institute of Technology, Pathumthani, Thailand.

CAI-Asia. 2010. Ambient Air quality standards in Asia survey report. CAI-Asia Center. http://www.cleanairnet.org/caiasia/

CAI-Asia. 2011. Road Map to Cleaner Fuels and Vehicles in Asia. CAI-Asia fact sheet No.17. CAI-Asia Center, Manila, Philippines. Accessed, 2012 from http://cleanairinitiative.org/portal/sites/default/files/documents/16_Roadmap_to_Cleaner_Fuels_and_Vehicle_in_Asia_0.pdf

Cao G., Zhang X., and Zheng F. 2006. Inventory of black carbon and organic carbon emissions from China, *Atmospheric Environment*, 40: 6516–6527.

CDM. 2004. *Clean Development Mechanism in China* (2nd ed.), *Taking a Proactive and Sustainable Approach*, World Bank, New York.

Center for Global and Regional Environmental Research (CGRER). 2006. Emission data. Retrieved from http://www.cgrer.uiowa.edu/EMISSION_DATA_new/index_16.html

Central Pollution Control Board (CPCB). 2011. Environmental Standards. Retrieved from http://cpcb.nic.in/

Chan C. K., and Yao X. 2007. Air pollution in mega cities in China, *Atmospheric Environment*, 42: 1–42.

Chang J., Sparks L., Guo Z., and Fortmann R. 1998. Evaluation of sink effects on VOCs from a latex paint, *Journal of Air and Waste Management Association*, 48: 953.

Corbett J. J., Winebrake J. J., Green E. H., Kasibhatla P., Eyring V., and Lauer A. 2007. Mortality from ship emissions: a global assessment, *Environmental Science and Technology*, 41: 8512–8518.

Department of Land Transport (DLT). 2009. Number of vehicle registered in Thailand as of 31 May 2009, Department of Land Transport. http://www.dlt.go.th.

Dockery D. W., and Pope C. A., III. 1994. Acute respiratory effects of particulate air pollution, *Annual Revision Public Health*, 15: 107–132.

Driejana I. R. 2002. Initial analysis of the ambient ozone concentration in the city of Bandung, Indonesia. Paper presented at the regional workshop on Better Air Quality in Asia and Pacific Rim cities. December 16–18. Hong Kong.

Driejana I. R. 2007. Identifying gaps and needs in the development of a nation air quality management policy in Indonesia. In *Proceedings of the International Conference on Air Quality Management in Southeast Asia*, Ho Chi Minh City, Vietnam, November 21–23, 2007. Canadian International Development Agency (CIDA)-Asian Institute of Technology (AIT), Bangkok. Thailand. ISBN 978-974-8257-45-7.

Emberson L. D., Bueker P., Ashmore M. R., Mills G., Jackson L., Agrawal M., Atikuzzaman M. D., Cinderby S., Engardt M., Jamir C., Kobayashi K., Kim Oanh N. T., Quadir F., and Wahid A. 2009. Dose-response relationships derived in North America underestimate the effects of ozone (O_3) on crop yields in Asia, *Atmospheric Environment*, 43: 1945–1953.

Energy Sector Management Assistance Programme (ESMAP). 2008. *Development Integrated Emission Strategies for Existing Land-Transport (DIESEL) Project*, Joint UNDP/World Bank Energy Sector Management Assistance Programme. Bangkok, Thailand.

Framework Convention on Tobacco Control in the South-east Asia Region, SEA/RC53/14 C. F. R. (2000).

Gadde B., Bonnet S., Menke C., and Garivait S. 2009. Air pollutant emissions from rice straw open field burning in India, Thailand and the Philippines, *Environmental Pollution*, 157: 1544–1558.

GB 3095. 2012. Ambient Air Quality Standards of China. Available and accessed 2012 from http://kjs.mep.gov.cn/hjbhbz/bzwb/dqhjbh/dqhjzlbz/201203/t20120302_224165.htm

Giang N. T. H. 2008. Roadside air quality in Ho Chi Minh City and vehicle emission estimation by back calculation, master's thesis, Asian Institute of Technology, Pathumthani, Thailand.

Godish T. 1991. *Indoor Air Pollution Control*, (3rd ed.), Lewis, Boca Raton, FL.

Godish T. 1997. *Air Quality* (3rd ed.), Lewis, Boca Raton, FL.

Gonsalves J. B. (2006). An assessment of biofuels industry in Thailand. United Nations Conference on Trade and Development. UNCTAD/DITC/TED/2006/7. Available online at 203.150.231.127/~arda/pdf/An%20Assessment.pdf

Gupta I., and Kumar R. 2006. Trends of particulate matter in four cities in India, *Atmospheric Environment*, 40: 2552–2566.

Hanoi People Committee (HPC). 2008. Hanoi Police Web site, Hanoi People Committee. Accessed 2011 from http://www.anninhthudo.vn/Xa-hoi/De-cao-viec-xu-ly-nghiem-va-triet-de/338099.antd

Health Effects Institute (HEI). 2010. *Outdoor Air Pollution and Health in the Developing Countries of Asia: A Comprehensive Review*, Health Effects Institute, Boston.

Hoff R., Mulr D., and Grift N. 1992. Annual cycle of PCBs and organohalogen pesticides in air in South Ontario. 1. Air concentration data, *Environmental Science Technology*, 26: 2266–2275.

Hopke P., Cohen D., Begum B., Biswas S., Ni B., Pandit G., Santoso M., Chung Y., Davy P., Markwitz A., Waheed S., Siddique N., Santos F., Pabroa P., Seneviratne M., Wimolwattanapun W., Bunprapob S., Vuong T., Duy Hien P., and Markowicz A. 2008. Urban air quality in the Asian region, *Science of the Total Environment*, 404: 1103–1112.

Huy T. N. Q. 2008. Photochemical smog modeling for ozone air quality management in Ho Chi Minh City, Vietnam, master's thesis, Asian Institute of Technology, Pathumthani, Thailand.

Intergovernmental Panel on Climate Change (IPCC). 2007. *The Physical Science Basis. Contribution of Working Group I to the Fourth Assessment Report of the Intergovernmental Panel on Climate Change*, edited by Solomon S., Qin D., Manning M., Chen Z., Marquis M., Averyt K. B., Tignor M., and Miller H. L., Cambridge University Press, Cambridge, UK.

International Institute for Applied Systems Analysis (IIASA). 2004. Regional air pollution information and simulation—Asia, International Institute for Applied Systems Analysis. http://www.iiasa.ac.at.

Iqbal A., and Kim Oanh N. T. 2011. Assessment of acid deposition over Dhaka division using CAMx-MM5 modeling system, *Atmospheric Pollution Research*, DOI 10.5094/APR.2011.05.

Iwata H., Tanabe S., Sakai N., Nishimura A., and Tatsukawa R. 1994. Geographical distribution of persistent organochlorine in air, water, and sediments from Asia and Oceania, and their implications for global redistribution from lower latitudes, *Environmental Pollution*, 85: 15–33.

Jenkin M. E., and Clemitshaw K. C. 2000. Ozone and other secondary photochemical pollutants: chemical processes governing their formation in the planetary boundary layer, *Atmospheric Environment*, 34: 2499–2527.

Jho W., and Lee H. 2009. The structure and political dynamics of regulating "yellow sand" in North-East Asia, *Asian Perspective*, 33: 41–72.

Kanabkaew T., and Kim Oanh N. T. 2010. Development of spatial and temporal emission inventory for crop residue field burning, *Environmental Monitoring and Assessment*, DOI 10.1007/s10666-010-9244-0.

Kim K. H., Choi G. H., Kang C. H., Lee J. H., Kim J. Y., Youn Y. H., and Lee S. R. 2003. The chemical composition of fine and coarse particles in relation with the Asian dust events, *Atmospheric Environment*, 37: 753–765.

Kim Oanh N. T., and Leelasakultum K. 2011. Analysis of meteorology and emission in haze episode prevalence over mountain-bounded region for early warning, *Science of the Total Environment*, 409: 2261–2271.

Kim Oanh N. T., Nghiem L. H., and Yin L. P. 2000. Emission of polycyclic aromatic hydrocarbons, toxicity and mutagenicity from domestic cooking using sawdust briquettes, wood and kerosene, *Environmental Science and Technology*, 36: 833–839.

Kim Oanh N. T., and Permadi D. A. 2009. Photochemical smog modeling for ozone air quality management. In *Modelling of Pollutants in Complex Environmental Systems* (Vol. 1), edited by Hanrahan G., ILM, Hertfordshire, UK.

Kim Oanh N. T., Pongkiatkul P., Upadhyay N., and Hopke P. 2009. Design a monitoring program of ambient particulate matter for source apportionment study by receptor modeling, *Atmospheric Environment*, 43: 3334–3344.

Kim Oanh N. T., Thiansathit W., Bond T. C., Subramanian R., Ekbordin W., and Ittipol P. 2010. Compositional characterization of $PM_{2.5}$ emitted from in-use diesel vehicles in Asia, *Atmospheric Environment*, 44(1): 15–22.

Kim Oanh N. T., Upadhyay N., Zhuang Y. H., Hao Z. P., Murthy D. V. S., Lestari P., Villarine J. T., Chengchua K., Co H. X., Dung N. T., and Lindgren E. S. 2006. Particulate air pollution in six Asian cities: spatial and temporal distributions, and associated sources, *Atmospheric Environment*, 40: 3367–3380.

Kim Oanh N. T., and Zhang B. 2004. Photochemical smog modeling for assessment of potential impacts of different management strategies on air quality of the Bangkok Metropolitan Region, Thailand, *Journal of Air and Waste Management Association*, 54(10): 1321–1338.

Klinmalee A. 2008. The development of assessment procedure for exposure to open rice straw burning emission: a case study of school children in Thailand, dissertation, Asian Institute of Technology, Pathumthani, Thailand.

Klinmalee A., Kasama S., and Kim Oanh N. T. 2008. Relationship between indoor and outdoor air pollution in selected public buildings in Thailand and potential exposure level, *Journal of Environmental Monitoring and Assessment*, 156: 581–594.

Koe L. C. C., Arellano A. F., and McGregor J. L. 2001. Investigating the haze transport from 1997 biomass burning in Southeast Asia: its impact upon Singapore, *Atmospheric Environment*, 35: 2723–2734.

Ladou J. 2004. The asbestos cancer epidemic, *Environmental Health Perspective*, 112: 285–290.

Large D. J., Kelly S., Spiro B., Tian L., Shao L., Finkelman R., Zhang M., Somerfield C., Plint S., Ali Y., and Zhou Y. 2009. Silica—volatile interaction and the geological cause of the Xuan Wei Lung cancer epidemic, *Environmental Science and Technology*, 43: 9016–9021.

Lee Y. C, Calori G. and Carmichael Gr. (2002). Ozone episodes in urban Hong Kong 1994–1999. *Atmospheric Environment*, 36 : 1957–1968.

Lestari P., and Mauliadi Y. D. 2009. Source apportionment of particulate matter at urban mixed site in Indonesia using PMF, *Atmospheric Environment*, 43: 1760–1770.

Liu X., Zhang G., Li J., Yu L.-L., Xu Y., Li X.-D., Kobara Y., and Jones K. C. 2009. Seasonal patterns and current sources of DDTs, chlordanes, hexachlorobenzene, and endosulfan in the atmosphere of 37 Chinese cities, *Environmental Science and Technology*, 43: 1316–1321.

Long D. H. 2006. Quality assurance and quality control for ambient air pollution data produced by automatic monitoring network in Hanoi, Vietnam, master's thesis, Asian Institute of Technology, Pathumthani.

Ma C. J., Tohno S., Kasahara M., and Hayakawa S. 2004. Properties of individual Asian dust storm particles collected at Kosan, Korea during ACE-Asia, *Atmospheric Environment*, 38: 1133–1143.

Ministry of Environment (MOE) Government of Japan. 2007. Background air monitoring of persistent organic pollutants in East Asian Countries 2004–2007. Paper presented at the Fifth Workshop on Environmental Monitoring of Persistent Organic Pollutants in East Asian Countries, November, Kyoto, Japan. Accessed 2011 from http://www.env.go.jp/en/chemi/pops/eaws.html

Ministry of Environment (MOE) Government of Japan. 2011. Environmental Quality Standards for Air. Retrieved from http://www.env.go.jp/en/

Molina M. J., and Molina L. T. 2004. Megacities and atmospheric pollution, *Air and Waste Management Association*, 54: 644–680.

Mumford J. L., Chapman R. S., Harris D. B., He X. Z., Cao S. R., Xian Y. L., and Li X. M. 1989. Indoor air exposure to coal and wood combustion emission associated with a high lung cancer rate in Xuan Wei, China, *Environment International*, 15: 315–320.

Mumford J. L., Harris D. B., Williams K., Chuang J. C., and M. C. 1987. Indoor air sampling and mutagenicity studies of emission from unvented coal combustion, *Environmental Science and Technology*, 21: 308–311.

Nghiem L. H. 2008. Photochemical modeling for simulation of ground-level ozone over the continental Southeast Asian region to assess potential impacts on rice crop, dissertation, Asian Institute of Technology, Pathumthani, Thailand.

Nghiem L. H., and Kim Oanh N. T. 2008. Evaluation of the mesoscale meteorological model (MM5)-community multi-scale air quality model (CMAQ) performance in hindcast and forecast of ground-level ozone, *Journal of Air and Waste Management Association*, 58: 1341–1350.

Novak K., Skelly J. M., Schaub M., Kräuchi N., Hug C., Landolt W., and Bleuler P. 2003. Ozone air pollution and foliar injury development on native plants of Switzerland, *Environmental Pollution*, 125: 41–52.

Pentamwa P. 2001. Levels and phase distribution of airborne polychlorinated biphenyls (PCBs) in the Bangkok metropolitan region, master's thesis, Asian Institute of Technology, Pathumthani, Thailand.

Pentamwa P., Kanaratanadilok N., and Kim Oanh N. T. 2010. *Indoor Pesticide Application Practices and Levels in Homes of Bangkok Metropolitan Region*, Environmental Monitoring and Assessment, doi: 10.1007/s10661-010-1834-4.

Pentamwa P., and Kim Oanh N. T. 2008a. Air quality in Southern Thailand during haze episode in relation to air mass trajectory, *Songklanakarin Journal of Science and Technology*, 30: 539–546.

Pentamwa P., and Kim Oanh N. T. 2008b. Levels of pesticides and PCBs in selected homes in Bangkok metropolitan region, Thailand, *Annals of the New York Academy of Sciences*, 1140: 91–112.

Permadi D. A. 2007. Surface O_3 air quality in Jakarta: current status, precursor emission inventory, and modeling efforts, master's thesis, Asian Institute of Technology, Pathumthani, Thailand.

Permadi D. A., and Kim Oanh N. T. 2008. Episodic ozone air quality in Jakarta in relation to meteorological conditions, *Atmospheric Environment*, 42(28): 6806–6815.

Pochanart P., Kreasuwun J., Sukasem P., Geeratithadaniyom W., Tabucanon M. S., Hirokawa J., Kajii Y., and Akimoto H. 2001. Tropical tropospheric ozone observed in Thailand, *Atmospheric Environment*, 35: 2657–2668.

Pollution Control Department (PCD). 2000. *Pollution Control Department: Estimated Emissions of Air Pollutants in Bangkok Metropolitan Region in 1997*. Pollution Control Department, Bangkok, Thailand.

Pollution Control Department (PCD). 2003–2008. *Annual Report 2003–2008*, Pollution Control Department, Bangkok, Thailand.

Pollution Control Department (PCD). 2004. *Thailand State of the Environment Report*, Pollution Control Department, Bangkok, Thailand.

Pollution Control Department (PCD). 2006. *Project on the Development of Environmental and Emission Standards of VOCs*, Pollution Control Department, Bangkok, Thailand.

Pollution Control Department (PCD). 2007. *Development of Environmental and Emission Standards of Volatile Organic Compounds (VOSs) in Thailand*, Pollution Control Deparment, Bangkok, Thailand.

Pollution Control Department (PCD). 2008. Annual air quality standards, Pollution Control Department. Accessed 2011 from http://infofile.pcd.go.th/law

Pollution Control Department (PCD). 2009. *Pollution Prevention and Mitigation Policy 1997–2016*, Pollution Control Department, Bangkok, Thailand.

Pongkiatkul P., and Kim Oanh N. T. 2007. Assessment of potential long-range transport of particulate air pollution using trajectory modeling and monitoring data, *Journal of Atmospheric Research*, 85: 3–17.

Pope C. A., III, Ezzati M., and Dockery D. W. 2009. Fine particulate air pollution and life expectancy in the United States, *The New England Journal of Medicine*, 360: 376–386.

Qiu X., Zhu T., Li J., Pan H., Li Q., Miao G., and Gong J. 2004. Organochlorine pesticides in the air around the Taihu Lake, China, *Environmental Science and Technology*, 38: 1368–1374.

Ramanathan V., and Crutzen P. J. 2003. New directions: Atmospheric brown clouds, *Atmospheric Environment*, 37: 4033–4035.

Reynolds C. C. O., and Kandlikar M. 2008. Climate impacts of air quality policy: switching to a natural gas-fueled public transportation system in New Delhi, *Environmental Science Technology*, 42: 5860–5865.

Robert I. K., Travis M. D., and Xiaofei Z. 2003. Octachlorodipropyl ether (S-2) mosquito coils are inadequately studied for residential use in Asia and illegal in the United States, *Environmental Health Perspective*, 111: 1439–1442.

Schwartz J., and Neas L. 2000. Fine particles are more strongly associated than coarse particles with acute respiratory health effects in schoolchildren, *Epidemiology*, 11: 6–10.

Sengkhammy B., and Kim Oanh N. T. 2007. Indoor air pollution in Vientiane, Laos, in relation to cooking fuel type and house design. In *Proceedings of the International Conference on Air Quality Management in Southeast Asia*, Ho Chi Minh City, Vietnam, November 21–23, 2007. Canadian International Development Agency (CIDA)-Asian Institute of Technology (AIT), Bangkok. Thailand. ISBN 978-974-8257-45-7.

Shah J. J., Nagpal T., Brandon C. J., Larssen S., Gronskei K. E., Hanegraaf M. C., Jansen H., Kuik O. J., Oosterhuis F. H., and Olsthoorn X. A. 1997. *Urban Air Quality Management Strategy in Asia-Guidebook* (Vol. 1), edited by Shah J. J., and Brandon C. J. E., World Bank, Washington, DC.

Shen L., Wania F., Lei Y. D., Teixeira C., Muir D. C. G., and Bidleman T. F. 2005. Atmospheric distribution and long-range transport behavior of organochlorine pesticides in North America, *Environmental Science and Technology*, 39: 409–420.

Shindell D., Kuylenstierna J. C. I., Vignati E., van Dingenen R., Amann M., Klimont Z., Anenberg S. C., Muller N., Janssens-Maenhout G., Raes F., Schwartz J., Faluvegi G., Pozzoli L., Kupiainen K., Höglund-Isaksson L., Emberson L., Streets D., Ramanathan V., Hicks K., Kim Oanh N.T, Milly G., Williams M., Demkine V., and Fowler D. (2012). Simultaneously mitigating near-term climate change and improving human health and food security. *Science*, January 2012: 183-189. doi:10.1126/science.1210026.

Shrestha R. M., Kim Oanh N. T., Shrestha R., Rupakheti M., Permadi D. A., Kanabkaew T., and Salony R. *Atmospheric Brown Cloud (ABC) Emission Inventory Manual (EIM)*, United Nation Environmental Programme (in press), Nairobi, Kenya.

Sillman S. 1999. The relation between ozone, NO_x and hydrocarbons in urban and polluted rural environments, *Atmospheric Environment*, 33: 1821–1845.

Smith K. R. 1987. *Biofuels, Air Pollution, and Health—A Global Review*, Plenum Press, New York.

State of Environment Report for Bangladesh (SOE-BLD). 2005. United Nations Environment Programme (UNEP), Nairobi, Kenya.

State of Environment Report for Delhi (SOE-Delhi). 2010. Department of Environment and Forests, Government of NCT of Delhi, New. Available online at http://www.indiaenvironmentportal.org.in/files/SoEDelhi2010.pdf

Streets D. G., Guttikunda S. K., and Carmichael G. R. 2000. The growing contribution of sulfur emissions from ships in Asian waters, *Atmospheric Environment*, 34: 4425–4439.

Subramanian R., Winijkul E., Bond T. C., Thiansathit W., Kim Oanh N. T., Paw-armart I., and Duleep K. G. 2009. Climate-relevant properties of diesel particulate emissions: results from a piggyback study in Bangkok, Thailand, *Environmental Science and Technology*, 43: 4213–4218.

Sun Y. L., Zhuang G. S., Ying W., Han L. H., Guo J. H., Mo D., Zhang W. J., Wang Z. F., and Hao Z. P. 2004. The air-borne particulate pollution in Bejing: concentration, composition, distribution and sources, *Atmospheric Environment*, 38: 5991–6004.

Supat W. 1999. Ambient air quality monitoring network in Thailand, paper presented at the National Workshop on Phasing Out Leaded Gasoline in Vietnam. November 28-December 1. Hanoi, Vietnam.

Tipayarom D., and Kim Oanh N. T. 2007. Effects from open rice straw burning emission on air quality in the Bangkok metropolitan region, *ScienceAsia*, 33: 339–345.

Torigoe K., Hasegawa S., Numata O., Yazaki S., Matsunaga M., Boku N., Hiura M., and Ino H. 2000. Influence of emission from rice straw burning on bronchial asthma in children, *Pediatrics*, 42: 143–150.

Truc V. T. Q., and Kim Oanh N. T. 2007. Roadside BTEX and other gaseous air pollutants in relation to emission sources, *Atmospheric Environment*, 41: 7685–7697.

United Nations Economic Commission for Europe (UNECE). 2004. *Handbook for the 1979 Convention on Long-Range Transboundary Air Pollution and Its Protocols*, United Nations, New York.

United Nations Environment Program (UNEP). 1991. *Urban Air Pollution*, UNEP/GEMS Environment Library, No. 4, United Nations Environment Program, Nairobi, Kenya.

United Nations Environment Program (UNEP). 2008. *Atmospheric Brown Clouds: Regional Assessment Report with Focus on Asia*, United Nations Environment Program, Nairobi, Kenya.

United Nations Environment Program (UNEP) and World Meteorology Organization (WMO). 2011. *Integrated Assessment of Black Carbon and Tropospheric Ozone*. Edited by Ullstein B., United Nations Environment Program, Nairobi, Kenya.

U.S. Environmental Protection Agency (USEPA). 1995. *The Inside Story: A Guide to Indoor Air Quality*, United States of Environmental Protection Agency. http://www.epa.gov/iaq/pubs/insidestory.html, accessed 2011.

U.S. Environmental Protection Agency (USEPA). 2009. *A Citizen's Guide to Radon: The Guide to Protecting Yourself and Your Family from Radon*, United States Environmental Protection Agency, Washington, DC.

Wanabongsea P., Tokonami S., and Bovornkitti S. 2005. Current studies on radon gas in Thailand, *International Congress Series*, 1276: 208–209.

Wang H. L., Zhuang Y. H., Hao Z. P., Cao M. Q., Zhong J. X., Wang X. K., and Kim Oanh N. T. 2008. Polycyclic aromatic hydrocarbons from rural household biomass burning in a typical Chinese village, *Journal of Science in China Series D: Earth Sciences*, 51: 1013–1020.

Warapetcharayut P., Paw-armart I., Pentakulchai T., Aransri, N., Charoenphonphanich C., 2007. Mobile source control strategies in Thailand. In *Proceedings of the International Conference on Air Quality Management in Southeast Asia*, Ho Chi Minh City, Vietnam, November 21–23, 2007. Canadian International Development Agency (CIDA)-Asian Institute of Technology (AIT), Bangkok. Thailand. ISBN 978-974-8257-45-7.

Watson J. G. 2002. Visibility: science and regulation, *Journal of Air and Waste Management Association*, 52: 628–713.

Weaver C., and Chan L. M. 1994. *Motorcycle Emission Standards and Emission Control Technology*, World Bank, Washington, DC.

Wingpud V. 2009. Level of particulate matter and volatile organic compounds in Map Ta Phut area, Rayong Province, Thailand, master's thesis, Asian Institute of Technology, Pathumthani, Thailand.

World Health Organization (WHO). 2000. Framework convention on tobacco control in the Southeast Asia Region, Document SEA/RC53/14, from http://w3.whosea.org/meeting/rc/rc53 /rc53-14.htm

World Health Organization (WHO). 2002a. *Specifications and Evaluations for Public Health Pesticides*, World Health Organization, Geneva, Switzerland.

World Health Organization (WHO). 2002b. The World Health Report 2002: reducing risks, promoting healthy life, World Health Organization. http://www.who.int/whr/1999/en/

World Health Organization (WHO). 2005. Indoor air pollution and health fact sheet no. 292, World Health Organization. Accessed 2011 from http://www.who.int/mediacentre/factsheets/fs292/en/

World Health Organization (WHO). 2006. Air quality guidelines: global update 2005, particulate matter, ozone, nitrogen dioxide, sulfur dioxide, WHO Regional Office for Europe, Copenhagen. Accessed 2011 from http://whqlibdoc.who.int/hq/2006/WHO_SDE_PHE_OEH_06.02_eng.pdf

World Health Organization (WHO). 2009. *WHO Handbook on Indoor Radon: A Public Health Perspective*, World Health Organization, Geneva, Switzerland.

World Health Organization (WHO). 2010. *WHO Guidelines for Indoor Air Quality: Selected Pollutants*, WHO Regional Office for Europe, Copenhagen, Denmark.

Yang H. H., Tsai C. H., Chao M. R., Su Y. L., and Chien S. M. 2006. Source identification and size distribution of atmospheric polycyclic aromatic hydrocarbons during rice straw burning period, *Atmospheric Environment*, 40: 1266–1274.

Zhang B., and Kim Oanh N. T. 2002. Photochemical smog pollution in the Bangkok metropolitan area in relation to O_3 precursor concentrations and meteorological conditions, *Atmospheric Environment*, 36: 4211–4222.

Zhang G., Chakraborty P., Li J., Sampathkumar P., Balasubramanian T., Kathiresan K., Takahashi S., Subramanian A., Tanabe S., and Jones K. C. 2008. Passive atmospheric sampling of organochlorine pesticides, polychlorinated biphenyls, and polybrominated diphenyl ethers in urban, rural, and wetland sites along the coastal length of India, *Environmental Science and Technology*, 42: 8218–8223.

Section I

Particulate Matter Monitoring and Source Apportionment Study

2

Receptor Modeling for Air Pollution Source Apportionment Study

Prapat Pongkiatkul and Nguyen Thi Kim Oanh

CONTENTS

2.1 Introduction

Accurate information on contributing sources to ambient air pollution in a location is the foremost requirement for development of efficient air quality management strategies. The top-down approach using emission inventories alone or combined with air quality dispersion modeling is usually used as a tool to generate information on source contributions to ambient air pollution. The dispersion modeling approach requires accurate emission input data to predict ambient air pollutant concentrations; hence, the lack of these data introduces much uncertainty to the top-down approach results. Besides, the source emission inventory usually does not include fugitive releases or provide direct information on secondary air pollutants that are formed in the atmosphere; thus, it is insufficient for air quality management in several situations.

To supplement the top-down approach, the so-called bottom-up approach is used, which assesses contributions from various sources based on observations at sampling sites/receptors by a receptor modeling tool (Gordon, 1998). Receptor models are based on mathematical procedures to identify sources and quantify the source contributions to ambient air pollution (Okamoto et al., 1990; Watson and Chow, 1989). Receptor models are widely used to estimate the source contributions for ambient particulate matter (PM), especially PM_{10} and $PM_{2.5}$ (particles with an aerodynamic diameter less than 10 and 2.5 μm, respectively), and volatile organic compounds (VOCs). To quantify source contributions to ambient PM, for example, extended PM compositional data are required, which normally include EC (elemental carbon), OC (organic carbon), elemental species, and ionic species (Dzubay et al., 1984). For VOCs, detailed speciation is required. In principle, the more complete the input of compositional data is, the better are the source apportionment results produced by the models.

This chapter describes fundamental principles of receptor modeling as well as the mathematical formulation of selected common receptor models applicable for PM source apportionment studies. Common tools for the ambient data screening, input data preparation, and model results evaluation are emphasized. A case study of ambient $PM_{2.5}$ source apportionment is illustrated with a stepwise application procedure of the positive matrix factorization (PMF) receptor model.

2.2 Receptor Modeling

Receptor models are a useful analytical tool that is based on the environmental measurement data to estimate the quantitative contributions from sources. Receptor models have been applied in source apportionment for various environmental studies, including wet deposition (Anttila et al., 1995; Hooper and Peters, 1989); dioxins and furans in sediment (Su and Christensen, 1997); and surface water quality (Reinikainen et al., 2001; Singh et al., 2005). In air pollution studies, receptor models are applied for air pollutants, such as PM and VOCs, that originate from a number of sources and do carry some markers of the original sources while in the ambient air. The source fingerprints or marker species are used to characterize respective source emissions, either by their abundance or their relative proportions (profile). The source profiles normally represent a category of sources in urban areas, such as diesel exhaust, rather than individual emitter, such as a single diesel vehicle (Kim Oanh et al., 2009).

Principally, receptor models are based on the measurement data of ambient air pollution (measured at a receptor) to estimate the contributions from sources without detailing the role of atmospheric processes. This is in contrast to dispersion modeling, which requires intensive source emission and meteorology data input to estimate ambient concentrations at different receptors. The receptor modeling approach thus complements the dispersion modeling approach, and a combination of both approaches would enhance understanding the quantitative linkage between emission sources and their impacts on ambient air quality. Some major features of receptor and dispersion models are listed in Table 2.1.

A general scheme of the receptor modeling approach for PM source apportionment is illustrated in Figure 2.1. In PM source apportionment studies, for example, ambient PM data, including mass and composition (elements, ionic species, etc.), are required to prepare the input files.

2.2.1 Mathematical Formulation of Receptor Models

The fundamental hypothesis in all mathematical receptor models is based on a simple mass conservation argument. A general assumption in receptor modeling is that source profiles do not change with time during the period from the emission until pollutants are measured at a receptor site. If p contributing sources exist and there is no interaction among air pollutants or any removal or formation processes, the total mass of an air pollutant measured at the receptor x would be a linear sum of the contributions of the individual sources g_k (Henry et al., 1984).

$$x = \sum_{k=1}^{p} g_k \tag{2.1}$$

Similarly, x_j, the mass concentration of species j, would be

$$x_j = \sum_{k=1}^{p} f_{kj} g_k \tag{2.2}$$

TABLE 2.1

Major Features of Receptor and Dispersion Models

Features	Receptor Models	Dispersion Models
Model formulation	Statistical analytical procedures, commonly based on the "trial-and-error" approach, to quantitatively apportion the measured ambient pollution levels at a site to potential contributing sources.	Deterministic models to solve a system of mathematical equations to estimate the ambient concentrations based on source emission and meteorology input data.
Source emission data	Information on the presence of major sources is required. Detailed source emission rates are not required, but source emission profiles with information on detailed variability of each source type are needed.	Detailed source emission rates and the spatial and temporal distribution are required to prepare the emission input file. Model results are highly dependent on accuracy of the source emission data.
Meteorological data	Commonly, meteorological data are used for interpretation of receptor model results. Use of meteorological data directly in the models is optional.	Meteorological data are required to prepare the meteorology input file.
Fugitive emissions	Models are able to take into account all unidentified and fugitive emissions that have contributed to the measured ambient pollution.	They cannot be directly applied to analyze emission scenarios.
Applicability	Receptor models are site specific, and results are relevant to the site and period for which measurements have been made; hence, this is not a forecast tool. They cannot identify and analyze the emission sources that are not included in the emission input file.	Dispersion models are universal, and in principle, they can be applied to any areas after successful evaluation, but complex terrains present a real challenge for dispersion modeling. Results can be produced for any time through the hindcast, nowcast, and forecast, depending on available input data. They can be used for analysis of emission scenarios.
Observation data	Receptor models require extended datasets generated by intensive physical and chemical analyses of samples.	Systematic observation data are required for dispersion model performance evaluation.
Determination of specific source contribution	Receptor models identify the contributions from source categories rather than individual source. They cannot deal efficiently with source colinearity.	Dispersion models are suitable for analyzing the contributions from specific sources or source categories. Their ability in predicting secondary particles varies and generally still needs improvement. No problems of colinearity exist.
Pollutants handled	Receptor models are applied for pollutants that exist in a mixture in ambient air and do carry markers of sources such as PM and VOCs.	Dispersion models can be applied for any pollutants included in the model formulation.

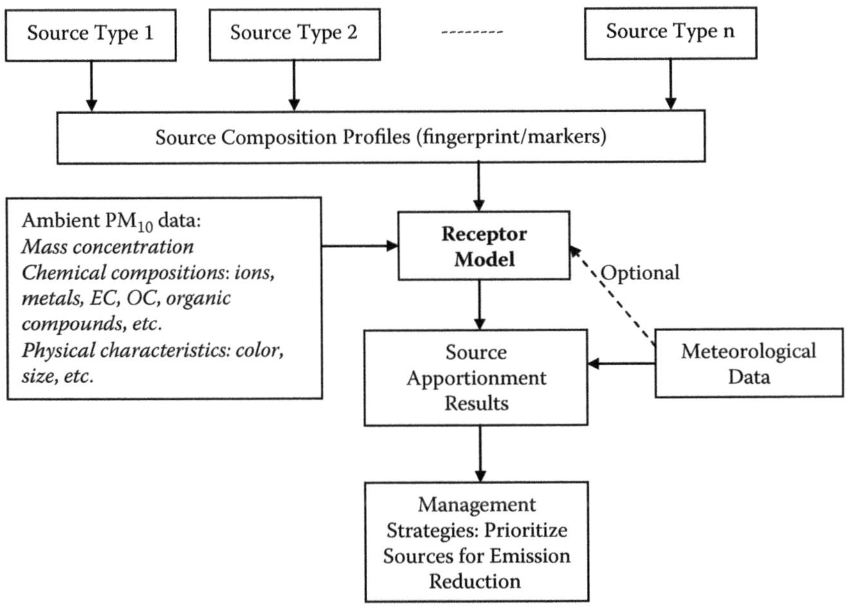

FIGURE 2.1
General scheme of receptor model application for PM source apportionment.

where f_{kj} is the mass fraction of species j in source k emission, generally called the *source composition profile*.

Receptor (mathematical) models are commonly classified into two main branches: chemical mass balance (CMB) and multivariate analysis (Henry et al., 1984). These two approaches have different advantages and disadvantages, which normally compensate each other. The basic idea of CMB is that composition patterns of emissions from various categories of sources are different enough that their contributions can be identified by measuring concentrations of many species in samples collected at a receptor site. CMB requires a priori information on the number of source categories and source emission composition/profiles f_{kj} in a given study region. CMB8 (version 8), developed by Watson et al. (1997), is a widely used receptor model.

The fundamental approach of multivariate models is to deal with a series of m measurements (samples) of species j during a sampling period at a receptor site. Thus,

$$x_{ij} = \sum_{k=1}^{p} g_{ik} f_{kj}, \qquad i = 1 \ldots m \tag{2.3}$$

The multivariate models identify the number of sources p and estimate source profiles f_{kj} associated with a matrix of source contribution g_{ik}. Hence, they are appropriate for use when source profiles are not available or are likely to be very unreliable. Factor analysis (FA) is one example of models in this category. The multivariate models, however, require a large number of ambient samples. Estimation of the minimum number of measurements m required was suggested by Henry et al. (1984):

$$m > 30 + \left(\frac{j+3}{2} \right) \tag{2.4}$$

where j is the number of species measured in m samples.

For example, if the number of species measured in each sample is 20, the minimum sample size required for multivariate model application is $m = 42$.

Among the multivariate receptor models, PMF, developed by Paatero and Tapper (1994), is widely applied. The model output, as a normal multivariate model, contains an indication of the number of sources and associated source profiles, although sometimes not all generated profiles are explainable. Recently developed multivariate models based on the multilinear engine (ME) provide a new conceptual approach for solving different multilinear and quasimultilinear mathematical problems to two-, three-, and many-dimensional data arrays (Paatero, 1999). The constrained physical receptor model (COPREM), developed by Wahlin (2003), is a hybrid physical receptor model that combines features from both CMB and the multivariate modeling approaches.

All receptor models are generally applied for source apportionment using measurement data collected at a single receptor site. When a multivariate model is applied for datasets collected from a few sites located in the same region, there is a strong pulling effect introduced by certain species in the process of calculation, which may result in poor (unexplainable) source profiles. Separated calculation for each receptor site, however, may produce different composition profiles for a similar source category. Attempts have been made to develop a dual-site receptor model (DUALM) that can be applied for PM source apportionment simultaneously using input data collected at two monitoring sites in a region where similar sources (profiles) exist (Pongkiatkul, 2006). DUALM can identify similar sources at both sites (e.g., secondary PM) as well as local/specific sources at each site. Auxiliary information, including wind data (speed and direction) simultaneously recorded at each monitoring site and the information on Julian date and seasons (four seasons), is included in the calculation to prevent rotational ambiguity and to add necessary constraints.

2.2.2 Source Fingerprints Used in Receptor Modeling

Typically, ambient PM are composed of both inorganic and organic chemical species that originated from natural and human-made sources. Each source emission contains specifically different species (elements, ions, etc.), which

TABLE 2.2
Typical Marker Elements and Ions Associated with PM Emission Sources

Sources	Marker Elements
Natural	
Soil	Al, Si, Ti, Fe, Sr, Ca, OC
Sea spray	Na, Cl
Secondary PM	NH_4^+, Na, NO_3^-, SO_4^{2-}, OC, WSOC
Antropogenic	
Road dust	Al, Si, Ti, Fe, Sr, Ca, K, OC, EC
Diesel vehicles	EC, OC, NO_3^-, SO_4^{2-}, Cu, Fe, Zn
Gasoline vehicles	OC, EC, Ca
Refuse incineration	K, Zn, Pb, EC, OC
Cement industry	Ca, Si, Al, Fe
Biomass burning	OC, EC, K, Cl^-, NH_4^+
Ferrous industry	Fe, Cr, Zn
Nonferrous industry	Zn, Cu, Pb, Al

Sources: Compiled from Chow et al. (2003); Chow et al. (2004); Dzubay et al. (1984); Garg et al. (2000); Garivait (1999); Han et al. (2011); Kim and Hopke (2006); Kim et al. (2003, 2004); Kim Oanh et al. (2010, 2011); Maykut et al. (2003); Ramadan et al. (2000); Schauer et al. (2002); Song et al. (2001); Vanderstraeten et al. (2008); Volckens et al. (2007); Watson and Chow (2001); Watson et al. (2001); and Zhao and Hopke (2006).

Note: EC, elemental carbon; OC, organic carbon; WSOC: water-soluble organic carbon.

can play the role of "source fingerprints" or source markers. The knowledge of marker species of a particular source is necessary for receptor modeling. In principle, the abundance of compositional species of PM and their relative proportions (source profiles) are used to recognize sources of ambient PM pollution. Similarly, for VOCs some specific species can serve as markers for original emission sources. Marker elements associated with various PM emission sources are summarized in Table 2.2. Examples of source profiles of typical PM sources that are used for the source apportionment studies in Asian cities are presented in Chapter 3 of this book.

2.2.3 Receptor Model Software

2.2.3.1 Chemical Mass Balance Version 8

The CMB8 model is formulated based on the effective variance least-squares method (Watson et al., 1997). The model consists of a solution of linear equations that express each receptor chemical concentration as a linear sum of products of source profile and source contributions (Equation 2.2). Chemical compositions of PM measured at emission sources (source profiles) and at a receptor site, both with appropriate uncertainty estimates, are the required input. The objective function of the CMB8 model is given in Equation 2.5. The model provides the output, which consists of the average quantitative contribution of each source and its range reflecting uncertainties (Henry et al., 1984; Watson et al., 1997). The newest version used at present is EPA-CMB version 8.2, which features Windows-based and menu-driven operation and is available for download (http://www.epa.gov/scram001/receptor_cmb.htm).

$$\chi^2 = \sum_{j=1}^{n} \frac{\left(x_j - \sum_{k=1}^{p} f_{kj} g_k \right)^2}{\sigma_{x_j}^2 + \sum_{k=1}^{p} \sigma_{f_{kj}}^2 \, g_k^2} \tag{2.5}$$

where σ_{x_j} is the uncertainty of the x_j measurement and $\sigma_{f_{kj}}$ is the uncertainty associated with the f_{kj} measurement.

Basic assumptions used in the CMB8 model formulation (Watson et al., 1997) are as follows:

1. Compositions of source emissions are constant over the period of ambient and source sampling.
2. Chemical species do not react with each other.
3. All potential sources contributing to the receptor site have been identified and have had their emissions characterized.
4. The number of sources or source categories is less than or equal to the number of species.
5. The source profiles are linearly independent of each other.
6. Measurement uncertainties are random and uncorrelated and have normal distribution.

In fact, not all these assumptions are always valid. In reality, the composition of a source emission (source profile) in a location can vary with time, raw materials, fuel types, and so on, which affect the validity of the first assumption. In addition, the changes in emission rates and in mean wind directions may also cause large uncertainties in modeling results produced for a particular receptor site. Further, sources with similar chemical and physical properties cannot be distinguished from each other by the CMB models (i.e., the colinearity problem).

CMB8 is applied in air pollution source apportionment studies for $PM_{2.5}$ and PM_{10} at different places and conditions, such as Taiwan (Chen and Lin, 2001); Qalabotjha, South Africa (Engelbrecht et al., 2002); Sihwa, Korea (Park et al., 2001); and Bangkok, Thailand (Upadhyay, 2002; Wangkiat et al., 2001). U.S. Environmental Protection Agency (USEPA) source profiles (SPECIATE

EPA) are normally used in these studies. CMB8 is also applied to identify source contributions to ambient VOCs (USEPA, 1998).

2.2.3.2 Positive Matrix Factorization

PMF assumes the nonnegativity of the factors, both loadings (source contribution) and scores (source profile). This nonnegativity principle built in the PMF is the major difference from other multivariate methods. The PMF does not rely on information from the correlation matrix but utilizes a point-by-point least-squares minimization scheme that takes the uncertainty of each data point into account. There are two versions of PMF programs, that is, PMF2 and PMF3. PMF2 is applicable for two-dimensional arrays, which are common for air pollution studies, while PMF3 is used in applications for solving three-dimensional arrays (Paatero, 2007). For example, PMF3 was applied to determine the source contributions to $PM_{2.5}$ from indoor, outdoor, and personal exposure sampling in Seattle, Washington (Larson et al., 2004).

In PMF2, if the number of samples is m and the number of species is n, the matrix X of data of i rows and j columns can be formed. The aim of PMF2 is to factorize matrix X into two matrices, namely, G $(i \times p)$ and F $(p \times j)$, and the unexplained part E of X, where p represents the number of factors extracted. Each row of F represents a single source of pollutants, and elements of the row show the concentrations of the species. The first column in matrix G shows how much concentration of a pollutant collected in each sample originated from the first source. The model can be written in the matrix form as shown in Equation 2.6.

$$X = GF + E \tag{2.6}$$

or in component form of

$$x_{ij} = \sum_{k=1}^{p} g_{ik} f_{kj} + e_{ij} \tag{2.7}$$

The objective of PMF2 analysis is to minimize Q, which is the sum of the squares of the residuals weighted inversely with the standard deviations of the data points (Equation 2.8).

$$Q = \sum_{i=1}^{m} \sum_{j=1}^{n} \frac{\left(x_{ij} - \sum_{k=1}^{p} g_{ik} f_{kj} \right)^2}{\sigma_{x_{ij}}^2} \tag{2.8}$$

where $\sigma_{x_{ij}}^2$ is the standard deviation of x_{ij}; $g_{ik} \geq 0$, and $f_{kj} \geq 0$; $k = 1, \ldots, p$.

PMF2 is available online, and the newest version of PMF is EPA-PMF version 3.0, which is based on the ME algorithm. It is Windows based, contains a user interface, and is downloadable (http://www.epa.gov/heasd/products/pmf/pmf.html).

PMF2 is widely applied for PM source apportionment studies in many places, such as Spokane, Washington (Kim et al., 2003); Hong Kong (Lee et al., 1999); and Bangkok, Thailand (Chueinta et al., 1999; Kittiruangpol, 2002; Wimolwattanapun et al., 2011).

2.2.3.3 Other Receptor Models

The COPREM, a hybrid physical receptor model, is based on the CMB approach but also incorporates the ability of multivariate mathematics to fit the chemical species in the source profiles (Wahlin, 2003). The fundamental hypothesis in COPREM is similar to the PMF model (Equation 2.7). However, the user can set up general constraints for the source profiles using knowledge of real source compositions at receptor locations (Wahlin, 2003).

Apart from the three receptor models described, there are other multivariate receptor models available. UNMIX, for example, is also a commonly applied model that is based on the FA approach (http://www.epa.gov/heasd/products/unmix/unmix.html). Principle component analysis (PCA) with varimax rotation is also often used to identify the contributing sources to ambient PM air pollution (Cheng et al., 2000; Lowenthal and Rahn, 1987) or VOCs (Truc and Kim Oanh, 2007). PCA, however, do not directly produce quantitative source contributions.

2.2.4 Strengths and Limitations of Some Commonly Used Receptor Models

CMB can be advantageous when the source composition profiles are available (Watson et al., 1997). However, difficulties may arise in CMB applications because the changes in source emission rates, emission compositions, and wind directions may cause large uncertainties in the source apportionment results. Also, there is the colinearity problem when some source compositions are almost similar (Henry et al., 1984). In contrast, multivariate models can identify unknown sources without intensive prior information on source compositions, but they need an ambient dataset produced by a large number of samples collected at a site. Simple multivariate models (e.g., FA), can give an indication of a number of sources and associated factor loadings (related to the source profiles) based on an eigenvector analysis of the matrix of correlation coefficients, but no direct quantitative source contributions (i.e., in percentage) are produced. Advance multivariate models (e.g., PMF and ME) can produce source profiles and information leading to quantification of source contributions. However, the number of factors (number of sources) should be estimated a priori. COPREM requires information on source profiles but allows some source profiles to be changed to obtain the best fit between the measured mass and the calculated mass. A summary of strengths and limitations of some commonly used receptor models is provided in Table 2.3.

TABLE 2.3

Strengths and Limitations of Some Common Receptor Models

Models	Strengths	Limitations
Chemical Mass Balance		
Chemical mass balance version 8 (CMB8)	Mathematically easier to implement Does not require a large number of samples Produces quantitative source contributions directly Has built-in performance-fitting measures User-friendly software	Requires comprehensive advance information on source profiles Faces colinearity problem when the source composition profiles are almost similar Cannot directly incorporate the time variation in ambient concentrations and source emissions Cannot identify sources with unknown profiles
Multivariate Models		
Positive matrix factorization (PMF)	Does not require comprehensive advance information on source composition profiles Incorporates the time variation Uses nonnegativity constraints, which is more physically realistic Rotation can be controlled by user	Requires a large number of ambient samples Employs the trial-and-error method Needs background knowledge of source composition profile for result interpretation Inherent rotational ambiguity of the factor analysis
Hybrid Model		
Constrained physical receptor model (COPREM)	Mathematically easier to implement Incorporates the time variation Allows user to set up constraints for selected source profiles Can produce quantitative source contributions	Requires comprehensive advance information on source composition profiles Cannot identify sources with unknown profiles Does not take source profile uncertainties into account

2.3 Receptor Modeling for PM Source Apportionment

2.3.1 Ambient Data Input Preparation

The principal input data for receptor modeling include ambient data (x_{ij} in Equation 2.3) and their uncertainty estimates. The ambient data contain the compositional concentrations of various ambient species, which are generated from laboratory measurements. Useful information on the generation of the ambient data for PM receptor modeling was presented by Kim Oanh et al. (2009 and references therein). The composition profiles (f_{kj} in Equation 2.2) of emission sources are also a required input file when the CMB model is used. This section describes the data screening and uncertainty estimates to prepare the ambient PM data input file for the receptor modeling purpose.

2.3.1.1 Data Screening

2.3.1.1.1 Raw Data Check

Raw data may contain some errors. Data screening is a process prior to a model run to minimize errors in the input data. The none-detected data points (in many cases are present as zero values in a dataset) should be recognized as the below detection limit (BDL) values. Polissar et al. (1998) suggested to replace the BDL data points in the PM ambient dataset by half of the detection limit. Missing values should be marked and may be replaced with, for example, the arithmetic mean of the respective species (Polissar et al., 1998) but should be done with caution. Seasonal averages are sometimes used to replace missing values in a large dataset.

2.2.3.1.2 Time Series Analysis

Time series plots for each species are useful for identification of extremely high data points. These data points, normally recognized as the values above fourfold of the standard deviation of a data series, should be thoroughly scrutinized as some of them may be the result of sampling and analytical errors or sometimes simply mistakes in the data entry process. The time series plot of several species can help to explain some extremely high values (e.g., to check whether they fluctuate in a similar manner, which in turn may be linked to changes in emission rates of a particular source). If there is no reasonable explanation, these data points should be treated as missing values.

2.2.3.1.3 Scatterplots

If the water-soluble and the total level of a particular species are both analyzed, then a good correlation of these two data series is generally expected. For example, if K^+ (water-soluble ion) by ion chromatography and K (element) by proton-induced x-ray emission (PIXE) are both produced, then a scatterplot between the K^+ and K series can provide useful information for data screening. Similarly, scatterplots between SO_4^{2-} and S are also normally analyzed.

2.2.3.1.4 Signal to Noise

The *signal-to-noise* ratio (S/N ratio), presented in Equation 2.9, is generally applied to screen out the species that have poor measurement results. The calculation should be made for each species. A small S/N ratio (less than 0.2) (Paatero and Hopke, 2003) indicates that the species has a high noise level (poor detection or a large number of BDL data points) and should be omitted from the dataset. Species with a large S/N ratio are well measured and thus of greater strength.

$$\text{S/N ratio} = \frac{\text{Sum of species concentration}}{\text{Average of detection limit} \times \text{Number of below detection limit}} \tag{2.9}$$

2.2.3.1.5 Reconstructed Mass

Checking the mass closure using the reconstructed mass approach is another efficient tool for data screening. Kim Oanh et al. (2006) and references therein, for example, suggested several groups to be estimated for the reconstructed mass, including the crustal matter, organic matter (OM), soot, sea-salt, ammonia (NH_4^+), sulfate (SO_4^{2-}), nitrate (NO_3^-), and the remaining trace elements. The reconstructed mass of each group can be calculated using the formula presented in Table 2.4. The OM is estimated by applying a multiplication factor to the OC mass to convert the latter into the organic compound mass. This factor commonly varies between 1.4 and 2.1 (Turpin and Lim, 2001), depending on the nature of the ambient PM measured at the receptor site. The crustal matter group includes oxides of principal elements of crustal origin. Chan et al. (1997) converted five elements to their oxides and applied a factor of 1.16 to compensate for the exclusion of MgO, Na_2O, K_2O, and H_2O from the crustal mass calculation (Table 2.4). Ideally, when all the species are measured and properly converted to the appropriate compounds, the reconstructed mass should be close to the measured mass. In principle, the remaining "unexplained" portion of the measured mass should be as small

TABLE 2.4

Groups for Reconstructed Mass Determination

Group	Formula	Description	References
Organic matter	$f \times$ OC (f = 1.4–2.1 depending on the nature of PM)	Mass of all organic compounds in PM	Turpin and Lim (2001)
Crustal matter	1.16 × (1.9 Al + 2.15 Si + 1.41 Ca + 1.67 Ti + 2.09 Fe)	Oxides of Al, Ca, Si, Ti, Fe, Na, Mg, and K that originated from earth (soil)	Chan et al. (1997)
K-biomass	Total K – 0.6Fe	Excess of K due to the contribution from biomass burning	Chan et al. (1997)
Sea salt	2.54 × Na^+	Contribution from the sea spray	Ho et al. (2003)
Soot	BC or EC	Emission from incomplete combustion	Kim Oanh et al. (2006)
Ammonia	NH_4^+	Secondary inorganinic PM	
Sulfate	SO_4^{2-}	Secondary inorganinic PM	
Nitrate	NO_3^-	Secondary inorganinic PM	
Trace elements	Remaining analyzed elements	From various sources	

as possible. However, in practice many species are not measured in the PM samples; hence, the reconstructed mass is sometimes only a small fraction of the measured mass.

The relative abundance of various groups in the reconstructed mass can also be used to identify contributing sources to ambient PM levels preliminarily as shown in Table 2.4. For example, high "soot" values indicate a high contribution of primary particles emitted from incomplete combustion, such as diesel vehicles or residential cooking. In regions where biomass burning is an important source, calculation of the K-biomass group provides a relevant indicator for the contribution of biomass burning to PM. Chan et al. (1997) suggested a method that assumes the crustal matter would have the average ratio K/Fe of 0.6 and the excess of K, that is (Total K – 0.6 Fe) can be assigned to the biomass burning contribution. It is noted that the reference ratio of K/Fe for the crustal matter may be location specific, which needs to be examined by local data. The K-biomass only indicates the abundance of K species (from biomass burning emission) in PM samples, so is just an indicator or surrogate of the biomass burning source and does not present its real contribution to ambient PM. K-biomass is therefore not accounted for in the reconstructed mass because the K content is already included in the crustal group.

After the data screening, the ambient composition data (matrix X with m number of samples and n number of species) with the measurement uncertainty should be prepared using a spreadsheet program (e.g., Microsoft Excel, Lotus123, or Apple iWork Numbers, etc.). Samples containing a significant number of missing species are normally omitted from the input dataset. If a species has more than one analytical result (by different instruments), then the more reliable results (e.g., with high S/N ratios and with fewer missing data points) should be used.

2.3.1.2 Estimation of Ambient Data Uncertainty

CMB, COPREM, and PMF receptor models require a measurement uncertainty of each compositional ambient data point in the input file. The uncertainty represents the measurement errors and gives the models the necessary confidence level in the validity of a data point. Measurement errors from both sampling (e.g., inadequate sampler calibration, fluctuation in airflow rates, uncharacterized natural variations in field conditions, etc.) and laboratory analyses (e.g., sensitivity and calibration of analytical methods, instrument noise, etc.) should be included in the uncertainty values (Kim Oanh et al., 2009). Thus, the uncertainty assigned to a specific data point should be based on both sampling and analytical errors. The total uncertainty can be determined by the square root of [(Sampling uncertainty)^2 + (Laboratory analytical uncertainty)^2], assuming that the two types of uncertainty are independent.

The sampling uncertainty is rarely available, mainly due to a large number of uncontrolled factors involved in the process and the common lack of colocated sampling data. To some extent, the sampling uncertainty can be estimated based on the difference in sampling flow rates measured before and after a sampling event. In many cases, only the laboratory analytical uncertainty is included in the input data. The laboratory analytical uncertainty includes the instrumental uncertainty that is readily provided by certain analytical equipment for the compositional measurements, such as x-ray fluorescence (XRF), PIXE, or instrumental neutron activation analysis (INAA). If sample pretreatment (filter cutting and extraction, sample cleanup) is involved, then a large uncertainty results from this process; hence, repeated analyses may be conducted for the uncertainty estimation. In principle, the laboratory analytical uncertainties are inversely proportional to the species concentrations in the samples; that is, the higher species concentration is, the lower analytical uncertainty will be. Kim Oanh et al. (2009) described several approaches to estimate laboratory uncertainties (instrumental and filter cut/extraction uncertainties or combined) for mass, water-soluble ions, and black carbon measurements. As a way of illustration, the relationships between the standard deviations (based on triplicate sample analyses) and concentrations of some elements analyzed by an inductively coupled plasma-optical emission spectrometer (ICP-OES) are shown in Figure 2.2. The lower levels of elements (e.g., Al and Si in the fine particles) were measured with higher uncertainty.

The data screening process, such as that suggested by Polissar et al. (1998), which did modify the measurement data points (e.g., replacement of the missing and BDL data points), would introduce certain errors in the model results. To minimize the errors, it is necessary to reduce the weight of those data points in the model calculation; hence, a new uncertainty (σ) has to be assigned. Polissar et al. (1998) suggested a method to calculate the uncertainty of different data point categories x_{ij} using Equation 2.10.

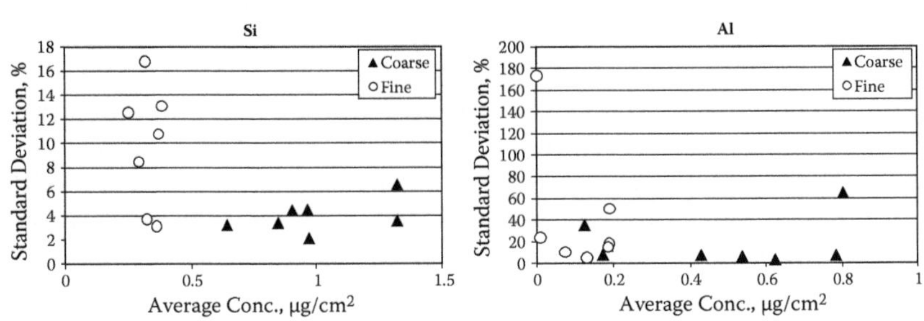

FIGURE 2.2
Examples of the standard deviation (%) of triplicate analyses versus element concentrations in fine and coarse PM samples, microgram per unit of filter area.

Assigned ambient concentration:

For determined values $x_{ij} = v_{ij}$

For BDL values $x_{ij} = \frac{d_{ij}}{2}$

For missing values $x_{ij} = \bar{v}_{ij}$

Assigned uncertainty:

For determined values $\sigma_{x_{ij}} = u_{ij} + \frac{d_{ij}}{3}$

For BDL values $\sigma_{x_{ij}} = \frac{d_{ij}}{2} + \frac{d_{ij}}{3}$

For missing values $\sigma_{x_{ij}} = 4\bar{v}_{ij}$ (2.10)

where v_{ij} is the measured concentration, u_{ij} is the analytical uncertainty, and d_{ij} is the analytical detection limit for the ith sample, j^{th} species.

2.3.1.3 Source Composition Profiles

Source composition profiles are the fractional abundance of species in the PM emitted from a source and are one of the input data files required for CMB modeling. For PMF modeling, available local source profile data are useful for modeling output interpretation.

A number of source composition profiles are available in the literature but mostly are based on studies conducted in the United States or Europe, such as those of Chow et al. (2003, 2004); Volckens et al. (2007); Watson and Chow (2001); and Vanderstraeten et al. (2008). A few emission profiles have also been developed for the major sources in Asian countries (e.g., Kim Oanh et al., 2010, 2011; Wang et al., 2008). USEPA SPECIATE 4.2 (USEPA, 2008) is the largest composition database for both PM and VOC emissions (http://www.epa.gov/ttnchie1/software/speciate/). Details of the source profiles used for the source apportionment in the six Asian cities under the AIRPET (Asian Regional Air Pollution Research Network) monitoring network are discussed in Chapter 3 of this book.

2.3.2 Model Run

2.3.2.1 Chemical Mass Balance Version 8

There are six input data files required for CMB8 modeling. The first file is named the control file, an optional file, that contains a list of the names of the other five input files. Three of these are the selection files, that is, source, species, and sample selection input files, which contain a list of the names of sources, species, and samples, respectively. The coordinates of the sampling site should also be included in the sample selection input file. The remaining two files are the ambient and source profile data files. The ambient data file contains the mass concentration, ions, elemental carbon, trace elements, and so on and their uncertainties, whereas the source profiles file contains the fractional abundance of each modeled species (same species of the ambient data file) and their uncertainties. Both data files should be prepared in a spreadsheet program (e.g., Microsoft Excel). Detailed input file formats were given by Watson et al. (1997).

As many as possible of the available source profiles relevant for a given study location should be included in the source profile input file. The CMB run should be done with different selections of the possible sources, in a trial-and-error way, to obtain the best results. CMB provides a set of performance measures to evaluate the validity of source contribution estimates, including the T-statistic (TSTAT), chi square, R square (R^2), and the percentage of mass explained. TSTAT is the ratio between the source contribution estimate and the standard error, which is suggested to be greater than 2.0. Low TSTAT values indicate that there might be some colinearity problems among the selected source profiles (Watson et al., 1997). Chi square is the weighted sum of squares of the differences between the calculated and measured fitting species concentrations. In principle, chi square values should be as small as possible and the values of between 1.0 and 2.0 are acceptable, whereas values greater than 4.0 indicate a fitting problem for some species (Pace and Watson, 1987; Watson et al., 1997). The R square is the fraction of the variance explained by the model, and its value should be closed to 1.0. Pace and Watson (1987) reported that the R-square values should be greater than 0.9. The percentage mass explained is the ratio (in percentage) of the sum of the model calculated source contribution estimates to the measured mass concentration. Values between 80% and 120% are acceptable. Values higher than 100% indicate an overestimation of PM mass. In principle, the percentage of the measured mass explained should be lower than 100%. The remaining unexplained mass may be recognized as "unknown" sources (source profiles are not included in the model calculation). The output of the CMB model contains source contribution estimates, eligible space colinearity (related to the potential of colinearity), and both measured and calculated species concentrations.

2.3.2.2 Positive Matrix Factorization

The PMF2 receptor model requires only the ambient PM data, which are separated into two input data files: the ambient PM concentration file and the uncertainty file. The input data files should not contain any letters, and the size of matrix should be the same for both files. The files should be prepared in a spreadsheet program and be converted to text files (.txt) for model input. The model run is done in Microsoft DOS with an executive, PMF2.exe, file. PMF2 also provides an initialization file (with .ini) as a control file that contains information on number of rows and columns and possible number of sources, along with the names of the input files. The EPA-PMF version 3.0 needs the same two input data files, but the number of rows and columns is determined automatically in the program.

An ideal ambient dataset should contain all species of interest (e.g., BC, OC, ions, elements, etc.) in every sample with their measurement uncertainties. However, generally some significant species (in terms of mass) may be missing in a dataset. In this case, Hopke (2007) suggested that the PM mass should be included as an input variable but with a large uncertainty (four times the measured mass values), and a normalization of "mean $G = 1$" should be set in the control file.

The main output of the PMF model consists of a source contribution matrix (G factor) and a source profile (F factor). Estimation of the number of factors (i.e., the number of possible contributing sources) is the most important task in PMF modeling. A trial-and-error process is performed with a series of PMF runs until the best explainable source profile (F factor) results are produced. Hopke (2007) suggested that the number of factors should be selected so that the Q value (Equation 2.8) produced is close to the theoretical Q value. The latter is approximately equal to the total number of data points in the ambient data file (number of samples × number of species). The output of PMF models also includes a "miscellaneous" file that contains the standard deviation of G, standard deviation of F, F explained variation, and scaled residues.

Rotations of PMF outputs are also necessary in the calculation, similar to any multivariate analysis, to produce the best G and F results. FPEAK is a built-in function in PMF that is used to transform the output matrices. The basic role of FPEAK is to subtract the values on one side (G factor or F factor) and simultaneously increase the values on another side without any change in the solution (Paatero and Tapper, 1994). The values of FPEAK range between -0.4 and 0.8 (in an arbitrary unit), but there are no rules to determine the best FPEAK. Hopke (2007) indicated that positive values of FPEAK would produce better results than negative ones.

If the mass is not included in the ambient input data file, then a linear regression (with slope of s_k) between the measured mass and source contribution matrix g_{ik} (i.e., mass rescaling) should be conducted for the final quantitative source apportionment results. In principle, the sum of all source contributions to each sample g_k should be equal to the measured PM mass concentration. Multiplication between source contribution g_{ik} and a factor s_k would produce the quantitative source contribution, as shown in Equation 2.11.

$$x_{ij} = \sum_{k=1}^{p} g_{ik} f_{kj} = \sum_{k=1}^{p} g_{ik} s_k \frac{f_{kj}}{s_k} = \sum_{k=1}^{p} g'_{ik} f'_{kj} \tag{2.11}$$

If the mass is already included in the ambient data input file, the *F* factor should be divided by the apportioned PM mass, whereas the G factor should be multiplied with the same apportioned PM mass values (Hopke, 2007). Background knowledge of local and long-range transport-contributing sources is essential for result interpretation.

2.3.3 Evaluation of Receptor Model Results

An explicit procedure for the evaluation of the receptor model results has yet been reported in the literature. The model outputs are normally comparatively analyzed with wind direction data and physical locations of contributing sources to justify the source contribution results, using, for example, the conditional probability function (CPF) for local sources and the potential source contribution function (PSCF) for regional sources.

2.3.3.1 Conditional Probability Function

A CPF plot is normally constructed to show the direction where a particular source is located. It is derived from the surface wind direction data and the source contribution estimates produced by a receptor model (Equation 2.12):

$$CPF = \frac{m_\theta}{n_\theta} \tag{2.12}$$

where n_θ is the total number of samples collected when wind directions are within sector θ; m_θ is the number of samples collected while the wind directions are within this wind sector (θ) and a particular source has a contribution greater than a selected percentile, normally the 75th, of all the observations at a given site.

The CPF is normally used to identify the local source contributions. It was successfully applied for the source apportionment study in Baltimore, Maryland (Ogulei et al., 2005); Houston, Texas (Xie and Berkowitz, 2006); and the Great Smoky Mountain area in the United States (Kim and Hopke, 2006).

2.3.3.2 Potential Source Contribution Function

The PSCF is used to identify the locations of regional sources that potentially contribute to the receptor site through long-range transport. Backward trajectories are normally determined using the hybrid single-particle Lagrangian integrated trajectory (HYSPLIT) model with gridded meteorological data (Begum et al., 2005; Kim and Hopke, 2006). $PSCF_{a,b}$ is the probability that air masses originate in the $(a,b)^{th}$ cell on the days that have a high contribution of the particular source to a receptor site and is determined using Equation 2.13:

$$PSCF_{a,b} = \frac{m_{a,b}}{n_{a,b}} \tag{2.13}$$

where

$n_{a,b}$ is the total cases that air masses originate in the $(a,b)^{th}$ cell with a the latitude and b the longitude coordinates of the grid.

$m_{a,b}$ is the number of cases that the air masses originate in the $(a,b)^{th}$ cell on a sampling day when a particular source has a contribution greater than a selected percentile, normally the 75th, of all the observations from a given site.

The PSCF was successfully applied for a source apportionment study in the Great Smoky Mountain area (Kim and Hopke, 2006); West Coast of the United States (Hwang and Hopke, 2007); eastern United States (Begum et al., 2005); New York State (Liu et al., 2003); China (Wang et al., 2006); and Thailand (Pongkiatkul and Kim Oanh, 2007). Note that this approach needs at least 300–500 data points (e.g., sampling days) to provide enough data points in a given grid for the analysis.

2.4 Examples of Receptor Model Applications

For illustration, the PMF model application for the source apportionment of $PM_{2.5}$ is demonstrated in this section using the ambient datasets produced by the Interagency Monitoring of Protected Visual Environments (IMPROVE) program (http://vista.cira.colostate.edu/IMPROVE/) at two sites, Cohutta (COHU) (Whitfield County, Georgia) and Great Smoky Mountain National Park (GRSM) of Tennessee. The GRSM station is located around 110 km northeast of the COHU station.

2.4.1 Input Data

Each dataset consists of 122 samples collected at a 3-day interval for the entire year of 2004. In total, 44 species were included in the database; these are summarized in Table 2.5. Data screening removed 6 samples from each dataset that contained significant missing compositional data points. Based on the S/N ratios (Equation 2.9), 27 (of 44) species and their uncertainties were finally included in the input data files: Al, As, Br, Ca, EC1, EC2, EC3, OC1, OC2, OC3, OC4, OP, Cu, H, Fe, Pb, Mn, NO_3^-, K, Se, Si, Na, Sr, SO_4^{2-}, Ti, V, and Zn. Prior to the model run, missing and BDL data points in each dataset were substituted with the mean and a half DL, as expressed previously

TABLE 2.5

Summary Statistics of Chemical Compositions of $PM_{2.5}$ from COHU and GRSM Stations (2004)

	COHU Station		GRSM Station	
Species	**Concentration, μg/m³**	**No. of BDL**	**Concentration, μg/m³**	**No. of BDL**
Mass	1.8–25.19	0	0.97–28.36	0
Al	0.02–0.63 (0.11)	60	0.02–0.46 (0.09)	65
NH_4^+	0.19–3.57 (1.34)	0	0.11–3.6 (1.18)	0
NH_4NO_3	0.09–2.75 (0.64)	0	0.09–4.95 (0.6)	0
$(NH_4)_2SO_4$	0.5–16.36 (5.85)	0	0.71–17.63 (5.9)	0
As	0.0002–0.0024 (0.0005)	27	0.0002–0.0017 (0.0005)	21
Br	0.0005–0.0104 (0.0027)	0	0.0002–0.011 (0.0029)	0
Ca	0.003–0.329 (0.038)	0	0.002–0.092 (0.022)	0
OM	0.52–14.07 (2.59)	0	0.29–7.62 (2.4)	0
EC1	0.09–3.27 (0.45)	0	0.07–1.49 (0.44)	1
EC2	0.02–0.24 (0.07)	11	0.01–0.24 (0.07)	12
EC3	0.003–0.04 (0.012)	73	0.003–0.036 (0.017)	76
EC	0.08–1.72 (0.31)	0	0.07–0.86 (0.32)	0
OC1	–0.02–0.82 (0.09)	45	–0.01–0.61 (0.09)	41
OC2	0.07–1.11 (0.38)	0	0.06–0.96 (0.34)	1
OC3	0.05–2.13 (0.59)	10	0.02–2.61 (0.59)	4
OC4	0.08–6.13 (0.58)	2	0.06–1.86 (0.51)	2
OP	–0.23–1.79 (0.24)	31	–0.33–0.75 (0.21)	27
OC	0.37–10.05 (1.85)	0	0.21–5.44 (1.71)	0
Cl^-	0.002–0.218 (0.03)	22	0.004–0.198 (0.035)	17
Cl	0.0005–0.0006 (0.0005)	118	0.0004–0.0011 (0.0008)	118
Cr	0–0.0076 (0.0002)	45	0–0.0006 0.0001)	47
Cu	0.0002–0.0022 (0.0007)	0	0.0001–0.003 (0.0008)	1
H	0.112–1.35 (0.539)	0	0.079–1.515 (0.549)	0
Fe	0.003–0.323 (0.040)	0	0.002–0.235 (0.027)	0
Pb	0.0002–0.0137 (0.0016)	2	0.0004–0.0043 (0.0017)	0
Mg	0.022–0.347 (0.096)	104	0.037–0.231 (0.102)	102
Mn	0.0001–0.0082 (0.0011)	0	0.0001–0.0039 (0.0009)	2
Ni	0.0001–0.0012 (0.0002)	28	0.0001–0.0011 (0.0002)	28
NO_3^-	0.07–2.13 (0.5)	0	0.07–3.84 (0.46)	0
NO_2^-	–0.02–0.07 (0.01)	84	–0.02–0.04 (0.01)	90
P	NA	NA	0.01–0.01 (0.01)	119
K	0.01–0.24 (0.05)	0	0.01–0.16 (0.05)	0
Rb	0.0001–0.0009 (0.0002)	45	0.0001–0.0006 (0.0002)	52
Se	0.0002–0.0055 (0.0011)	0	0.0002–0.0047 (0.0011)	0
Si	0.02–2.05 (0.23)	7	0.01–1.01 (0.17)	8
Na	0.05–1.01 (0.24)	79	0.04–1.64 (0.22)	80

TABLE 2.5 ***(Continued)***

Summary Statistics of Chemical Compositions of $PM_{2.5}$ from COHU and GRSM Stations (2004)

	COHU Station		GRSM Station	
Species	**Concentration, μg/m³**	**No. of BDL**	**Concentration, μg/m³**	**No. of BDL**
Soil	0.03–7.88 (0.83)	0	0.04–4.3 (0.61)	0
Sr	0.0001–0.0024 (0.0004)	11	0–0.0027 (0.0003)	17
SO_4^{2-}	0.41–11.7 (4.14)	0	0.57–14.66 (4.31)	0
S	0.12–3.97 (1.42)	0	0.17–4.27 (1.43)	0
Ti	0.0003–0.0348 (0.0039)	1	0.0002–0.0288 (0.0026)	1
V	0.0001–0.002 (0.0006)	8	0.0001–0.0017 (0.0005)	10
Zn	0.001–0.043 (0.006)	0	0.001–0.022 (0.006)	0
Zr	0.0001–0.0013 (0.0003)	107	0–0.001 (0.0003)	107

Note: In parentheses are the average values. BDL, below detection limit; NA, not available.

TABLE 2.6

Summary of PMF Model Results at COHU and GRSM

	Source Contribution at COHU		Source Contribution at GRSM	
Sources	**μg/m³**	**%**	**μg/m³**	**%**
Secondary SO_4^{2-}	7.2	73.2	7.5	67.5
Secondary NO_3^-	0.1	1.0	0.2	2.2
Diesel exhaust	0.1	1.4	0.6	5.1
Gasoline exhaust	0.3	2.8	1.9	17.3
Industry	0.5	4.9	0.1	1.1
Aged sea salt, mixed	1.4	14.3	0.3	2.7
Airborne soil	0.2	2.4	0.5	4.2
Calculated mass	9.9		11.1	
Measured mass	10.0		10.5	
R^2	0.976		0.982	

(Equation 2.10). Average number of missing values in the datasets was between four and five for each species.

2.4.2 Source Profiles and Source Contribution

A summary of PMF model results at COHU and GRSM is given in Table 2.6. The PMF produced seven factors at both COHU and GRSM but with some differences in the source profiles. At both sites, the mass explained was reasonable, but a better result (99% of total measured mass was explained) was obtained for COHU than for GRSM, which had an overestimation (106% of total mass). The first factor found at both COHU and GRSM was called the "secondary sulfate," which accounted for the highest contribution to the

$PM_{2.5}$ mass (73.2% at COHU and 67.5% at GRSM). It was characterized by a high concentration of SO_4^{2-} in the profile (Figure 2.3). Sulfur dioxide (SO_2) emitted from a number of coal-fired power plants surrounding the sites may be a potential precursor for these secondary particles. Secondary nitrate particles were identified by a high concentration of NO_3^-, and they accounted for 1% of the $PM_{2.5}$ mass at COHU and 2% at GRSM.

A high concentration of carbon fraction (EC and OC) and significant amounts of Cu, Fe, and Zn were observed in factor 3, which may be linked to diesel exhaust. Small amounts of Cu were also presented in this source (Figure 2.3), which may have been released from brake pads (Kim and Hopke, 2006), hence appearing in the diesel exhaust profile. Iron was likely formed from muffler ablation, while Zn would be linked to motor oil additives (Garg et al., 2000; Kim et al., 2004; Maykut et al., 2003; Zhao and Hopke, 2006). The diesel exhaust emission contributes about 1.5% and 5% to the PM mass concentration at COHU and GRSM, respectively.

Gasoline-powered vehicle exhaust PM emission can be distinguished from diesel exhaust based on its higher concentrations of OC than EC. This source profile has the highest OC fractions among all sources extracted at both COHU and GRSM (Figure 2.3). Calcium found in the gasoline emission profile may be associated with catalytic converters (Schauer et al., 2002). Gasoline exhaust emission accounted for 3% of the $PM_{2.5}$ mass concentration at COHU and 17% at GRSM (Table 2.6).

A mixture of markers with high EC and OC concentrations and with high amounts of Fe, Pb, Mn, and Zn could be linked to emissions from several industries in the area. An aged sea salt mixed-particle profile was identified by the high concentrations of Na. Note that chlorine (Cl) was excluded from the input data due to the high number of BDL values. The low Cl levels found in the PM samples may be linked to the long transport distance of air masses from the sea (470 km from the Atlantic Ocean) to the sites that allowed the conversion from NaCl to Na_2SO_4 (Zhao and Hopke, 2006). These aged sea salt particles may be mixed with particles emitted from several industries around the sites. This source contributed 14% to the total $PM_{2.5}$ mass concentration at COHU and 3% at GRSM. Airborne soil particles profiles were identified at both COHU and GRSM (Figure 2.3) based on the abundance of Al, Ca, Si, and Ti (Dzubay et al., 1984). This source contributed 2% to the $PM_{2.5}$ mass at COHU and 4% at GRSM.

The temporal variations in the source contributions are presented in Figure 2.4. The contributions of the secondary sulfate had similar temporal patterns at both COHU and GRSM sites and feature higher values during summer and autumn. Secondary nitrate PM contributions had a significant seasonal variation, with higher values during winter, which may be explained by the enhanced formation of secondary nitrate particles under lower temperature and higher relative humidity in the season (Kim and Hopke, 2006). Higher contributions of diesel exhaust were also observed during summer and autumn but with different patterns between these two

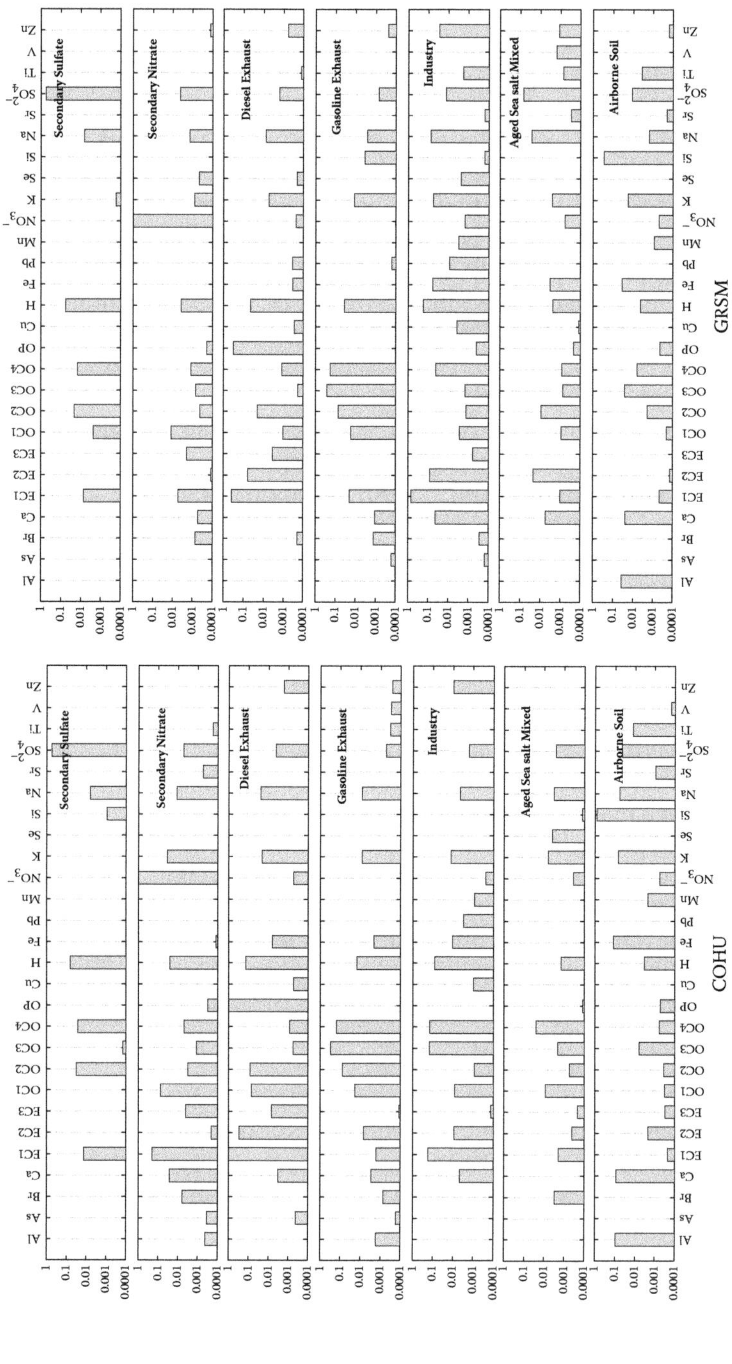

FIGURE 2.3
Composition profiles of contributing sources to $PM_{2.5}$ at COHU and GRSM sites provided by PMF.

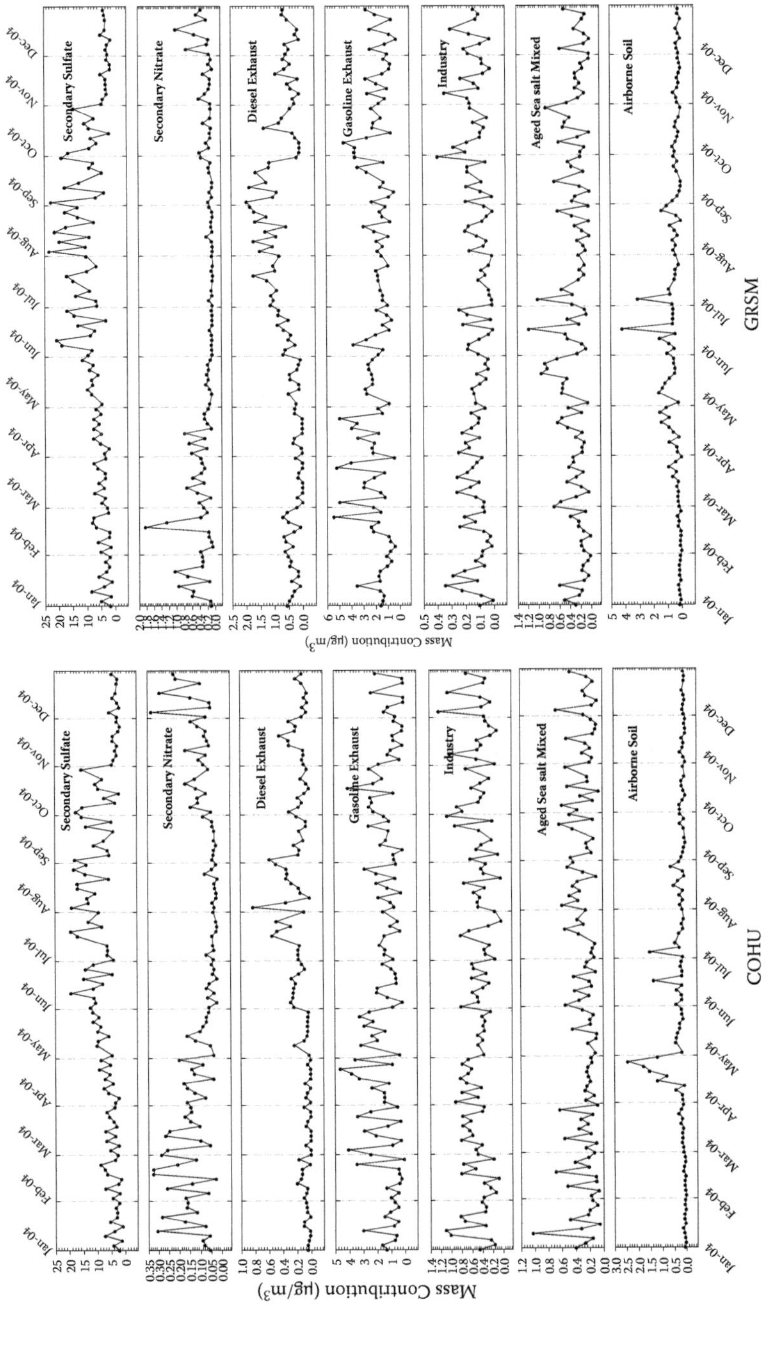

FIGURE 2.4
Time series of potential source contributions to $PM_{2.5}$ at COHU and GRSM sites provided by PMF.

sites. This may be due to a difference in temporal variations in the traffic volume in these two areas. High contributions of gasoline-powered vehicle exhaust PM were observed during spring at both COHU and GRSM sites, which were different from the diesel exhaust contributions. Industrial emission contributions were highly fluctuating among the samples collected during the year but did not show any clear seasonal patterns. The contributions of aged sea salt mixed particles did not show a clear seasonal variation (Figure 2.4). Airborne soil PM contributions were small but quite stable during the year, except for a few peaks in April and the summer months (June–July) at both sites, which may be linked to long-range transport sources. The analysis of air mass backward trajectories (HYSPLIT) on the days with high contributions from airborne soil PM showed that the peaks in April were likely linked to long-range transport from the north and northwest of the United States, whereas the peaks during summer may possibly be linked to that from the south and southwest of the United States. The findings were consistent with those from Kim and Hopke (2006).

2.4.3 Evaluation of Model Results

The evaluation of PMF results is illustrated using the correlation between measured and calculated mass and CPF plots. A good correlation between calculated mass and measured mass was obtained for each site ($R^2 = 0.97$ for COHU and $R^2 = 0.98$ for GRSM), as seen in Figure 2.5.

The CPF plots at both stations were constructed to identify the direction of a particular local source surrounding the sites. Clearer associations of the source contributions to the wind direction were observed for other identified sources than the secondary sulfate and nitrate contributions (Figure 2.6). This was expected as there were a number of large coal-fired power plants located around the two sites (Commission for Environmental Cooperation

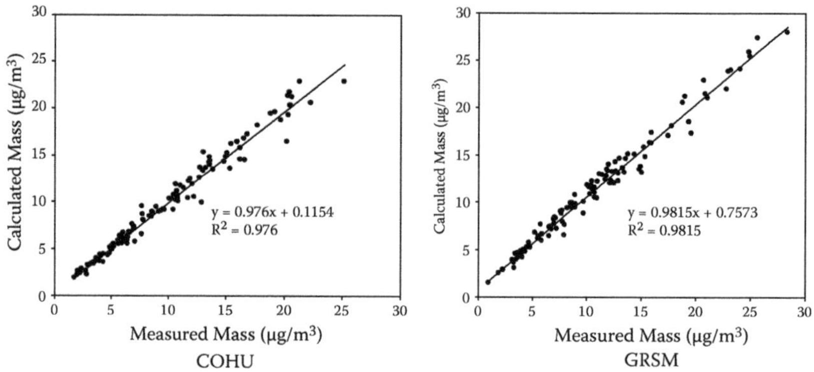

FIGURE 2.5
Comparison between measured and PMF calculated $PM_{2.5}$ mass at COHU and GRSM sites.

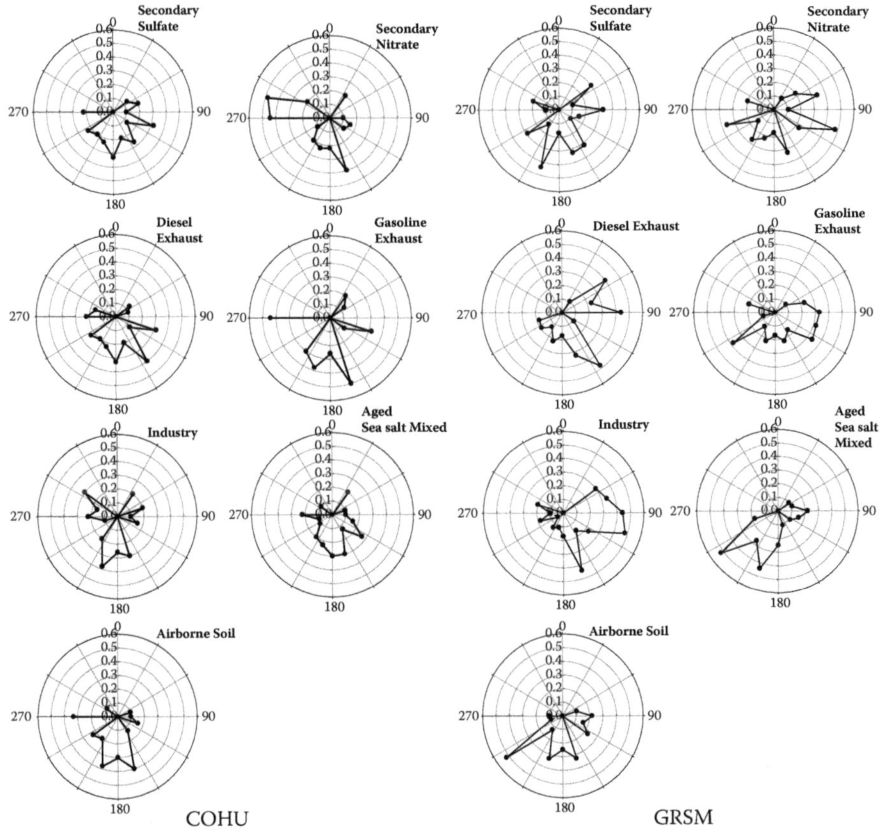

FIGURE 2.6
Conditional probability function plots for contributing sources to $PM_{2.5}$ at COHU and GRSM provided by PMF.

of North America [CECNA], 2004); hence, emissions of the SO_2 and NO_x precursors would be spread in all directions. Besides, the secondary PM needs a relatively long time period to form, and during that the particles are sufficiently mixed over a large region. Nevertheless, the secondary PM still appeared to be associated to the south rather than to the north areas of both sites, which appeared to be linked to the prevalent wind directions at these sites. The CPF plots for diesel exhaust and gasoline exhaust PM contributions consistently pointed to certain directions (i.e., to the south of COHU and to the south and southeast of GRSM), which can be linked to the road locations. The plots also showed a strong contribution of the industry source from the south to COHU and mainly from the east to GRSM, which points to the directions of some large lead industries in the area (USEPA, 2004). For airborne soil particles, the CPF plots consistently showed strong contributions from the south and southeast, which were somewhat similar to the traffic contribution directions and suggested the local road dust was a potential source. The aged sea salt mixed PM and the peaked airborne soil PM contribution, for example, may also have originated from long-range transport pollution. To further reveal these contributions, PSCF plots should be used. In particular, the HYSPLIT backward trajectories also provided useful information to analyze the potential contribution of long-range transport as mentioned for the case of airborne soil PM.

2.5 Summary

Receptor models complement dispersion models to reveal the casual links between source emissions and consequent effects at receptor sites. Receptor models use mathematical procedures, commonly with a trial-and-error approach, to quantify the source contributions based on specific source maker species present in ambient samples. The models are commonly applied in air pollution research for the source apportionment studies of PM and VOCs. The quantitative source apportionment results produced by receptor models can be used to identify the most important sources to prioritize emission control efforts. Several types of receptor models are available, which can be divided into two groups: CMB and multivariate models. Based on the method used to solve the general equation, CMB requires both source profiles and ambient measurement data, while the multivariate models, such as PMF, require only ambient PM data input. COPREM is a hybrid model that combines features from both model groups. Each model type has its own limitations and advantages.

Preparation of the input data of required quality is a crucial task for the success of receptor model applications. This includes raw data screening, reconstructed mass, data uncertainty estimates, and preparation of source composition profiles. The major steps of running a receptor model generally include (a) identifying potential sources in a study area, (b) selecting the number of important sources (and source profiles for CMB), (3) running the model, and (4) interpreting the results. Evaluation of receptor modeling results is not straightforward, but some supplementary tools, such as CPF and PSCF, may be used in examining the source apportionment results.

References

Anttila P., Paatero P., Tapper U., and Jarvinen O.A.E. 1995. Source identification of bulk wet deposition in Finland by positive matrix factorization, *Atmospheric Environment*, 29: 1705–1718.

Begum B.A., Kim E., Jeong C.H., Lee D.W., and Hopke P.K. 2005. Evaluation of the potential source contribution function using the 2002 Quebec forest fire episode, *Atmospheric Environment*, 39: 3719–3724.

Chan Y.C., Simpson R.W., McTainsh G.H., Vowles P.D., Cohen D.D., and Bailey G.M. 1997. Characterisation of chemical species in $PM_{2.5}$ and PM_{10} aerosols in Brisbane, Australia, *Atmospheric Environment*, 31: 3773–3785.

Chen K.S., and Lin C.F. 2001. Determination of source contributions to ambient $PM_{2.5}$ in Kaohsiung, Taiwan, using a receptor model, *Journal of Air and Waste Management Association*, 51: 489–498.

Cheng Z.L., Lam K.S., Chan L.Y., Wang T., and Cheng K.K. 2000. Chemical characteristics of aerosols at coastal station in Hong Kong I. Seasonal variation of major ions, halogens, and mineral dusts between 1995 and 1996, *Atmospheric Environment*, 34: 2771–2783.

Chow J.C., Watson J.G., Ashbaugh L.L., and Magliano K.L. 2003. Similarities and differences in PM_{10} chemical source profiles for geological dust from the San Joaquin Valley, California, *Atmospheric Environment*, 37: 1317–1340.

Chow J.C., Watson J.G., Kuhns H., Etyemezian V., Lowenthal D.H., Crow D., Kohl S.D., Engelbrecht J.P., and Green M.C. 2004. Source profiles for industrial, mobile, and area sources in the Big Bend Regional Aerosal Visibility and Observational study, *Chemosphere*, 54: 185–208.

Chueinta W., Hopke P.K., and Paatero P. 1999. Investigation of sources of atmospheric aerosal at urban and suburban residential areas in Thailand by positive matrix factorization, *Atmospheric Environment*, 34: 3319–3329.

Commission for Environmental Cooperation of North America (CECNA). 2004. *North American Power Plant Air Emissions*, Commission for Environmental Cooperation of North America, Montreal, Canada.

Dzubay T.G., Stevens R.K., Balfour W.D., Williamson H.J., Cooper J.A., Core J.E., De Cesar R.T., Crutcher E.R., Dattner S.L., Davis B.L., Heisler S.L., Shah J.J., Hopke P.K., and Johnson D.L. 1984. Interlaboratory comparison of receptor model results for Houston aerosol, *Atmospheric Environment*, 18: 1555–1556.

Engelbrecht J.P., Swanepoel L., Chow J.C., Watson J.G., and Egami R.T. 2002. The comparison of source contributions from residential coal and low-smoke fuels, using CMB modeling, in South Africa, *Environmental Science and Policy*, 5: 157–167.

Garg B.D., Cadle S.H., Mulawa P.A., Groblicki P.J., Laroo C., and Parr G.A. 2000. Brake wear particulate matter emissions, *Environmental Science and Technology*, 34: 4463–4469.

Garivait H. 1999. A study on air pollution by airborne polycyclic aromatic hydrocarbons (PAHs) in Bangkok urban atmosphere, doctoral dissertation, Asian Institute of Technology, Pathumthani, Thailand.

Gordon E.G. 1998. Receptor models, *Environmental Science and Technology*, 22: 1132–1142.

Han S., Youn J.S., and Jung Y.W. 2011. Characterization of PM_{10} and $PM_{2.5}$ source profiles for resuspended road dust collected using mobile sampling methodology, *Atmospheric Environment*, 45(20): 3343–3351.

Henry R.C., Lewis C.W., Hopke P.K., and Williamson H.J. 1984. Review of receptor model fundamentals, *Atmospheric Environment*, 18: 1507–1515.

Ho K.F., Lee S.C., Chan C.K., Yu J.C., Chow J.C., and Yao X.H. 2003. Characterization of chemical species in $PM_{2.5}$ and PM_{10} aerosols in Hong Kong, *Atmospheric Environment*, 37: 31–39.

Hooper R.P., and Peters N.E. 1989. Use of multivariate analysis for determining sources of solutes found in wet deposition in the United States, *Environmental Science and Technology*, 23: 1263–1268.

Hopke P.K. 2007. *A Guide to Positive Matrix Fractorization*. Department of Chemistry, Clarkson University. http://www.epa.gov/ttnamti1/files/ambient/pm25/workshop/laymen.pdf.

Hwang I.J., and Hopke P.K. 2007. Estimation of source apportionment and potential source locations of $PM_{2.5}$ at a west coastal IMPROVE site, *Atmospheric Environment*, 41: 506–518.

Kim E., and Hopke P.K. 2006. Characterization of fine particle sources in the Great Smoky Mountains area, *Science of the Total Environment*, 368: 781–794.

Kim E., Hopke P.K., and Edgerton E.S. 2004. Improving source identification of Atlanta aerosol using temperature resolved carbon fractions in positive matrix factorization, *Atmospheric Environment*, 38: 3349–3362.

Kim E., Larson T.V., Hopke P.K., Slaughter C., Sheppard L.E., and Claiborn C. 2003. Source identification of $PM_{2.5}$ in an arid northwest U.S. city by positive matrix factorization, *Atmospheric Environment*, 66: 291–305.

Kim Oanh N.T., Pongkiatkul P., Upadhyay N., and Hopke P. 2009. Design a monitoring program of ambient particulate matter for source apportionment study by receptor modeling, *Atmospheric Environment*, 43: 3334–3344.

Kim Oanh N.T., Thiansathit W., Bond T.C., Subramanian R., Ekbordin W., and Ittipol P. 2010. Compositional characterization of $PM_{2.5}$ emitted from in-use diesel vehicles in Asia, *Atmospheric Environment*, 44(1): 15–22.

Kim Oanh N.T., Thuy L.B., Tipayarom D., Manadhar B.R., Pongkiatkul P., Simpson C.D., and Liu L.S. 2011. Source characterization of aerosol emission from field burning of rice straw, *Atmospheric Environment*, 45: 493–502.

Kim Oanh N.T., Upadhyay N., Zhuang Y.H., Hao Z.P., Murthy D.V.S., Lestari P., Villarine J.T., Chengchua K., Co H.X., Dung N.T., and Lindgren E.S. 2006. Particulate air pollution in six Asian cities: Spatial and temporal distributions, and associated sources, *Atmospheric Environment*, 40: 3367–3380.

Kittiruangpol P. 2002. Source apportionment of fine and coarse particulate matters in Bangkok metropolitan region using positive matrix factorization receptor model, master's thesis, Asian Institute of Technology, Pathumthani, Thailand.

Larson T., Gould T., Simpson C., Liu L.J., Claiborn C., and Lewtas J. 2004. Source apportionment of indoor, outdoor, and personal $PM_{2.5}$ in Seattle, Washington, using positive matrix factorization, *Journal of Air and Waste Management Association*, 54: 1175–1187.

Lee E., Chan C.K., and Paatero P. 1999. Application of positive matrix factorization in source apportionment of particulate pollutants in Hong Kong, *Atmospheric Environment*, 33: 3201–3212.

Liu W., Hopke P.K., Han Y.J., Yi S.M., Holsen T.M., Cybart S., Kozlowski K., and Milligan M. 2003. Application of receptor modeling to atmospheric constituents at Potsdam and Stockton, NY, *Atmospheric Environment*, 37: 4997–5007.

Lowenthal D.H., and Rahn K.A. 1987. Application of the factor-analysis receptor model to simulated urban- and regional-scale data sets, *Atmospheric Environment*, 21: 2005–2013.

Maykut N.N., Lewtas J., Kim E., and Larson T.V. 2003. Source apportionment of $PM_{2.5}$ at an urban IMPROVE site in Seattle, Washington, *Environmental Science and Technology*, 37: 5135–5142.

Ogulei D., Hopke P.K., Zhou L., Paatero P., Park S.S., and Ondov J.M. 2005. Receptor modeling for multiple time resolved species: The Baltimore supersite, *Atmospheric Environment*, 39: 3751–3762.

Okamoto S., Hayashi M., Nakajima M., Kainuma Y., and Shiozawa K. 1990. A factor analysis-multiple regression model for source apportionment of suspended particulate matter, *Atmospheric Environment*, 24: 2089–2097.

Paatero P. 1999. The multilinear engine-a table-driven, least squares program for solving multilinear problems, including the n-way parallel factor analysis model, *Computational and Graphical Statistics*, 8(4): 854–888.

Paatero P. 2007. *User's Guide for Positive Matrix Factorization Programs PMF2 and PMF3*, Part 1: Tutorial, Part 2: Reference, University of Helsinki, Finland.

Paatero P., and Hopke P.K. 2003. Discarding or downweighting high-noise variables in factor analytic models, *Analytica Chimica Acta*, 490: 277–289.

Paatero P., and Tapper U. 1994. Positive matrix factorization: A non-negative factor model with optimal utilization of error estimates of data values, *Environmetrics*, 5: 111–126. doi:10.1002/env.3170050203.

Pace T.G., and Watson J.G. 1987. *Protocol for Applying and Validating the CMB Model*, United States Environmental Protection Agency, Research Triangle Park, NC.

Park S.S., Bae M.S., and Y.J. K. 2001. Chemical composition and source apportionment of $PM_{2.5}$ particles in the Sihwa Area, Korea, *Journal of Air and Waste Management Association*, 51: 393–405.

Polissar A.V., Hopke P.K., Paatero P., Malm. W.C., and Sisler J.F. 1998. Atmospheric aerosol over Alaska 2. Elemental composition and sources, *Geophysical Research*, 103(D15): 19045–19057.

Pongkiatkul P. 2006. Development of hybrid receptor models for air pollution source apportionment study, doctoral dissertation, Asian Institute of Technology, Pathumthani, Thailand.

Pongkiatkul P., and Kim Oanh N.T. 2007. Assessment of potential long-range transport of particulate air pollution using trajectory modeling and monitoring data, *Atmospheric Research*, 85(1): 3–17.

Ramadan Z., Song X.H., and Hopke P.K. 2000. Identification of sources of Phoenix aerosol by positive matrix factorization, *Journal of Air and Waste Management Association*, 50: 1308–1320.

Reinikainen S.P., Laine P., Minkkinen P., and Paatero P. 2001. Factor analytical study on water quality in lake Saimaa, Finland, *Fresenius Journal of Analytical Chemistry*, 369: 727–732.

Schauer J.J., Kleeman M.J., Cass G.R., and Simoneit B.R.T. 2002. Measurement of emissions from air pollution sources. 5. C1–C32 organic compounds from gasoline-powered motor vehicles, *Environmental Science and Technology*, 36: 1169–1180.

Singh K.P., Malik A., and Sinha S. 2005. Water quality assessment and apportionment of pollution sources of Gomti River (India) using multivariate statistical techniques—a case study, *Analytica Chimica Acta*, 538: 355–374.

Song X.H., Polissar A.V., and Hopke P.K. 2001. Source of fine particle composition in the northeastern US, *Atmospheric Environment*, 35: 5277–5286.

Su M.C., and Christensen E.R. 1997. Apportionment of sources of polychlorinated dibenzo-p-dioxins and dibenzofurans by a chemical mass balance model, *Water Research*, 31: 2935–2948.

Turpin B.J., and Lim H.J. 2001. Species contributions to $PM_{2.5}$ mass concentrations: Revisiting common assumptions for estimating organic mass, *Aerosol Science and Technology*, 35: 602–610.

Truc V.T.Q., and Kim Oanh N.T. 2007. Roadside BTEX and other gaseous air pollutants in relation to emission sources, *Atmospheric Environment*, 41: 7685–7697.

Upadhyay N. 2002. Source apportionment of fine and coarse fractions of particulate matters in Bangkok metropolitan region by receptor modeling, master's thesis, Asian Institute of Technology, Pathumthani, Thailand.

U.S. Environmental Protection Agency (USEPA). 1998. *CMB8 Application and Validation Protocol for PM2.5 and VOC*, United States Environmental Protection Agency, Research Triangle Park, NC.

U.S. Environmental Protection Agency (USEPA). 2004. 2004 TRI Public data release, toxics release inventory (TRI) program. http://www.epa.gov/tri/tridata/tri04/index.htm, accessed 2011.

U.S. Environmental Protection Agency (USEPA). 2008. SPECIATE 4.2. EPA's repository of total organic compound (TOC) and particulate matter (PM) speciated profiles for a variety of sources for use in source apportionment studies. http://www.epa.gov/nrmrl/news/012010/news012010.html.

Vanderstraeten P., Lenelle Y., Meurrens A., Carati D., Brenig L., Delcloo A., Offere Z.Y., and Zaady E. 2008. Dust storm originate from Sahara covering Western Europe: A case study, *Atmospheric Environment*, 42: 5489–5493.

Volckens J., Braddock J., Snow R.F., and Crews W. 2007. Emissions profile from new and in-use handheld, 2-stroke engines, *Atmospheric Environment*, 41: 640–649.

Wahlin P. 2003. A multivariate receptor model with a physical approach, *Atmospheric Environment*, 37: 4861–4867.

Wang Y.Q., Zhang X.Y., and Arimoto R. 2006. The contribution from distant dust sources to the atmospheric particulate matter loadings at XiAn, China during spring, *Science of the Total Environment*, 368: 875–883.

Wang H.L., Zhuang Y.H., Hao Z.P., Cao M.Q., Zhong J.X., Wang X.K., and Kim Oanh N.T. 2008. Polycyclic aromatic hydrocarbons from rural household biomass burning in a typical Chinese village, *Journal of Science in China Series D: Earth Sciences*, 51: 1013–1020.

Wangkiat A., Garivait H., Harvey N.W., and Okamoto S. 2001. Application of CMB8 receptor model for source apportionment in Bangkok metropolitan area. http://www.iic.tuis.ac.jp/edoc/journal/ron/r5-1-10/r5-1-10i.html.

Watson J.G., and Chow J.C. 1989. Receptor in air resources management: A summary of APCA international specialty conference, *Journal of Air and Waste Management Association*, 39: 419–426.

Watson J.G., and Chow J.C. 2001. Source characterization of major emission sources in the Imperial and Mexicali Valleys along the U.S. Mexico border, *Science of the Total Environment*, 276: 33–47.

Watson J.G., Chow J.C., and Houck J.E. 2001. $PM_{2.5}$ chemical source profiles for vehicle exhaust, vegetative burning, geological material, and coal burning in northwestern Colorado during 1995, *Chemosphere*, 43: 1141–1151.

Watson J.G., Robinson N.F., Lewis C., and Coulter T. 1997. *Chemical Mass Balance Receptor Model Version 8 (CMB8) User's Manual*, Desert Research Institute, Reno, NV.

Wimolwattanapun W., Hopke P.K., and Pongkiatkul P. 2011. Source apportionment and potential source locations of $PM_{2.5}$ and $PM_{2.5-10}$ at residential sites in metropolitan Bangkok, *Atmospheric Pollution Research*, 2: 172–181.

Xie Y., and Berkowitz C.M. 2006. The use of positive matrix factorization with conditional probability functions in air quality studies: An application to hydrocarbon emissions in Houston, Texas, *Atmospheric Environment*, 40: 3070–3091.
Zhao W., and Hopke P.K. 2006. Source identification for fine aerosols in Mammoth Cave National Park, *Atmospheric Environment*, 80: 309–322.

3

Monitoring and Source Apportionment for Particulate Matter Pollution in Six Asian Cities

Nguyen Thi Kim Oanh, Prapat Pongkiatkul, Melliza Templonuevo Cruz, Nghiem Trung Dung, Ligy Phillip, Guoshun Zhuang, and Puji Lestari

CONTENTS

3.1 Introduction

Atmospheric particulate matter (PM) has multiple effects on human health and the environment. Major emission sources, highlighted in Chapter 1 of this book, release not only a huge amount of primary PM but also precursor gases, which subsequently form secondary PM in the atmosphere. High levels of total suspended PM (TSP) and PM_{10} (particles with an aerodynamic diameter less than 10 μm) in populated urban areas of Asia have been reported (Gupta et al., 2006; Health Effects Institute [HEI], 2004; Molina and Molina, 2004; Tsai and Chen, 2006). Fragmented available data of $PM_{2.5}$ (particles with an aerodynamic diameter less than 2.5 μm) also showed very high levels of this toxic air pollutant, often exceeding the WHO guideline by a factor of 2 or more (Hopke et al., 2008; Kim Oanh et al., 2006).

The extent of health and atmospheric effects of particles depends on their size and compositions. Smaller particles can penetrate the lungs deeper than large particles and hence are more damaging. PM_{10} have long been connected to adverse health effects, including increased mortality, whereas fine particles ($PM_{2.5}$) pose even higher risks (Dockery and Pope, 1994; Donaldson et al., 1998; Pope et al., 2009; Schwartz and Neas, 2000). These fine particles also have large surface areas for adsorption of toxic organic compounds, such as polycyclic aromatic hydrocarbons (PAHs). Furthermore, due to their ability to absorb and scatter solar radiation, the atmospheric particles, especially fine ones, effectively reduce visibility (Kim et al., 2001; Watson, 2002) and affect Earth's radiation energy balance (Bond et al., 2004; Intergovernmental Panel on Climate Change [IPCC], 2007). Black carbon (BC) particles, for example, strongly absorb the solar radiation and hence can also lead to warming of Earth's atmosphere (Ramanathan and Carmichael, 2008).

There have been growing research interests on characterization of levels and compositions of size-segregated PM. These data help understand the nature of the particulate pollution in relation to local sources, long-range transport, and atmospheric transformation processes, which are essentially required in formulation of effective air quality management strategies. In particular, the data can be analyzed by receptor modeling to identify the contributing sources and to quantify their contributions to PM pollution, as detailed in Chapter 2 of this book.

There are still insufficient systematic, regionally compatible, and long-term PM data records for the Asian developing countries. Until recently, only the TSP was the most monitored and is still a regulated pollutant in many countries (refer to detail in Chapter 1). Scarcity of information on ambient levels of fine particles and their physical and chemical properties in large urban areas of Asia prevents researchers and policy makers from gaining insights into the nature of PM pollution. To fill the gap partly, the Asian Regional Air Pollution Research Network or AIRPET had designed and implemented a monitoring program to collect such data. One of the main research objectives of the program was to provide a comprehensive assessment of the PM pollution in the region. The focus was on $PM_{2.5}$ and PM_{10} or $PM_{10-2.5}$ (coarse fraction) levels and composition in six cities or metropolitan regions in Asia, namely, the Bangkok metropolitan region (BMR, Thailand); Bandung (Indonesia); Beijing (China); Chennai (India); metro Manila (Philippines); and Hanoi metropolitan region (Vietnam). The network was mainly funded by the Swedish International Development Cooperation Agency (Sida) through the Asian Regional Research Program in Environmental Technology (ARRPET; http://www.arrpet.ait.ac.th/index.html) and was coordinated by the Asian Institute of Technology (AIT). This chapter summarizes the findings on the PM levels, composition, and source contributions in the six AIRPET cities based on the monitoring results for the 2002–2008 period. The focus is on the design of the regional PM monitoring program with necessary quality assurance/quality control (QA/QC) to generate data of suitable quality.

3.2 Design of Regional PM Monitoring

3.2.1 Overview of AIRPET PM Monitoring Program Design

The six selected cities or regions are different in topography, meteorology, population, primary energy use, industry, vehicular mix, and population density. The climate of the region is dominated by monsoons, with two distinct seasons, dry and wet. Although the dry and wet seasons cover different months of the year in different countries, only these two broad season categories were used for consistency in intercity comparison of the monitoring results. In all cities located in the Northern Hemisphere, the wet season is within the summer months (May/June to October/December), and the dry season is the rest of the year. The opposite is true for Bandung (Indonesia) due to its location in the Southern Hemisphere. During the wet season, AIRPET samples were collected only when there was no intensive and continuous rain at the monitoring sites.

3.2.1.1 Sampling

3.2.1.1.1 Monitoring Site Selection

Siting is a challenging task in crowded and congested urban areas in Asian developing countries, where mixed land use patterns are common, especially when the limit in resources prohibits having a large number of sites. Sampling sites with known strong impacts from certain sources such as roadside or industrial estates are expected to have higher contributions from nearby sources and hence can be used to characterize the source emission impacts on ambient air quality. However, they would not provide an overall picture of the ambient pollution levels in an airshed. Because of complex urban activities forming different air bubbles (a bubble is a subairshed affected by some common sources) in a large urban area, more than one monitoring site is normally required (Kim Oanh et al., 2009). These sites would capture common urban air bubbles, such as upwind, downwind, city center, commercial, residential, industrial, and so on. Thus, an air bubble should at least be characterized by a monitoring site that would reflect the representative mixture of pollution from major contributing sources to that air bubble. A careful consideration of the emission source distribution and typical meteorological conditions (wind pattern, rain pattern, etc.) in a study area during a year should be made when selecting monitoring sites.

For PM source apportionment studies in particular, a few but representative mixed sites are generally desirable in an urban airshed. Sites located upwind and downwind of a city help in revealing the air pollution transport across the study area. If remote sites are added, regional-scale phenomena such as long-range transport can also be reflected. Auxiliary information such as meteorology and gaseous pollutants at selected monitoring sites can greatly aid in the evaluation of the source apportionment results. The information can be directly included in the analysis by some advanced receptor models, such as those based on the multilinear engine (ME) algorithm (Zhou et al., 2005).

The AIRPET PM monitoring program was designed to provide harmonious data in the six cities or metropolitan regions. The sampling sites were classified into six categories characterizing an urban airshed: upwind, traffic, residential, commercial, industrial, and urban mixed (downwind) sites. Each city or metropolitan region commonly had five to six sites out of these six categories. The classification of sites was based on the major influencing emission sources or activities in a given urban air bubble and was done to achieve maximum consistency between all six cities. The upwind site refers to an upwind location of the city in the dry season, and the traffic site was located in the city central areas and close to traffic lanes (within 3–10 m). The commercial site was located in urban central areas, the residential site was located in a crowded part of a city, while the industrial site was located in areas with concentrated industrial activities within the metropolitan regions. Finally, the urban mixed site was located in the downwind direction during the dry season and was affected by most of the important sources in the selected urban areas. The sampling intake was generally placed at a height of 3–5 m. Overall characteristics of the AIRPET sites are given in Table 3.1. Based on the number of samples generated, the AIRPET urban monitoring sites were divided into two types: the intensive sites, with a large number of PM samples collected, and the minor sites, for which the total number of samples was around 30 during the entire period. Each city had one or two intensive sites that provided a necessary number of samples for statistical analyses, including receptor modeling studies. In AIRPET phase 2 (2004–2008), a remote site located 35–250 km upwind of each selected city was included, and samples were also collected at these remote sites in both dry and wet seasons.

TABLE 3.1

Information on Sampling Sites in Six Asian Cities and Remote Areas

Country	City/Area	Site Characterization	Periods	PM Size
Thailand	BMR	6 sites (upwind, traffic, mix, residential, suburban, agricultural)	Mar 02–Feb 08	$PM_{10\text{-}2.5}$; $PM_{2.5}$
	KhaoYai	Remote mountain background (120 km NE from Bangkok)	Jan 05–Jun 05	$PM_{10\text{-}2.5}$; $PM_{2.5}$
China	Beijing	5 urban sites (upwind, traffic, residential, industry, suburban)	Jul 01–Jan 07	PM_{10}; $PM_{2.5}$
	Duolun	Remote rural background (180 km N from Beijing)	Jul 06–Jan 07	PM_{10}; $PM_{2.5}$
India	Chennai	5 urban sites (upwind, traffic, industry, mixed traffic, urban background)	Nov 02–Nov 06	PM_{10}; $PM_{2.5}$
	Kavalur	Remote mountain background (250 km NW from Chennai)	Feb 06–Jan 07	PM_{10}; $PM_{2.5}$
Indonesia	Bandung	5 urban sites (upwind, traffic, mix, residential, industry)	Jun 01–Jul 07	$PM_{10\text{-}2.5}$; $PM_{2.5}$
	Rancabali	Remote mountain background (35 km SW from Bandung)	Aug 05–May 05	$PM_{10\text{-}2.5}$; $PM_{2.5}$
Philippines	Manila	6 urban sites (urban background, traffic, mix, industry, commercial, urban-mixed)	Aug 00–Oct 07	$PM_{10\text{-}2.5}$; $PM_{2.5}$; PM_{10}
	Gabaldon	Remote mountain background (150 km N from Manila)	Jan 06–Jun 08	PM_{10}; $PM_{2.5}$
Vietnam	Hanoi	5 urban sites (upwind, traffic, mix, residential, commercial)	Oct 02–Mar 07	$PM_{10\text{-}2.5}$; PM_{10}; $PM_{2.5}$
	TamDao	Remote mountain background (70 km NW from Hanoi)	Sep 05–Jan 06	$PM_{10\text{-}2.5}$; $PM_{2.5}$

Note: BMR, Bangkok metropolitan region, including Bangkok City and surrounding provinces.

3.2.1.1.2 Sampling Equipment

An important requirement of a sampling system for receptor modeling studies is the ability to collect enough PM mass on filters for subsequent chemical and physical analyses required to produce a suitable PM compositional dataset. The distinction between the coarse and fine fractions is of fundamental importance for receptor modeling because of the differences in the nature, origin, and environmental fates of these two fractions (Chow, 1995). In addition, a separation between the fine (normally more acidic) and the coarse particles (more alkali) helps minimize the alteration of the chemical composition of PM resulting from acid-base reactions during the sampling and storage (Kim Oanh et al., 2009). Some PM samplers provide both coarse and fine PM samples simultaneously (i.e., dichotomous samplers) (California Air

Resources Board [CARB], 2001) or Gent samplers (Hopke et al., 1997), which are particularly suitable for PM sampling for the subsequent source apportionment studies. Organic compounds such as PAHs or aliphatic compounds are desirable markers for a number of combustion sources, such as vehicle exhaust and biomass burning but normally require a larger amount of PM deposit than those provided by dichotomous or Gent samplers. A Hi-Vol Puff sampler that has a volumetric flow of a few hundred liters per minute may be used in parallel to collect these semi-VOC compounds (Kim Oanh et al., 2009).

In all AIRPET cities, monitoring for $PM_{2.5}$ and PM_{10} or $PM_{10-2.5}$ was conducted using dichotomous or dichot (CARB, 2001), Minivol (Airmetrics, 2012), and other equivalent samplers depending on the availability. A dichot, available in each city of the AIRPET network, was colocated with other samplers to collect about 30 samples simultaneously for the performance evaluation. The fine ($PM_{2.5}$) and coarse ($PM_{10-2.5}$) fractions collected by dichots were summed to get the PM_{10} mass concentration.

3.2.1.1.3 *Frequency of Filter-Based Sampling versus Online Monitoring*

Filter-based PM sampling for mass concentration measurements is normally performed over 24 h, which provides a convenient value for comparison with 24-h ambient air quality standards. However, shorter sampling intervals provide better insight into the temporal variations of contributing sources, meteorology, as well as the atmospheric chemistry (affecting secondary particle formation). The measurements should be performed as frequently as possible, preferably semi-continuously or online. Advanced techniques that can measure online mass/mass equivalent or particle number concentrations in different size ranges are available (Chow et al., 2008). Online monitoring techniques for PM composition are also available, such as those for water-soluble ions, elements, and semicontinuous analyzer for elemental carbon/organic carbon (EC/OC), as highlighted in the work of Kim Oanh et al. (2009). A range of other particle properties (light scattering, light absorption) is also monitored online and can be used as the PM mass surrogate/equivalent.

The filter-based measurements are still more commonly used as they are less expensive and more readily available in most developing countries; therefore, these measurements were performed in the AIRPET network. The AIRPET PM sampling was mainly conducted over a 24-h period, but in some instances, samples were also collected separately for daytime and nighttime. For consistency, in this chapter only the 24-h $PM_{2.5}$ and PM_{10} data are discussed.

3.2.1.1.4 *Filter Selection, Filter Pretreatment, and Weighing*

Selection of filters should be made to suit the intended subsequent analytical parameters and methods (Chow, 1995). In the AIRPET monitoring program, several types of filters were used, with the most common being Teflon, quartz fiber, and mixed cellulose ester. Before sampling, quartz filters were prefired at 550°C for 5–6 h to remove carbonaceous/organic contaminants.

The treatment of quartz filters at this temperature is recommended for samples to be analyzed for organic components.

3.2.1.2 Sample Analysis

The AIRPET PM samples were analyzed for mass and composition, including water-soluble ions, BC, and elements. Analyses for EC and OC were also done for samples collected in Bangkok and Beijing. The analytical methods used in the AIRPET network remained basically the same as those used in the early phase of the project, which are discussed in Kim Oanh et al. (2006).

3.2.1.2.1 Gravimetric Measurement

The filters/samples should be conditioned for an extended period before weighing. QA/QC blank reference filters should be weighed before and after weighing real samples to ensure that the filter mass was unchanged over time under the weighing conditions (U.S. Environmental Protection Agency [USEPA], 1998). The interferences in gravimetric analysis of filters are those related to the nature of filters and those related to the electrostatic effect. The latter is commonly removed by exposing the filters to a low-level radioactive source (500 pCi of polonium-210) prior to and during sample weighing (Engelbrecht et al., 1980).

In the AIRPET monitoring network, prefired quartz filter samples or Teflon filters were used for mass determination. The PM samples were conditioned for about 24 h (temperature 22°C ± 5°C, relative humidity 40% ± 10%) before weighing. For measurement of the fine fraction mass ($PM_{2.5}$), a calibrated six-digit analytical balance was commonly used. Being susceptible to relative humidity, the mixed-cellulose filters were not used for determination of the PM mass.

3.2.1.2.2 Composition Analysis

Generally, for PM apportionment studies by the receptor modeling, PM samples should be analyzed for markers of major emission sources in a study area. In Asian countries major sources of primary PM include vehicle exhaust, residential cooking, biomass burning, construction activity, soil/road dust, and industry. In addition, the secondary particles also constitute a large portion of ambient PM (Kim Oanh et al., 2006). Thus, markers such as BC or EC, sulfate, and various fuel additives (elements) should be analyzed to identify the contribution of diesel exhaust. The ratio between EC and OC is also useful to distinguish the PM emitted from biomass burning and those from diesel vehicle exhaust (Kim Oanh et al., 2010, 2011). Likewise, sulfate, nitrate, and ammonium should be included to identify the secondary inorganic aerosols, while potassium is a useful inorganic marker for the contribution of biomass burning smoke. Gasoline vehicles are also expected to emit a considerable amount of PM. However, the markers for gasoline vehicle PM have not yet been readily established after lead additives were phased out. A full range of

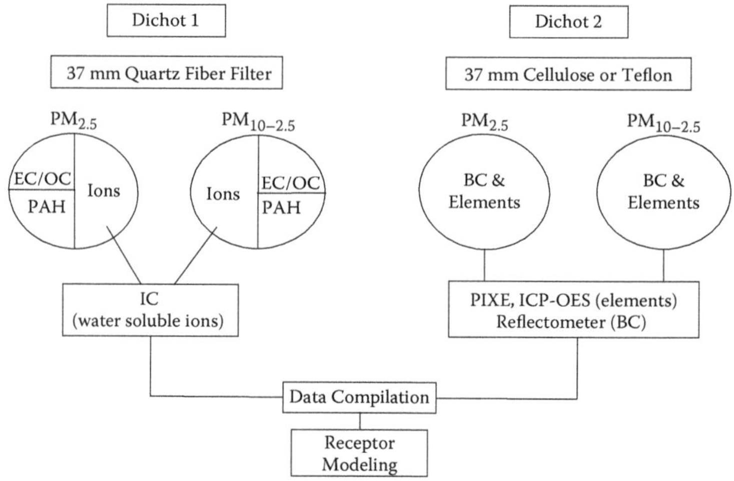

FIGURE 3.1
Segregation of PM filter samples collected by AIT using colocated dichots for different composition analyses. (*Source:* Adapted from Kim Oanh N.T., Pongkiatkul P., Upadhyay N., and Hopke P. 2009. Design a monitoring program of ambient particulate matter for source apportionment study by receptor modeling, *Atmospheric Environment*, 43: 3334–3344, courtesy of AE.)

organic compounds can be included for better source apportionment studies (Chowdhury et al., 2007), such as levoglucosan and methoxyphenols for biomass burning (Kim Oanh et al., 2011; Simoneit, 2002) or hopanes, steranes, and other compounds for vehicle emission (Schauer et al., 1999, 2002), but the cost would increase considerably. Markers of the coarse particle fraction may include, among others, crustal elements (soil particles), Na^+ and Cl^- (sea spray), and cement-related elements (construction activity). Road dust, for example, would contain the tire and brake wear (e.g., barium in brake linings) or resuspended lead particles and hence may be identified using these markers. More detail on the markers of major emission sources is presented in Chapter 2.

Samples collected by the AIRPET network were segregated for multiple analyses to generate as complete a PM composition dataset as possible for the purpose of receptor modeling. For illustration, the segregation of filters for different analyses at AIT is presented in Figure 3.1.

The analyzed chemical compositions included about 20 elements (Fe, Ni, Pb, Sr, Ti, V, Zn, Si, Al, Ca, Cd, Co, Cr, Cu, Mg, Mn, Mo, Tl, etc.), 8 water-soluble ions (Na^+, NH_4^+, K^+, Ca^{2+}, Mg^{2+}, NO_3^-, SO_4^{2-}, and Cl^-), and BC. The elements were analyzed by different equipment available for each laboratory, including atomic absorption spectroscopy (AAS; Indonesia); ICP-OES (inductively coupled plasma-optical emission spectrometry; AIT); ICP-AES (inductively coupled plasma-atomic emission spectroscopy; China); ICP-MS (inductively coupled plasma-mass spectrometry; India, Vietnam); PIXE (proton-induced x-ray emission; AIT, Philippines); and XRF (x-ray fluorescence; AIT,

Vietnam). The water-soluble ions were mainly analyzed by ion chromatography using the procedure given by Chow (1995). BC was measured using an EEL M43D digital smokestain reflectometer in each participating laboratory. In addition, EC/OC analysis was done for Bangkok samples using the thermal optical transmittance (TOT) method following the NIOSH (National Institute for Occupational Safety and Health) 5040 protocol and for some samples in Beijing by a C/H/N elemental analyzer (Dan et al., 2004).

3.2.2 Receptor Modeling

Several receptor models are available, each having specific advantages and disadvantages as detailed in Chapter 2. In this chapter, the chemical mass balance version 8 (CMB8) model was applied for source apportionment of $PM_{2.5}$ collected in six cities. The profiles of major PM sources used for the analysis are summarized in Table 3.2 and in the Appendix. The source profiles have been compiled from various data sources, both experimentally produced (Kim Oanh et al., 2010, 2011) and produced by the constrained physical receptor model (COPREM) for BMR (Upadhyay, 2002). Source profiles presented in Table 3.2 contain both EC and BC, which are significantly different for diesel exhaust. In the model calculation, either EC or BC is selected, depending on the availability of EC or BC measurements for ambient PM samples.

3.3 Quality Assurance and Quality Control in PM Monitoring

Data quality is of the foremost concern, and implementation of a harmonized QA/QC procedure is especially important when many laboratories in different cities are involved. Use of common monitoring methodologies or comparison of data generated by different sampling and analytical techniques in the network is important to obtain a harmonized regional database. The QA/QC of a monitoring network should be followed at all possible stages from sampling (siting, sampling method selection, sampler calibration, colocated sampling, selection of filter media, filter preparation, sampling execution, sample preservation), to chemical analyses (filter segregation, weighing, trip blanks, laboratory blanks, blind samples, split samples, standard reference material [SRM], etc.), and finally to data analysis and data management.

3.3.1 Sampling QA/QC

In the AIRPET network, a QA/QC procedure was established and implemented to ensure adequate data quality. Besides the siting issue discussed, the sampling QA/QC included filter selection and preparation, periodical sampler calibration, and colocated sampling. All samplers were periodically

TABLE 3.2

Selected Source Composition Profiles (in Percentage of PM Mass, %) Used in PM Apportionment by CMB8 (Other Profiles Are Presented in the Appendix)

	Field Rice Straw Burning[a]	Diesel Composite[b]	Biomass Burning	Soil	Sea Salt	Steel Industry	$(NH_4)_2SO_4$	Construction
	Size							
Name	$PM_{2.5}$	$PM_{2.5}$	PM	PM	PM	PM	PM	PM
Na^+	0.29 ± 0.28	0.28 ± 0.63	0.2 ± 0.36	2.37 ± 0.71	32.44 ± 9.73	0 ± 0	0 ± 0	0 ± 0.01
K^+	5.0 ± 3.4	0.13 ± 0.42	5.9 ± 3.54	7.12 ± 2.14	1.17 ± 0.35	5 ± 0.86	0 ± 0	1.51 ± 0.3
Mg^{2+}	0.03 ± 0.03	0.24 ± 0.69	0 ± 0.01	1.75 ± 0.18	3.91 ± 1.17	0 ± 0	0 ± 0.01	0 ± 0.01
Ca^{2+}	0.12 ± 0.04	0.81 ± 1.66	0.4 ± 0.4	3.04 ± 0.91	1.23 ± 0.37	0.55 ± 0.09	0 ± 0.01	23.51 ± 4.7
NH_4^+	2.38 ± 1.17	0.13 ± 0.32	0 ± 0.1	0 ± 0.1	0 ± 0.1	0 ± 0	27.3 ± 5.46	0 ± 0.1
NO_3^-	0.34 ± 0.38	0.34 ± 0.89	0.9 ± 1.19	0 ± 0.01	0 ± 0.01	0.55 ± 0.09	0 ± 0	0 ± 0.01
SO_4^{2-}	0.98 ± 0.72	1.87 ± 2.95	5 ± 5.5	0 ± 0.01	0 ± 0.01	39.34 ± 6.74	72.7 ± 21.81	0 ± 0.01
Cl^-	6.93 ± 3.16	0.22 ± 3.89	3.1 ± 0.46	0.04 ± 0.01	58.3 ± 17.49	0 ± 0	0 ± 0	0.02 ± 0
BC[c]	—	67.58 ± 32.86	2.38 ± 0.95	0 ± 0.1	0 ± 0.1	0 ± 0	0 ± 0.1	0 ± 0.1
OC	33.54 ± 8.8	19.15 ± 13.91	38.63 ± 11.59	0 ± 0.1	0 ± 0.1	10 ± 2.74	0 ± 0.1	0 ± 0.1
EC	5.77 ± 2.79	46.71 ± 19.26	2.38 ± 0.71	0 ± 0.1	0 ± 0.1	0 ± 0.1	0 ± 0.1	0 ± 0.1
Mg	0.127 ± 0.195	—	—	—	—	—	—	—
Al	0.06 ± 0.03	0.10 ± 0.18	0.3 ± 0.27	11.97 ± 3.59	0 ± 0.01	0 ± 0	0 ± 0.01	2.54 ± 0.51
Si	0.04 ± 0.05	—	0.32 ± 0.27	23.25 ± 6.98	0 ± 0.01	4.95 ± 1.36	0 ± 0.01	10.34 ± 2.07
S	0.12 ± 0.04	—	1.5 ± 0.45	0.04 ± 0.01	2.72 ± 0.82	11.96 ± 2.09	0 ± 0	0.02 ± 0.01

Cl	—	—	3.1 ± 0.46	0.04 ± 0.03	58.3 ± 17.49	0 ± 0	0 ± 0	0.02 ± 0.6
K	—	—	5.9 ± 3.54	7.12 ± 2.14	1.17 ± 0.35	0 ± 0	0 ± 0	1.51 ± 0.3
Ca	0.15 ± 0.15	—	0.4 ± 0.4	3.04 ± 0.91	1.23 ± 0.37	0 ± 0	0 ± 0	23.51 ± 4.7
Ti	0.01 ± 0.01	0.002 ± 0.004	0.02 ± 0.02	1.34 ± 0.4	0 ± 0.01	0 ± 0	0 ± 0	0.28 ± 0.08
V	0.01 ± 0.01	0.0008 ± 0.01	0 ± 0.01	0 ± 0.01	0 ± 0.01	0.55 ± 0.1	0 ± 0	0.01 ± 0.01
Cr	0.02 ± 0.02	0.03 ± 0.06	—	—	—	—	—	—
Mn	0.002 ± 0.003	0.01 ± 0.01	0.02 ± 0.03	0.15 ± 0.1	0 ± 0.1	0.55 ± 0.1	0 ± 0	0.16 ± 0.1
Fe	0.02 ± 0.02	0.20 ± 0.28	0.23 ± 0.42	4.19 ± 1.26	0 ± 0.02	11.66 ± 2.08	0 ± 0	8.15 ± 1.63
Ni	0.01 ± 0.01	0.01 ± 0.02	0 ± 0.1	0.01 ± 0.1	0 ± 0.1	0.55 ± 0.1	0 ± 0	0 ± 0.1
Cu	0.04 ± 0.06	0.06 ± 0.15	0.01 ± 0.05	0.01 ± 0.01	0 ± 0.01	0.55 ± 0.1	0 ± 0	0 ± 0.02
Zn	0.01 ± 0.02	0.06 ± 0.17	0.04 ± 0.01	0.01 ± 0	0 ± 0	0 ± 0	0 ± 0	0 ± 0.01
As	0.0003 ± 0.0003	—	0.01 ± 0.02	0 ± 0.02	0 ± 0.02	0.05 ± 0.01	0 ± 0	0 ± 0.02
Se	0 ± 0	—	0.01 ± 0.01	0 ± 0.01	0 ± 0.01	0 ± 0	0 ± 0	0 ± 0.03
Br	0.005 ± 0.001	—	0.05 ± 0.08	0 ± 0.01	0.2 ± 0.1	0.05 ± 0.1	0 ± 0	0 ± 0.01
Rb	0.004 ± 0.001	—	0.02 ± 0.01	0.03 ± 0.01	0 ± 0.01	0.05 ± 0.1	0 ± 0	0.01 ± 0.01
Sr	0.003 ± 0.004	0.003 ± 0.023	0 ± 0.1	0.04 ± 0.1	0.04 ± 0.1	0.05 ± 0.1	0 ± 0.1	0.04 ± 0.1
Zr	0 ± 0	—	0 ± 0.1	0.02 ± 0.4	0 ± 0.1	0.5 ± 0.1	0 ± 0.1	0.02 ± 0.1
Cd	0.006 ± 0.002	0.001 ± 0.007	—	—	—	—	—	—
Sn	nd	—	0 ± 0.1	0 ± 0.1	0 ± 0.1	0.05 ± 0.1	0 ± 0.1	0 ± 0.1
Pb	0.01 ± 0.013	0.02 ± 0.04	0 ± 0.1	0 ± 0.1	0 ± 0.1	0.55 ± 0.1	0 ± 0.1	0 ± 0.1

Sources: [a]Kim Oanh et al. (2011); [b]Kim Oanh et al. (2010); other profiles produced by COPREM for BMR (Upadhyay, 2002) for both PM_{10} and $PM_{2.5}$. [c]Either BC or EC was selected for calculation to be compatible with the ambient datasets.

Note: nd, not detected; —, not available; 0 (zero), values were generally below 10^{-4}% of PM mass.

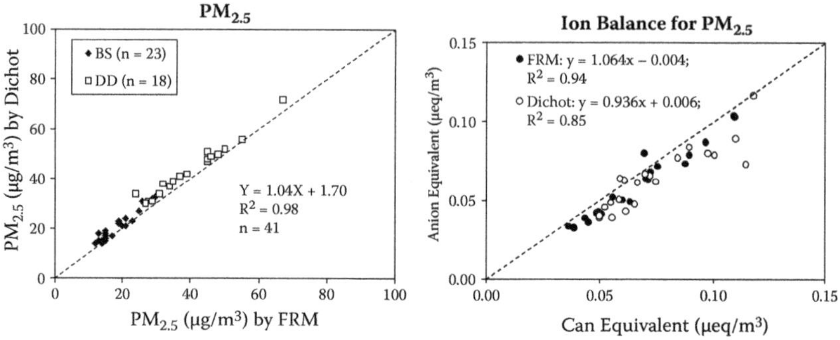

FIGURE 3.2
$PM_{2.5}$ mass and ion charge balance of colocated FRM and dichot samplers conducted by AIT, Bangkok. (*Source:* Adapted from Kim Oanh N.T., Pongkiatkul P., Upadhyay N., and Hopke P. 2009. Design a monitoring program of ambient particulate matter for source apportionment study by receptor modeling, *Atmospheric Environment*, 43: 3334–3344, courtesy of AE.)

calibrated for designated flow rates at actual temperature and pressure at the samplings sites. Each city was provided with one Sierra Anderson dichot (series 241), which was used as a reference sampler for the network and was collocated with other samplers used in each city. The dichot results agreed with those of most of the other samplers used in six AIRPET cities, yielding a slope of 0.93–1.06 for $PM_{2.5}$ mass and a slope of 0.97–1.05 for $PM_{10}/PM_{10-2.5}$ mass (R^2 of 0.86–0.99), as detailed in the work of Kim Oanh et al. (2006). The dichot performance was also compared with the USEPA Federal Reference Method (FRM) samplers (RASS and Partisol®-Plus), which showed good agreement for both $PM_{2.5}$ mass and ionic composition (Figure 3.2).

After sampling, all filters were examined visually to detect any visible damages. Only those samples collected continuously over a sampling period of 24 ± 1.0 h and having no visible damage on filters were considered valid. The AIRPET protocol also included filter handling, which specified that each preweighed clean filter must be placed in a labeled petri dish and kept in an airtight plastic ziplock until the sampling start. After sampling, on removal from the filter holder, each filter was immediately placed in a clean petri dish, bagged in an airtight plastic ziplock, and kept in an icebox during transportation to the laboratory for analysis.

3.3.2 Analytical QA/QC

The QA/QC for the chemical analyses of the AIRPET samples included several steps. First, the sample storage period was kept as short as possible to avoid loss of analytes and potential cross contamination during the storage (e.g., Bangkok samples for ionic species were extracted at the AIT laboratory within 24 h after collection). A large number of laboratory blanks and trip blanks were made; that is, 1 in every 20–30 filters in a filter lot was used as a laboratory blank, and a trip blank was made for every sampling period at a site. The blanks were treated in the same way as the actual samples, and the reported compositional data were blank corrected. At the AIT laboratory, a QA/QC (reference) filter was used during the pre- and postsampling weighing and was reweighed after every 5–10 sample filters on each day. If the readings of the QA/QC filter preweight and postweight differed by more than 30 μg/filter, then the filter weighing results of that particular day were discarded, and reweighing was done on another day when the condition was satisfied. The EEL M43D digital smokestain reflectometer was calibrated against the BC mass by the experiments conducted at AIT (Kim Oanh et al., 2009). BC results of PM samples collected in some cities were also cross-checked with the measurements by the AIT equipment.

Repeated and split-sample analyses were conducted to check the consistency of analytical equipment and methods. At the AIT laboratory, for PM mass determination, each filter was weighed at least three times. Repeated BC measurements were done at five or six different points on each sampled filter. In addition, split samples were prepared using 30 pairs of $PM_{2.5}$ and $PM_{10-2.5}$ filters; each was segregated into three to six parts for separate analyses of ions and elements. The standard deviation of each species concentrations produced by these analyses was required for the estimation of measurement uncertainty (Polissar et al., 1998) to prepare the ambient data input file for receptor models as detailed in Chapter 2.

The analytical methods for elements used by different laboratories within the AIRPET network were checked using SRM. The average analytical result produced by different laboratories of the network agreed reasonably with the corresponding certified value, with a difference between the two values less than 30% for most elements and their variation ranges largely overlapped (Kim Oanh et al., 2006).

3.3.3 Data Quality Check

The data quality was checked in several steps. The coarse fraction of dichot samples was corrected for the inherent carried-over fine fraction (CARB, 2001). The samples taken at shorter intervals (daytime and nighttime) were summed to produce the 24-h average concentrations and are reported at the standard conditions of temperature and pressure (25°C, 1 atmosphere). The internal consistency of PM composition data series at each site was checked using several methods, including simple mass closure, reconstructed mass, consistency of chemical compositions (sulfate vs. total S, soluble K^+ vs. total K, soluble Cl^- vs. total Cl), and anion and cation balance. The last was checked for those sites where sufficient ionic data were available to calculate the ratio between the sum of the cation equivalents (NH_4^+, Na^+, K^+, Mg^{2+}, Ca^{2+}) and that of the anion equivalents (Cl^-, NO_3^-, SO_4^{2-}).

A simple mass closure may be estimated as the ratio between the gravimetric PM mass and the sum of all analyzed chemical species (without conversion to their corresponding compounds that potentially exist in the air). Similar to the reconstructed mass discussed in Chapter 2, all the analyzed species should be checked against double counting; for example, soluble Ca^{2+}, Cl^-, and K^+ are normally excluded if Ca, Cl, and K are already accounted for in the summation. Principally, the simple mass closure ratio should be considerably below 100% because there are nonquantified components of the PM samples, including, for example, oxygen associated with elements in the metal oxides, hydrogen and oxygen associated with OC in the organic compounds, water, and the unknowns. However, for very low levels of the $PM_{2.5}$ mass concentration (consequently also its components), in some cases the mass closure determined using average component concentrations may be highly fluctuating and may unrealistically reach close to or even above 100%. The main cause of this problem is the high uncertainty of the analytical results for low levels of analytes in samples, as discussed in Chapter 2.

3.4 Levels and Temporal Variation of Particulate Matter

3.4.1 Seasonal Variations

3.4.1.1 Urban Areas

In total, during the period 2002–2008, the AIRPET network produced a large number of valid samples, 3,300 $PM_{2.5}$ and 2,500 PM_{10} or $PM_{10-2.5}$ in six cities and over 215 pairs of $PM_{2.5}$ and PM_{10} samples at the remote sites. $PM_{2.5}$ and PM_{10} mass concentrations averaged over all sampling sites in each city during the period are presented in Table 3.3 together with the $PM_{2.5}/PM_{10}$ ratios (slope of the $PM_{2.5}$ and PM_{10} regression line). Note that only $PM_{2.5}$ and $PM_{10-2.5}$ or PM_{10} samples available in pairs are presented in this table.

The results presented in Table 3.3 for this 8-year period of sampling are consistent with those presented for the first 4-year sampling (2001–2004) in the work of Kim Oanh et al. (2006). Certainly, large numbers of samples collected over a longer period in each city should provide more representative PM levels. The dry season was characterized with significantly higher PM levels than the wet season in every city except for Manila. Manila had more or less the same average PM levels in both seasons, although the upper values of the PM ranges measured in the dry season were higher, which is also featured by a larger standard deviation (SD) of the $PM_{2.5}$ and PM_{10} data series presented in Table 3.3. In Chennai, the $PM_{2.5}$ levels were also almost the same in both seasons, but PM_{10} levels in the dry season were remarkably higher.

Higher PM levels in the dry season may be explained by less-efficient wet removal, additional source emissions (more biomass open burning and construction activity), enhanced secondary PM formation through more intensive photochemistry (dry and hot conditions in most of the cities), more

TABLE 3.3

Summary of PM Mass Concentrations (Average ± 1 SD) (μg/m³) (25°C, 760 mmHg) during 2002–2008 in Six Urban Areas

	Stations	No. of Sample Pairs	$PM_{2.5}$, Avg ± SD	PM_{10}, Avg ± SD	$PM_{2.5}/PM_{10}$[a] Slope	R^2
Dry Season	BMR (Thailand)	401	48 ± 22	72 ± 30	0.66	0.92
	Beijing (China)	308	132 ± 87	215 ± 174	0.48	0.59
	Chennai (India)	192	32 ± 15	104 ± 64	0.28	0.62
	Bandung (Indonesia)	237	51 ± 15	78 ± 22	0.68	0.80
	Manila (Philippines)	661	44 ± 26	56 ± 33	0.65	0.46
	Hanoi (Vietnam)	171	115 ± 50	144 ± 82	0.70	0.83
Wet Season	BMR (Thailand)	259	18 ± 14	31 ± 20	0.61	0.93
	Beijing (China)	205	106 ± 56	164 ± 84	0.60	0.32
	Chennai (India)	106	32 ± 14	89 ± 39	0.37	0.33
	Bandung (Indonesia)	116	36 ± 13	58 ± 13	0.65	0.83
	Manila (Philippines)	572	45 ± 22	57 ± 23	0.75	0.80
	Hanoi (Vietnam)	57	40 ± 27	67 ± 39	0.68	0.87

[a] Ratio between $PM_{2.5}$ and PM_{10} is the slope of the linear regression line.

soil dust emission, potential long-range and regional transport (continental origins and pathways of air masses), and so on. It is noted that Manila had the largest number of samples, mainly taken at one mixed site (Manila Observatory, MO). The site was influenced by a number of sources, with a significant share from traffic emissions since it is in the vicinity of two major roads. Traffic on these roads tends to build up, especially on rainy days, and is only lighter during the school summer break (April–May), which may explain the similar PM levels measured at this site in both seasons.

Variations in PM levels between the cities can be attributed to the differences in local emission sources, in the contribution of regional/long-range transport air pollution, and the natural conditions of the cities, such as location (e.g., continental vs. coastal). In both seasons, the highest $PM_{2.5}$ and PM_{10} levels were recorded in Beijing, with a large day-to-day variation. In the dry season, the second-highest PM concentrations were observed in Hanoi. In the wet season, the second-highest $PM_{2.5}$ levels were observed in Manila, although the levels in Hanoi were almost in the same range.

The high PM levels in Beijing and Hanoi during the dry season may have been caused by both natural and human-made factors. During winter, these two cities have more stable atmospheric conditions typical for the high-pressure systems frequently observed over China, with a ridge extending to Northern Vietnam. Poor atmospheric dispersion conditions in this synoptic pattern enhance a high buildup of air pollutants, including PM in the dry season. Other important reasons for the high PM pollution in Beijing include cold winters (emission from fuel combustion for space heating), low annual

TABLE 3.4

PM Mass Concentrations (Average ± 1 SD) (μg/m³) (25°C, 760 mmHg) Collected at Remote Sites in Six Countries

	Stations	No. of Sample Pairs	$PM_{2.5}$, Avg ± SD	PM_{10}, Avg ± SD	$PM_{2.5}/PM_{10}$[a] Slope	R^2
Dry Season	Khao Yai (Thailand)	39	46 ± 10	57 ± 11	0.80	0.91
	Duolun (China)	32	60 ± 32	97 ± 52	0.53	0.15
	Kavalur (India)	16	24 ± 4	41 ± 14	0.54	0.04
	Ranca Bali (Indonesia)	7	10 ± 4	13 ± 5	0.78	0.95
	Gabaldon (Philippines)	10	11 ± 5	16 ± 7	0.72	0.86
	Tam Dao (Vietnam)	30	51 ± 29	67 ± 31	0.78	0.91
Wet Season	Khao Yai (Thailand)	9	5 ± 2	11 ± 3	0.48	0.15
	Duolun (China)	29	54 ± 41	145 ± 112	0.21	0.03
	Kavalur (India)	12	10 ± 5	26 ± 9	0.35	0.0003
	Ranca Bali (Indonesia)	10	8 ± 3	11 ± 4	0.70	0.87
	Gabaldon (Philippines)	5	8 ± 2	14 ± 4	0.59	0.97
	Tam Dao (Vietnam)	15	25 ± 12	31 ± 11	0.83	0.93

[a] Ratio between $PM_{2.5}$ and PM_{10} is the slope of the linear regression line.

average precipitation (as compared to other cities, as shown in Chapter 1), and dust storms (Sun et al., 2004). The four remaining cities (Bangkok, Bandung, Chennai, and Manila) have no cold winter and are located along the coastal lines, hence benefitting from an enhanced dispersion of air pollution.

3.4.1.2 Remote Sites

PM levels measured at remote sites are summarized in Table 3.4. Relatively large numbers of samples (~30) were collected at remote sites in China (both seasons) and Thailand and Vietnam (dry season). Higher PM levels during the dry season than the wet season were also recorded at the remote sites. Exceptional were the PM_{10} levels measured in Duolun, China, which were quite high and comparable in both seasons, indicating high background levels and therefore a potential contribution from the regional transport of PM to the city. The $PM_{2.5}$ levels at the remote mountain background sites of Thailand and Vietnam were also high in the dry season, which, for example, were comparable with those measured in the city of Bangkok (Table 3.3). The regional/long-range transport of PM pollution to these remote sites during the dry season following the northeast monsoon may be significant and requires further research. At other remote sites, the number of samples was quite small (Table 3.4) and hence fully representative data may not yet be provided. Longer sampling periods are required at all the sites to gain more meaningful insight into the processes/sources contributing to PM levels at the remote areas and the potential regional transport of PM to the cities.

3.4.2 $PM_{2.5}/PM_{10}$ Ratio

3.4.2.1 Urban Areas

$PM_{2.5}/PM_{10}$ ratios (fraction of $PM_{2.5}$ in PM_{10}) ranked from the lowest values of 0.28–0.37 for Chennai to the highest values of 0.70–0.75 for Hanoi and Manila (Table 3.3) in both seasons. The lower ratio for the PM samples collected in Beijing during the dry season as compared to the wet season may be attributed to more prevalent dust storms during this time of the year, which should contribute more to the coarse particle fraction than to the fine one. The higher $PM_{2.5}/PM_{10}$ ratios in BMR during the dry season compared to the wet season suggests an increase in the contribution of combustion sources (rice straw field open burning) and secondary particles (higher photochemistry and ozone in the dry season, as highlighted in Chapter 1).

Large variations in the ratios for the day-to-day PM samples observed in some cities resulted in low R^2 values (0.33–0.46), for instance, for Beijing (wet season), Chennai (wet season), and Manila (dry season). This also potentially indicates the changes in the predominant contributing sources to daily PM samples. In particular, the large values of R^2 for Bangkok, Bandung, and Hanoi in both seasons and Manila (wet season) showed a strong correlation between $PM_{2.5}$ and PM_{10} and further suggest a presence of common source types contributing proportionally to both fine and coarse PM levels. Examples of correlation between $PM_{2.5}$ and PM_{10} at the AIT site (BMR) are given in Figure 3.3.

3.4.2.2 Remote Sites

In the dry season, $PM_{2.5}/PM_{10}$ ratios at remote sites were significantly higher than the urban areas in all countries (Table 3.4), which showed more dominant processes/sources contributing to the fine particles than the coarse particles (e.g., the secondary formation and the long-range transport). In the wet

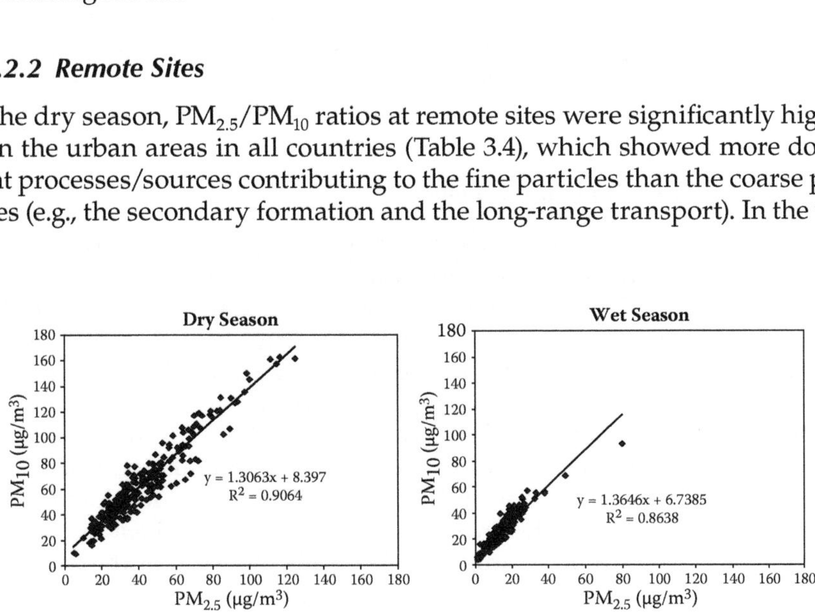

FIGURE 3.3
Correlation between $PM_{2.5}$ and PM_{10} at the AIT site of BMR (2003–2008).

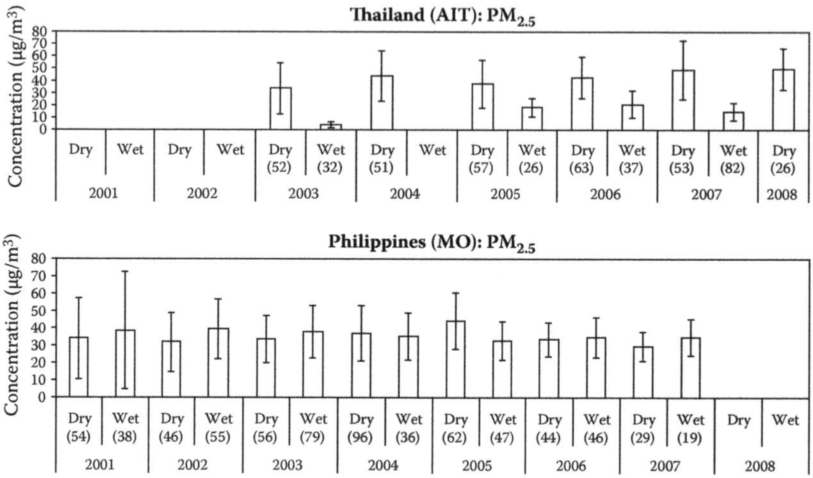

FIGURE 3.4
Seasonal average $PM_{2.5}$ measured at AIT (BMR) and MO (Manila) during 2001–2008; whisks represent the standard variation (±1 SD).

season, the ratios at the remote sites were lower than those in the cities only for Thailand, China, India, and Philippines. The R^2 values at some sites were small, which may indicate variations in contributing sources to daily PM levels. However, the small number of samples collected in most of the remote sites may be a reason for these low R^2 values. Larger numbers of samples are required to produce representative results at the remote sites.

3.4.3 Yearly Variations in PM Levels

Among the urban sampling sites, only two sites generated continuous and consistent PM measurement records during the 8-year monitoring period: the AIT site of BMR and the MO site of Manila. The seasonal average $PM_{2.5}$ levels at AIT during the monitoring period clearly showed higher levels in the dry season as compared to the wet season (Figure 3.4). However, there was no clear year-to-year trend during the 8 years except for a slight increase in PM levels at AIT during the last 3 years (2006–2008). This increase should be examined in relation to, for example, changes in emission sources and meteorological conditions during the period. The urban development that has recently expanded to the areas surrounding AIT may be a major cause. Neither was a clear trend in annual $PM_{2.5}$ levels observed at the MO site in Manila. The differences in $PM_{2.5}$ levels between dry and wet seasons at MO (Manila) were not as pronounced as at AIT (Bangkok), which may be due to the influence of local traffic emission around the MO site, as mentioned previously.

3.5 Chemical Compositions and Source Apportionment of $PM_{2.5}$ in Six Urban Areas

3.5.1 PM Chemical Composition

Chemical composition data were produced for each sample at every site in the six cities in the ready format for receptor model applications. In general, in urban areas during the dry season, the most abundant chemical components of $PM_{2.5}$ included OC, BC (or EC), and water-soluble ions (SO_4^{2-}, NH_4^+, NO_3^-, and K^+). Remarkable levels of Na^+ and Cl^- were also found in the samples from Chennai. High BC levels in $PM_{2.5}$ were observed at all sites and were quite uniform across each city, which indicates the prevalence of incomplete combustion (e.g., vehicles and biomass/refuse burning). The elemental levels found in $PM_{2.5}$ were generally at lower levels and were also quite uniform across all sampling sites in a city.

For the coarse fraction ($PM_{10-2.5}$), OC was also the highest fraction, followed by water-soluble ions (Cl^-, Ca^{2+}, Na^+, NO_3^-, and SO_4^{2-}). Generally, low BC levels were found in the coarse fraction ($PM_{10-2.5}$); hence, its levels were almost the same in both $PM_{2.5}$ and PM_{10}. The most abundant elements were those typically present in the crustal material (e.g., Si, Ca, Fe, Al, and Ti). High concentrations of Na^+ and Cl^- were commonly observed at those sites located near the coastal lines and were likely linked to the influence of sea spray. In addition, high levels of Cl^- at the suburban agricultural sites in BMR may be attributed to the fresh smoke of agroresidue field-burning activities.

During the wet season, the major constituents of fine and coarse fraction PM in percentages were almost the same as those in the dry season, but the absolute levels of these components (in $\mu g/m^3$) were significantly lower in most cities.

3.5.2 PM Reconstructed Mass

The chemical composition data obtained for PM samples in each site of the AIRPET cities were used to establish the reconstructed mass, which also provides information on major contributing sources. For the AIRPET network samples, eight composition mass groups were used, similar to those presented in the work of Kim Oanh et al. (2006) and discussed in Chapter 2 of this book. Note that a factor of 1.7 was used to convert OC to organic matter (OM = 1.7 OC).

The reconstructed mass results for $PM_{2.5}$, averaged over all monitoring sites for each city, are presented in Figure 3.5. The average mass closure (mass explained) generally ranged between 43% and 86% for both seasons. Overall, the "unexplained" mass (i.e., the remaining unexplained portion of $PM_{2.5}$) was still high, especially for the samples without OC data (e.g., for Hanoi in the dry season and Manila in both seasons). With OC data available, the mass closure was improved significantly, that is, 85% or more for samples in BMR (both seasons) and Beijing (dry season). However, although

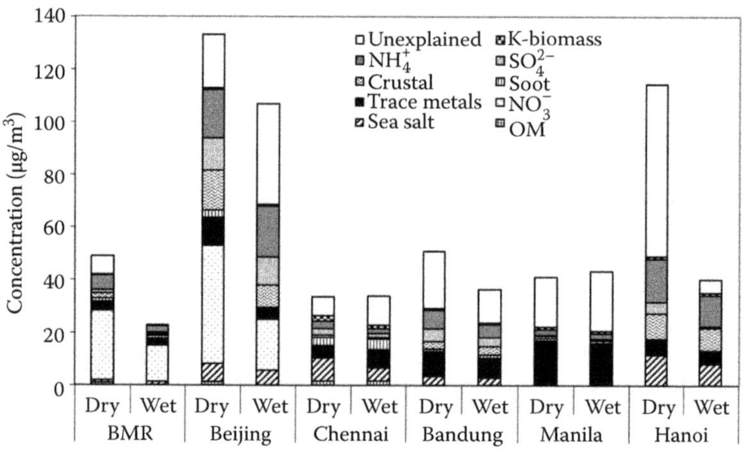

FIGURE 3.5
Average composition data of $PM_{2.5}$ in six urban areas in dry and wet seasons (2002–2008). *Note:* OC data were only available for BMR and Beijing samples.

a number of samples had OC measurements, the mass closure for Beijing samples in the wet season was still low (~64%). Additional chemical species such as organic compounds, carbonate carbon, and others may be analyzed to improve the $PM_{2.5}$ mass closure in the cities.

For BMR and Beijing, where OC data were available, the OM group was the major component of $PM_{2.5}$ at all sites in both seasons. Other significant composition groups included sulfate, ammonium, and soot (BC). In all the cities, the soot levels were considerably high, with the highest values recorded for Manila during both seasons, which indicates a predominant contribution of incomplete combustion (e.g., diesel vehicles). The crustal group was higher in the dry season as compared to the wet season. The highest crustal levels were shown in the $PM_{2.5}$ samples of Chennai, Hanoi, and Beijing, suggesting a remarkable contribution of particles of geological origin. During the dry season, the K-biomass group had higher values, especially in Beijing, BMR, and Chennai, indicating a higher contribution from biomass burning emission. Note that as previously discussed in Chapter 2, the K-biomass group only indicates the abundance of K species from biomass burning emission in PM samples and does not present the absolute contribution of biomass burning smoke to ambient PM.

3.5.3 CMB Source Apportionment Results

The quantitative source apportionment results for $PM_{2.5}$ in the six urban areas, produced by CMB8 using the source profiles presented, are shown in Figure 3.6. The results presented here are the average of the source contributions in each city. The secondary PM presented in

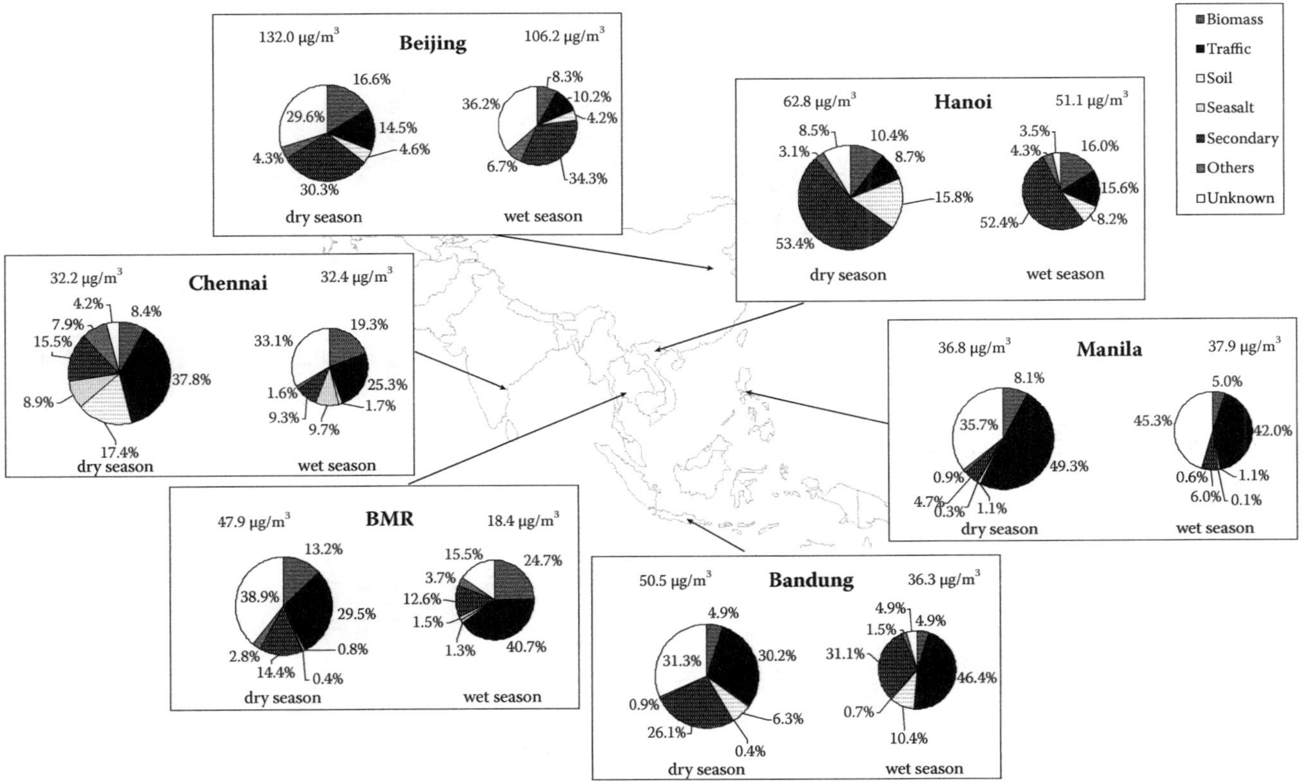

FIGURE 3.6
Average contributions to $PM_{2.5}$ in six Asian cities in percentages. The total average mass is given for dry and wet seasons.

Figure 3.6 is the sum of the inorganic particles of $NaNO_3$, $(NH_4)_2SO_4$, and NH_4NO_3, whereas the "others" include construction, oil burning, and industrial emissions.

The most important contributors to $PM_{2.5}$ were traffic (diesel vehicles), secondary PM, and biomass burning, although their relative importance was different between the cities. Collectively, these three contributors accounted for 54–84% of $PM_{2.5}$ mass in these cities. In BMR, Manila, Chennai, and Bandung, traffic was the largest contributor to $PM_{2.5}$ (25–49%) in both seasons. The secondary inorganic PM was the largest contributor to $PM_{2.5}$ in Hanoi, Beijing, and Bandung (26–52%), while biomass burning was the second-largest source in BMR, Chennai, and Vietnam (10–25%).

The relative contributions of other sources in Figure 3.6 varied significantly between the cities. For example, sea spray was an important $PM_{2.5}$ source in Chennai (9–10%), which may be explained by its coastal location, but it had only a minor contribution in other cities (generally below 1% of $PM_{2.5}$ mass). The construction activities and industrial emission ("others" in Figure 3.6) generally contributed less than 8% of the $PM_{2.5}$ mass and had a higher contribution in Beijing, Chennai, and Hanoi than other cities. Overall, the "unknown" (representing the unidentified sources) was still high (reaching 30–45%) in most of the cities. Lack of the source profiles (e.g., for coal-burning emission or motorcycle emission) is the main reason for this large unknown portion. Note that only the diesel vehicle emission source profile was used to present the traffic emission in the CMB analysis; hence, emission of other vehicle types with different PM profiles could not be quantified. For example, the motorcycle PM emission, which may be important in many Asian cities, especially Hanoi, is also listed under unknown. In addition, some common profiles were used for all cities, so the local specific features of the emission sources (different industries, different diesel fleet compositions, etc.) were not considered. Further improvement in the city-specific source profiles is required to obtain better receptor modeling results by CMB. In addition, applications of multivariate models such as positive matrix factorization (PMF) (detailed in Chapter 2) would help to reveal the source contributions without the need for prior information on detailed source emission profiles.

3.6 Summary

High levels of PM reported for Asian cities are the results of both local and regional emissions of primary particles and the precursors. Systematic monitoring data of PM levels and composition can provide insights into the nature of the pollution problems and the potential effects on health and the environment. This chapter highlighted a design of a regional monitoring network to produce harmonious data with necessary QA/QC and illustrated it with the case study of the AIRPET project involving six cities in Asia (Bangkok, Bandung, Beijing, Chennai, Manila, Hanoi). The PM monitoring was conducted over an 8-year period (2002–2008) at five or six urban sites in each city and at one remote site in each country. The cities are different in terms of natural conditions (geological positions, population, and meteorology) and in the intensity of the human-made and natural emission sources, which is reflected in the different PM levels. The highest PM levels during the period were recorded in Beijing, followed by Hanoi. Overall, the PM levels during the dry season were higher than for the wet season. Based on the available monitoring data, there was no reducing trend observed in yearly $PM_{2.5}$ over the 8-year period in Bangkok and Manila. At the remote sites, the PM levels were also high, thus featuring the potential regional and long-range transport of PM to the urban areas.

The PM composition data, reconstructed mass, and CMB receptor model results indicated the largest contributions to $PM_{2.5}$ in the six cities were from traffic, secondary inorganic PM, and biomass burning. Collectively, these three source categories contributed about 54–84% to $PM_{2.5}$ mass in the cities. The source apportionment study should be further improved by including additional source profiles (coal combustion, motorcycle emission, etc.) and city-specific source profiles into the CMB analysis, which would lower the fraction of the unknown $PM_{2.5}$ sources in the receptor model results. Applications of multivariate receptor models may help improve source apportionment results, especially when no sufficient detailed source emission profiles are available.

APPENDIX TO CHAPTER 3

Additional Source Composition Profiles (Percentage of Mass, %) Used for PM Apportionment by CMB

Species	Oil Burning	Lead-Rich Source	$NaNO_3$	NH_4NO_3	Zinc-Rich Industry	Tin-Rich Industry	Diesel
Na^+	0 ± 0.01	0 ± 0.01	22.58 ± 4.52	0 ± 0	0 ± 0.01	0 ± 0.01	0 ± 0.01
K^+	0 ± 0.01	0 ± 0.01	0 ± 0	0 ± 0	0 ± 0.01	0 ± 0.01	0 ± 0.01
Mg^{2+}	0 ± 0.01	0 ± 0.01	2.72 ± 0.54	0 ± 0.01	0 ± 0.01	0 ± 0.01	0 ± 0.01
Ca^{2+}	0 ± 0.01	0 ± 0.01	0.86 ± 0.17	0 ± 0.01	0 ± 0.01	0 ± 0.01	0 ± 0.01
NH_4^+	0 ± 0.1	0 ± 0.1	0 ± 0	22.6 ± 4.52	0 ± 0.1	0 ± 0.1	0.1 ± 0.1
NO_3^-	0 ± 0.01	0 ± 0.01	71.11 ± 14.22	77.5 ± 7.75	0 ± 0.01	0 ± 0.01	0 ± 0.01
SO_4^{2-}	0 ± 0.01	0 ± 0.01	0 ± 0	0 ± 0	0 ± 0.01	0 ± 0.01	0 ± 0.01
Cl^-	0 ± 0.01	0 ± 0.01	0 ± 0	0 ± 0	0 ± 0.01	0 ± 0.01	0 ± 0.01
BC	0 ± 0.1	0 ± 0.1	0 ± 0.1	0 ± 0.1	0 ± 0.1	0 ± 0.1	64.89 ± 12.98
OC	0 ± 0.1	0 ± 0.1	0 ± 0.1	0 ± 0.1	0 ± 0.1	0 ± 0.1	26.75 ± 8.03
EC	0 ± 0.1	0 ± 0.1	0 ± 0.1	0 ± 0.1	0 ± 0.1	0 ± 0.1	64.89 ± 19.47
TC	—	—	—	—	—	—	—
Li	—	—	—	—	—	—	—
Be	—	—	—	—	—	—	—
Mg	—	—	—	—	—	—	—
Al	0 ± 0.01	0 ± 0.01	0 ± 0.01	0 ± 0.01	0 ± 0.01	0 ± 0.01	0 ± 0.01
Si	0 ± 0.01	0 ± 0.01	0 ± 0.01	0 ± 0.01	0 ± 0.01	0 ± 0.01	0 ± 0.01
S	0 ± 0.1	0 ± 0.1	1.89 ± 0.38	0 ± 0	0 ± 0.1	0 ± 0.1	0.3 ± 0.05
Cl	0 ± 0.03	0 ± 0.03	0 ± 0	0 ± 0	0 ± 0.6	0 ± 0.6	0 ± 0.6

K	0 ± 0.06	0 ± 0.06	0 ± 0	0 ± 0	0 ± 0.06	0 ± 0.06	0 ± 0.05
Ca	0 ± 0.03	0 ± 0.03	0.86 ± 0.17	0 ± 0	0 ± 0.01	0 ± 0.01	0 ± 0.01
Ti	0 ± 0.01	0 ± 0.01	0 ± 0	0 ± 0	0 ± 0.01	0 ± 0.01	0 ± 0.03
V	0.9 ± 0.2	0 ± 0.01	0 ± 0.01	0 ± 0	0 ± 0.01	0 ± 0.01	0 ± 0.01
Cr	—	—	—	—	—	—	—
Mn	0 ± 0.1	1.85 ± 1.85	0 ± 0	0 ± 0	3.7 ± 1.11	0 ± 0.1	0 ± 0.1
Fe	0 ± 0.02	35.27 ± 24.69	0 ± 0	0 ± 0	21.68 ± 6.5	0 ± 0.02	0 ± 0.08
Co	—	—	—	—	—	—	—
Ni	0.62 ± 0.19	0 ± 0.1	0 ± 0	0 ± 0	0 ± 0.1	0 ± 0.1	0 ± 0.1
Cu	0 ± 0.01	1.51 ± 0.6	0 ± 0	0 ± 0	1.98 ± 0.59	2.55 ± 0.77	0.1 ± 0.04
Zn	0.04 ± 0.01	0 ± 0	0 ± 0	0 ± 0	72.28 ± 21.68	0 ± 0.01	0.1 ± 0.01
As	0.01 ± 0.02	0 ± 0.02	0 ± 0	0 ± 0	0 ± 0.02	1.64 ± 0.49	0 ± 0.02
Se	0.06 ± 0.06	0.09 ± 0.04	0 ± 0	0 ± 0	0.35 ± 0.11	0 ± 0.03	0 ± 0.03
Br	0 ± 0.01	0 ± 0.01	0 ± 0	0 ± 0	0 ± 0.01	0 ± 0.01	0 ± 0.01
Rb	0 ± 0.01	0.08 ± 0.03	0 ± 0	0 ± 0	0 ± 0.01	1.04 ± 0.42	0 ± 0.01
Sr	0 ± 0.1	0.1 ± 0.1	0.03 ± 0.01	0 ± 0.1	0 ± 0.1	0.22 ± 0.09	0 ± 0.1
Zr	0 ± 0.1	0.1 ± 0.1	0 ± 0.1	0 ± 0.1	0 ± 0.1	0.42 ± 0.21	0 ± 0.1
Cd	—	—	—	—	—	—	—
Sn	0 ± 0.1	0 ± 0.1	0 ± 0.1	0 ± 0.1	0 ± 0.1	19.6 ± 3.92	0 ± 0.1
Ba	—	—	—	—	—	—	—
Pb	0 ± 0.1	60.99 ± 18.3	0 ± 0.1	0 ± 0.1	0 ± 0.1	74.52 ± 22.36	0 ± 0.1

Sources: Upadhyay (2002), with the profiles produced by COPREM for BMR used for both PM_{10} and $PM_{2.5}$. Either BC or EC was selected for calculation to be compatible to the ambient datasets.

Note: —, not available; 0 (zero), values were generally below 10^{-4}% of PM mass.

References

Airmetrics. 2012. MiniVol portable air sampler: Basic operation. Accessed 2012 from http://www.airmetrics.com/products/minivol/basicop.html

Bond T.C., Venkataraman C., and Masera O. 2004. Global atmospheric impacts of residential fuels, *Energy for Sustainable Development*, 8(3): 20–32.

California Air Resources Board (CARB). 2001. Air monitoring quality assurance. Volume II: Standard operating procedures for air quality monitoring. Appendix M: Dichotomous sampler. http://www.arb.ca.gov/airwebmanual/, accessed 2011.

Chow J.C. 1995. Measurements methods to determine compliance with ambient air quality standards for suspended particles, *Journal of Air and Waste Management Association*, 45: 320–382.

Chow J.C., Doraiswamy P., Watson J.G., Chen L.W., Ho S.S., and Sodeman D.A. 2008. Advances in integrated and continuous measurements for particle mass and chemical, composition, *Journal of Air and Waste Management Association*, 58: 141–163.

Chowdhury Z., Zheng M., Schauer J.J., Sheesley R.J., Salmon L.G., Cass G.R., and Russell A.G. 2007. Speciation of ambient fine organic carbon particles and source apportionment of $PM_{2.5}$ in Indian cities, *Journal of Geophysical Research*, 112, doi: 10.1029/2007JD008386.

Dan M., Zhuang G., Li X., Tao H., and Zhuang Y. 2004. The characteristics of carbonaceous species and their sources in $PM_{2.5}$ in Beijing, *Atmospheric Environment*, 38(21): 3443–3452.

Dockery D.W., and Pope C.A., III. 1994. Acute respiratory effects of particulate air pollution, *Annual Revision Public Health*, 15: 107–132.

Donaldson K., Li X.Y., and MacNee W. 1998. Ultrafine (nanometer) particle mediated lung injury, *Journal of Aerosol Science*, 29(5–6): 553–560.

Engelbrecht D.R., Cahill T.A., and Feeney P.J. 1980. Electrostatic effects on gravimetric analysis of membrane filters, *Journal of the Air Pollution Control Association*, 30(4): 391–392.

Gupta P., Christopher S.A., Wang J., Gehrig R., Lee Y., and Kumar N. 2006. Satellite remote sensing of particulate matter and air quality assessment over global cities, *Atmospheric Environment*, 40(30): 5880–5892.

Health Effects Institute (HEI). 2004. *Health effects of outdoor air pollution in developing countries in Asia. Special report 15 executive summary*. Health Effects Institute, Boston.

Hopke P., Cohen D., Begum B., Biswas S., Ni B., Pandit G., Santoso M., Chung Y., Davy P., Markwitz A., Waheed S., Siddique N., Santos F., Pabroa P., Seneviratne M., Wimolwattanapun W., Bunprapob S., Bac V.T., Hien P.D., and Markowicz A. 2008. Urban air quality in the Asian region, *Science of the Total Environment*, 404: 1103–1112.

Hopke P.K., Xie Y., Raunemaa T., Biegalski S., Landsberger S., Maenhaut W., Artaxo P., and Cohen D. 1997. Characterization of the Gent stacked filter unit PM_{10} sampler, *Aerosol Science and Technology*, 27: 726–735.

Intergovernmental Panel on Climate Change (IPCC). 2007. *The Physical Science Basis. Contribution of Working Group I to the Fourth Assessment Report of the Intergovernmental Panel on Climate Change*, edited by Solomon S., Qin D., Manning M., Chen Z., Marquis M., Averyt K. B., Tignor M., and Miller H. L., Cambridge University Press, Cambridge, UK.

Kim Y.J., Kim K.W., and Oh S.J. 2001. Seasonal characteristics of haze observed by continuous visibility monitoring in the urban atmosphere of Kwangju, Korea, *Environmental Monitoring and Assessment*, 70: 35–46.

Kim Oanh N.T., Pongkiatkul P., Upadhyay N., and Hopke P. 2009. Design a monitoring program of ambient particulate matter for source apportionment study by receptor modeling, *Atmospheric Environment*, 43: 3334–3344.

Kim Oanh N.T., Thiansathit W., Bond T.C., Subramanian R., Ekbordin W., and Ittipol P. 2010. Compositional characterization of $PM_{2.5}$ emitted from in-use diesel vehicles in Asia, *Atmospheric Environment*, 44(1): 15–22.

Kim Oanh N.T., Thuy L.B., Tipayarom D., Manadhar B.R., Pongkiatkul P., Simpson C.D., and Liu L.S. 2011. Source characterization of aerosol emission from field burning of rice straw, *Atmospheric Environment*, 45: 493–502.

Kim Oanh N.T., Upadhyay N., Zhuang Y.H., Hao Z.P., Murthy D.V.S., Lestari P., Villarine J.T., Chengchua K., Co H.X., Dung N.T., and Lindgren E.S. 2006. Particulate air pollution in six Asian cities: Spatial and temporal distributions, and associated sources, *Atmospheric Environment*, 40: 3367–3380.

Molina M.J. and Molina L.T. 2004. Critical review: Megacities and atmospheric pollution, *Journal of Air and Waste Management Association*, 54: 644–680.

Polissar A.V., Hopke P.K., Paatero P., Malm. W.C., and Sisler J.F. 1998. Atmospheric aerosol over Alaska 2. Elemental composition and sources, *Geophysical Research*, 103(D15): 19045–19057.

Pope C.A., III, Ezzati M., and Dockery D.W. 2009. Fine particulate air pollution and life expectancy in the United States, *The New England Journal of Medicine*, 360: 376–386.

Ramanathan V., and Carmichael G.R. 2008. Global and regional climate changes due to black carbon, *Nature Geoscience*, 1: 221–227.

Schauer J.J., Kleeman M.J., Cass G.R., and Simoneit B.R.T. 1999. Measurement of emissions from air pollution sources 2: C1 through C30 organic compounds from medium duty diesel trucks, *Environmental Science and Technology*, 38: 1578–1587.

Schauer J.J., Kleeman M.J., Cass G.R., and Simoneit B.R.T. 2002. Measurement of emissions from air pollution sources. 5. C1–C32 organic compounds from gasoline-powered motor vehicles, *Environmental Science and Technology*, 36: 1169–1180.

Schwartz J., and Neas L. 2000. Fine particles are more strongly associated than coarse particles with acute respiratory health effects in schoolchildren, *Epidemiology*, 11: 6–10.

Simoneit B.R.T. 2002. Biomass burning—a review of organic tracers for smoke from incomplete combustion, *Applied Geochemistry*, 17: 129–162.

Sun Y., Zhuang G., Yuan H., Zhang X., and Guo J. 2004. Characteristics and sources of 2002 super dust storm in Beijing, *Chinese Science Bulletin*, 49(7): 698–705.

Tsai Y.I., and Chen C.L. 2006. Atmospheric aerosol composition and source apportionments to aerosol in southern Taiwan, *Atmospheric Environment*, 40: 4751–4763.

Upadhyay N. 2002. Source apportionment of fine and coarse fractions of particulate matters in Bangkok metropolitan region by receptor modeling, master's thesis, Asian Institute of Technology, Pathumthani, Thailand.

U.S. Environmental Protection Agency (USEPA). 1998. *$PM_{2.5}$ Mass Weighting Laboratory Standard Operating Procedures for the Performance Evaluation Program*, United State Environmental Protection Agency, Washington, DC.

Watson J.G. 2002. Visibility: Science and regulation, *Journal of Air and Waste Management Association*, 52: 628–713.

Zhou L., Kim E., Hopke P.K., Stanier C.O., and Pandis S.N. 2005. Mining airborne particulate size distribution data by positive matrix factorization (PMF), *Journal of Geophysical Research*, 110, doi: 10.1029/2004JD004707.

4

Satellite Tools for Air Quality Management with Focus on Particulate Matter

Jill A. Engel-Cox, Amy K. Huff, Thongchai Kanabkaew, and Nguyen Thi Kim Oanh

CONTENTS

4.1 Introduction: Satellite Remote Sensing for Air Quality Monitoring

Satellite remote sensing provides information about regional air quality and long-distance transport of pollutants, which complements available ground-based data. While ground-based monitors provide the most accurate measurements of air quality at a specific location in the vicinity of the monitor, satellite sensors provide data with global, consistent coverage; these measurements are combined with science-based algorithms for observation of regional pollutant concentrations. Satellite sensor data are particularly relevant to regions in Asia with few monitors or where transport of pollutants may occur across oceans or remote regions. Satellite datasets are typically available at minimal or no cost to end users, making them a cost-effective resource for understanding air pollution at regional, country, and administrative levels. This chapter presents an overview of satellite sensors relevant to air quality, more detailed data for the measurement of particulate matter (PM), and case studies of quantitative and qualitative applications in Asia.

4.1.1 Brief Description of Satellite Sensors

Space agencies, such as the U.S. National Aeronautics and Space Administration (NASA), the European Space Agency (ESA), and the Indian Space Research Organization (ISRO), have launched dozens of satellites for Earth observation since the mid-1990s. Earth-observing satellites include several components. The satellite platform is the infrastructure of the satellite, which generally includes solar panels for power and hardware for instruments. Each Earth-observing satellite platform typically has one to four instruments or sensors, which in turn measure Earth's reflectance in specific spectral wavelength bandwidths; these measurement ranges are called *channels.* Each sensor measures multiple (4 to 40+) channels. These wavelength data are combined through scientific algorithms to provide a value that represents the environment being measured. Satellite sensors and their associated algorithms measure the full suite of environmental parameters, such as land cover, sea surface temperature and color, solar radiation, meteorological characteristics, and air quality.

Two defining features of satellite data are spatial resolution and temporal resolution. Spatial resolution is the smallest area of Earth's surface that a satellite sensor can observe. Spatial resolution varies by the type of instrument. High or fine spatial resolution measurements (e.g., 10 m or less) permit observations of detailed features, such as individual roads, trees, and buildings. Low or coarse spatial resolution measurements (e.g., 1 km or more) permit observation of large regional features, such as cities and forests. Figure 4.1 shows examples of images with high and low spatial resolution. Typical spatial resolutions for air quality satellite data range from 250 m to 25 km.

Temporal resolution is the time period required for a sensor to observe the same location on Earth, and it depends on the orbit of the satellite. There are two main types of satellite orbits: polar orbiting and geostationary. Polar-orbiting satellites are in a low-altitude (700- to 800-km) orbit over the poles. Earth rotates under the satellite as it orbits, so each time the satellite makes an overpass, it observes a new area. As a result, polar-orbiting satellites typically have global coverage and high spatial resolution, but lower temporal resolution (from 1 to 16 days). Polar-orbiting satellites have a wide variety of applications, such as air quality, land cover, water quality, and vegetation studies. Figure 4.2(a) shows a representation of a polar-orbiting satellite. Geostationary satellites are in a high-altitude (35,000- to 36,000-km)

(a) (b)

FIGURE 4.1
Examples of (a) low and (b) high spatial resolution satellite images. (Image 4.1a is a black and white image converted from a 1-km MODIS true color image, courtesy of NASA's MODIS Rapid Response Subsets, http://rapidfire.sci.gsfc.nasa.gov/subsets/; image 4.1b is from Google Earth's default imagery, http://www.google.com/earth/index.html.) Color image can be viewed at http://www.crcpress.com/product/isbn/9781439862254

(a) (b)

FIGURE 4.2
(a) Representation of a polar-orbiting satellite. The narrow arrow represents the path of the satellite around the North and South Poles, and the wide arrow represents the rotation of the Earth. (b) Representation of a geostationary satellite. The narrow arrow represents the path of the satellite around the Equator, and the wide arrow represents the rotation of Earth. Note that the geostationary satellite is revolving around Earth at the same rate that Earth is rotating. Color image can be viewed at http://www.crcpress.com/product/isbn/9781439862254

orbit; at this altitude, the orbital speed of the satellite matches the rotation rate of Earth. Consequently, geostationary satellites make continuous measurements of the same area on Earth, so they have high temporal resolution (typically minutes to hours). Geostationary satellites are primarily used to monitor meteorological conditions. Figure 4.2(b) shows a representation of a geostationary satellite. There is often a trade-off between the spatial and temporal resolution of satellite measurements; sensors that have high temporal resolution, like geostationary satellites, tend to make measurements with lower spatial resolution than analogous measurements from polar-orbiting satellites.

Air quality has strong temporal variations, and ambient air quality measurements are normally presented as hourly, daily, or multiday averages. Spatial variability of most air pollution monitoring is on a neighborhood-to-regional scale, except for highly local effects, such as street-level pollution. Since polar-orbiting satellites tend to have higher spatial resolution, they have typically been most often used for air quality monitoring, particularly satellites that have daily coverage and contain instruments with many spectral bands for high-accuracy measurements. Geostationary satellites have hourly or less-temporal resolution, which makes them well suited to observe the evolution of air quality events; however, most current geostationary satellites carry instruments with only four or five spectral channels and have a lower spatial resolution than analogous polar-orbiting instruments. As a result, geostationary air quality measurements tend to be less accurate than analogous measurements from polar-orbiting satellites, although in some applications the higher temporal resolution of geostationary satellite measurements outweighs their lower accuracy and spatial resolution. The next generation of geostationary satellites, such as the U.S. Geostationary Operational Environmental Satellite-R Series (GOES-R) (Schmit et al., 2005), will include instruments that contain a higher number of spectral bands and higher spatial resolution, which promise to provide a new source of air quality observations with high temporal resolution, high spatial resolution, and high accuracy.

4.1.2 Examples of Satellite Sensors that Measure Air Quality

Several satellite sensors have been used by researchers for different aspects of global air quality monitoring. The NASA Moderate Resolution Imaging Spectroradiometer (MODIS) sensor is useful for monitoring particulate pollution (aerosols). MODIS has 36 channels, so it is versatile for monitoring air quality as well as other environmental parameters. MODIS is located on two satellite platforms: Terra, which passes over any part of Earth at about 10:30 a.m. local time, and Aqua, which passes over any part of Earth at about 1:30 p.m. local time. Each MODIS sensor has 1-day temporal resolution; thus, up to two images per day are available from MODIS, one from each satellite. The spatial resolution of MODIS measurements is 250 m to 1 km, depending on the channel used to make the measurement.

From these measurements, a variety of MODIS data products are available. True-color images combine the red, green, and blue spectral channels to create images that are similar to photographs at resolutions as fine as 250 m. The aerosol optical depth (AOD) product is a measure of the scattering and absorption of light by aerosols in a vertical column from the surface of Earth to the top of the atmosphere. Correlation studies with ground-based measurements have found AOD to be a good proxy for fine particles. The application of satellite data to the monitoring of PM, particularly regarding AOD measurements, is discussed in more detail in Section 4.2 of this chapter. Several other sensors also have aerosol products, such as NASA's Ozone Monitoring Instrument (OMI) on the Aura platform and NASA's Multi-angle Imaging SpectroRadiometer (MISR) on the Terra platform.

Gaseous pollutants can also be measured using defined spectral channels based on the pollutant. For example, OMI has products for SO_2, NO_2, and O_3. NO_2 and SO_2 can be measured at regional scales from space, and OMI measurements have been used to monitor urban areas, extremely large industrial plumes, and volcanic events (SO_2) (Krotkov et al., 2006; Wenig et al., 2008). CO and CO_2 have also been measured from space at larger spatial scales, particularly to study biomass burning (Deeter et al., 2003; McMillan et al., 2005; Warner et al., 2007). However, one of the pollutants of significant interest to human health, tropospheric O_3, is particularly challenging to monitor from space due to the stratospheric ozone layer. Residual analysis techniques are used to separate tropospheric and stratospheric levels of O_3 (e.g., Fishman et al., 2003) since 90% of the total O_3 column is in the stratosphere. Studies using the Total Ozone Mapping Spectrometer (TOMS) instrument, a precursor to OMI, indicated that only 20% of the variances in the TOMS total column data were from variations in tropospheric O_3, with the remainder possibly because of long-range transport of biomass-burning emission (Combrink et al., 1995). Similar results have been found for tropospheric ozone data measured by OMI (e.g., Ziemke et al., 2006), indicating that these data can be used for qualitative regional air pollution studies but not ozone monitoring. Parameters of many recent sensors relevant to air quality are summarized in the work of Martin (2008).

4.1.3 Key Air Quality Applications of Satellite Sensors

Given consistent spatial coverage, satellite sensor data are particularly relevant to two key types of air quality monitoring: (a) regional near-real-time monitoring and historical studies of air pollutant distribution and intensity and (b) long-range transport of pollutants. Where ground-based monitors exist, they tend to be clustered in urban locations, as appropriate for monitoring direct impacts to human health. However, pollutants can be broadly dispersed across a region, and concentrations can be high in rural areas due to biomass burning, clusters of industry, natural sources, or transport of pollutions from other locations. Satellite data permit mapping of pollutants across regions over various time periods to understand the distribution of pollutants and their relative intensity. In this way, satellite measurements can create a synoptic picture of air quality in an urban airshed. Distribution information and patterns can be used for setting air quality policy, planning monitoring programs, mitigation planning, near-real-time monitoring, and public communication. When satellite measurements are integrated with ground-based data, the regional distribution of pollutants can be quantified (e.g., Engel-Cox et al., 2004).

Medium- to long-range transport of air pollutants can cover large areas and significant distances (hundreds to thousands of kilometers), and it is difficult to assess such large areas using sparsely distributed ground-based monitor networks. Complex three-dimensional (3D) models can be used to study the long-range transport of air pollution (Fairlie et al., 2009), but the lack of adequate spatial distribution of ground-monitoring data for model evaluation is an issue. In such cases, a remote sensing-based approach is essential. Satellite products can provide the overall picture of pollutant distributions for semiquantitative model evaluation. They can also be used, in combination with back trajectory analysis, to document specific cases of long-range transport of pollutants from wildfires, dust storms, and regional industrial pollution, such as the Asian/atmospheric brown cloud.

4.2 Satellite Monitoring for Particulate Matter

The monitoring of particulate matter ($PM_{2.5}$ and PM_{10}, particles with an aerodynamic diameter less than 2.5 and 10 μm, respectively) has been the most successful and well studied of the air pollutants by satellite sensors. This is because particles in the atmosphere can be relatively easily detected as they scatter and absorb visible light, producing a change in the surface reflectance of visible solar radiation. Fine particles such as sulfates are pollutants of particular importance to human health; they also effectively scatter visible light and thus are easy to detect. PM pollution is relatively well mixed in the troposphere and concentrated in the boundary layer, so it is less easily confused with stratospheric effects (as is the case for ozone).

As a proxy for measuring fine particles using remote sensors such as satellites, scientists developed the parameter AOD (also called aerosol optical thickness, or AOT). It is derived from multiple spectral channels using algorithms based on radiative transfer equations and surface reflectance of radiation at multiple wavelengths. A detailed description of the AOD algorithm can be found in the work of Kaufman et al. (2002), King et al. (2002), and Levy et al. (2007).

AOD measurements are unitless; typical values of AOD range from 0 to about 4, with values most commonly between 0 and 1. The MODIS AOD dataset is 10-km resolution at nadir, meaning each pixel is 10 km × 10 km when the satellite is looking straight down. It is important to note that, since the scattering by particles is largely in the visible part of the spectrum, no AOD values can be retrieved under cloudy conditions or at night. In addition, no AOD values can be derived where there is sunlight glinting off a water surface, corresponding to a narrow part of the middle of each swath.

Because AOD is a vertical column measurement, it is not necessarily equivalent to PM concentrations at the surface. However, numerous studies in North America, Europe, and Asia have shown that AOD is well correlated with fine PM measured at the ground. These studies have largely found a linear relationship, although with variations in correlation coefficients. Hoff and Christopher (2009) presented a tabular summary of many of the key publications on the relationship of PM to AOD.

The relationship of AOD to $PM_{2.5}$ or PM_{10} varies by location and season. Several factors influence both the specific linear regression equation and the correlation. First, the MODIS algorithm seems to be most sensitive to fine particles and sulfates and less sensitive to nitrates and dust. Other AOD algorithms, such as the OMI aerosol index, are more sensitive to absorbing aerosols. Thus, some components of PM measured by ground monitors may not be accurately measured by a satellite sensor. Second, satellite AOD algorithms perform best over consistent dark surfaces, such as oceans or unbroken forests. Bright surface reflectance, found in deserts and arid regions and over snow, can prevent retrieval of AOD; therefore, measurements over these types of surfaces increase variability and error in the measurement of AOD. Third, different seasons or atmospheric conditions can vary the relationship between PM and AOD. There tend to be fewer AOD retrievals during rainy seasons due to increased cloud cover, and AOD values are generally lower. Winter in higher latitudes may have bright or highly variable surface reflectance due to snow cover, which limits measurement of AOD. High humidity may cause hydroscopic growth of particles, which causes a higher AOD measurement for a given concentration of $PM_{2.5}$ than would occur under conditions of low humidity. Pollutant type may vary during the seasons, from sulfate-based industrial pollution during the summer, to nitrate and carbon from combustion sources during the winter. Mixing height will vary during the day, by season, and by location, changing the concentration of pollutants measured by ground monitors compared to the total column measurements of AOD. These variable effects are magnified when there are low aerosol concentrations since the error range of low AOD values statistically makes determining a linear regression difficult.

Finally, satellite data are spatial averages (e.g., 10 km × 10 km for MODIS) yet represent a single point in time during the satellite overpass. Ground-based PM data are made for a single location but are time averaged, typically hourly or daily (24 h). The mismatch of time and space averaging between ground-based and satellite data introduces another uncertainty factor when comparing the two datasets.

Ground-based datasets also have limitations that hinder their comparison to satellite data. In some cases, consistent long-term data are needed to develop significant statistical relationships between satellite data and ground-monitoring data. For some global locations, these long-time series of data are not available or have significant data gaps. The practice of making one measurement of the 24-h time average of PM concentration every 3 to 6 days, which is common in many areas of the world, creates major data gaps when comparing monitor data to daily satellite data. Monitoring programs in some countries place monitors in locations of highly localized pollution (e.g., roadsides), which do not accurately represent the ambient levels of regional pollution measured by satellites. Also, in Asia, the most common particulate pollution datasets are total suspended particles (TSP) and PM_{10}; few $PM_{2.5}$ time series datasets are available. Since light is scattered most efficiently by finer particles, the majority of the published studies have compared AOD to $PM_{2.5}$. The relationship of AOD to TSP is not well understood, and the relationship to PM_{10} is based on a more limited set of analyses, which makes these datasets less useful for comparison to AOD measurements.

Given these limitations, the value of satellite data for monitoring PM comes from the integration of both space-based and ground-based data. Each dataset has its strengths and weaknesses, and the two complement each other to make an effective set of tools for air quality monitoring. Once the relationship between AOD and $PM_{2.5}$ has been established for a region, AOD measurements can be used to evaluate the regional distribution and intensity of pollutants quantitatively, estimate exposure, monitor events in real time, and analyze historical events in detail. When AOD data are combined with trajectory models, long-distance transport can be tracked, documented, and visualized. These methods have been used extensively in the United States and Europe, but there is perhaps more potential for using satellite sensors to monitor air quality in Asia, considering its low density of ground-based monitors, large land mass, high pollutant levels in industrial and highly populated areas, and significant potential for transport events from biomass burning in Southeast Asia, dust storms in the Gobi, and regional industrial haze in South and East Asia.

4.3 Applications in Asia

The integration of satellite imagery and surface-based data provides significant information on the spatial, temporal, and chemical characteristics of aerosol events. These data have been used by policy makers and as early detection warning systems (Falke et al., 2001). High air pollution in Asia, especially PM (see Chapter 1) has been difficult to assess due to a lack of monitoring data for a comprehensive assessment. Thus, satellite data can be used to supplement sparse monitoring networks to assess, analyze, and even predict pollution events. This section gives several examples of applications of air quality data integration in Asia.

4.3.1 Quantitative Analysis of AOD and PM_{10} Relationship: Case Study in Thailand

There are still a limited number of automatic/permanent ground-level PM_{10} observations in Thailand. Of a total of more than 50 stations across the country, 17 stations are located in the capital city of Bangkok (Pollution Control Department [PCD], 2010). Other stations are located sparsely and asymmetrically in the city centers of provinces with high residential and industrial density, such as Chiangmai, Nakhonratchasima, Khonkaen, or Songkla. Some highly populated and large provinces are still not covered by permanent air quality monitoring networks, such as Tak, Nan, Ubonratchathani, Nongkhai, and others (PCD, 2008). PM monitoring using satellite-derived AOD would provide useful information to capture an overall picture with larger spatial coverage for the air quality management in the country as well as pollutants transported into the country.

To realize this possibility, the MODIS AOD Terra data (at 550-nm wavelength from the level 2 MOD04 aerosol products) with a horizontal resolution of 10×10 km^2 over the country were retrieved over a 6-year period (2002–2007) from the NASA LAADS Web site (http://ladsweb.nascom.nasa.gov/). The AOD data were further extracted for three grids containing three stations in Bangkok, where PM_{10} data were available, for two grid sizes (10×10 and 30×30 km^2) with each station located at the grid's center, respectively. The results presented here include (a) spatial distribution of AOD and PM_{10} during a PM pollution episode over the country and (b) linear regressions between AOD and hourly PM_{10} at the selected stations with consideration of the grid sizes of AOD data, the cloud screening, and other local meteorological conditions (relative humidity and wind speed).

4.3.1.1 Spatial Distribution of AOD and PM_{10}

The MODIS AOD Terra and PM_{10} levels measured at different stations in Thailand were comparatively analyzed for a selected PM episode February 3–5, 2007 (Figure 4.3). During this episode, 24-h PM_{10} were reported to be high, exceeding the National Ambient Air Quality Standard (NAAQS; 120 μg/m^3) at a number of locations. High AOD values were also observed over the country, with the maximum reached nearly 1.0 in Bangkok, Chiangmai, and in some locations in the Gulf of Thailand. High PM_{10} (the period average and 1 SD [standard deviation] are shown in brackets) were recorded in Bangkok and at several other locations where AOD were also high. In northeastern

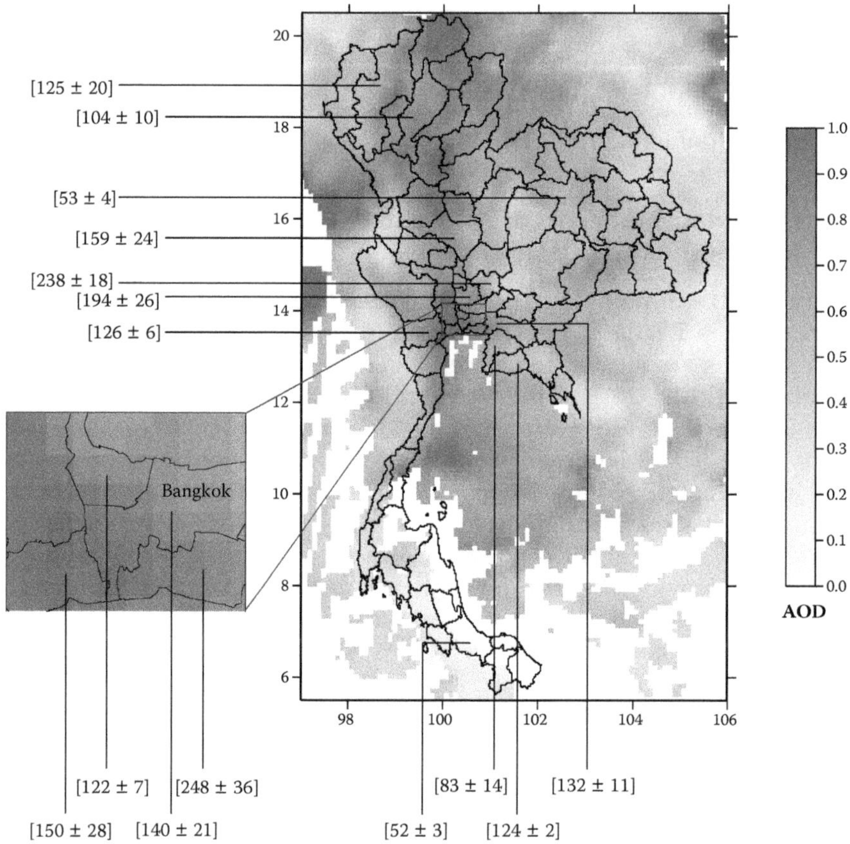

FIGURE 4.3
AOD over Thailand in comparison with average PM_{10} during February 3–5, 2007 (brackets indicate the 3-day period average PM_{10} and 1 SD, μg/m³). Color image can be viewed at http://www.crcpress.com/product/isbn/9781439862254

Thailand, where the average PM levels were relatively low, around 50 μg/m³, low AOD values were also observed. The satellite failed to retrieve AOD values in some areas during the period, resulting in missing values (blank in Figure 4.3) seen over the sea off the coast of southern Thailand, which may be due to a combination of cloudiness and effects of sun glint over the ocean (Levy et al., 2009). Generally, satellite AOD data can qualitatively provide an overview picture of PM air quality. However, the difference in the AOD values is not large enough to reflect the corresponding difference in the measured PM levels at different locations, such as over the Bangkok metropolitan region (BMR; inset in Figure 4.3). The averaging effects, which are different for MODIS AOD values (spatially over 10 × 10 km² and temporally over a short period per day) and the PM ground measurements (spatially at one point and temporally over 3 days of the episode), may be a reason.

The true color image on February 3, 2007, shows many hot spots observed mainly over Cambodia and in some places over northeastern Thailand, as shown in Figure 4.4(a) featuring open burning in Southeast Asia. The HYSPLIT (hybrid single-particle Lagrangian integrated trajectory) backward trajectories (3 days, starting at 1,000 m) show the possibility of the transport of regional emission from surrounding Southeast Asian countries to the Gulf of Thailand, as seen in Figure 4.4(b), which may justify some high AOD values observed over the sea.

4.3.1.2 Regression Analysis of AOD-PM_{10}

To investigate further the linear regression relationships between AOD and PM, the hourly PM_{10} data, averaged for the period of 10:00–11:00 a.m., which overlaps the MODIS Terra satellite overpass time (about 10:30 a.m., local standard time), were retrieved from the continuous hourly record at three ambient stations (10T, National Housing; 11T, Huai Khwang; 12T, None Tree Vitaya), with the locations shown in Figure 4.5.

The accuracy of using satellite data to estimate ground mass concentrations depends on the strength of the regression relationship between AOD and PM mass, which depends on the accuracy of retrieved AOD and of PM measurements and other factors, such as local meteorological conditions. Since AOD data measure light scattering of an integrated columnar aerosol load from the ground up to the top of the atmosphere and PM mass concentration measures the particulate mass concentration at the ground surface, the relationship between PM mass concentrations and AOD may be partially dependent on meteorological variables (Dinoi et al., 2010). This case study examined the linear regressions between AOD and hourly PM_{10} for each station, taking into account the effects of the meteorological conditions, including cloud cover factor (CF), relative humidity (RH), and wind speed (WS). The cloudiness data were collected at a meteorology station in BMR, while RH and WS were measured also at the ambient air quality monitoring stations.

The AOD-PM_{10} relationships in this study (for hourly PM_{10} measured at each station and the average hourly PM_{10} at three stations) are comparable to other studies presented in Table 4.1. Note that the grid size of AOD data used by different studies indicated in Table 4.1 were different (i.e., 50×50 km^2 in Gupta and Christopher, 2008, and 1×1 km^2 in Wang et al., 2010), and no cloud screening was reported in the studies except for the standard cloud screen done by the NASA product.

4.3.1.2.1 *Effects of Cloud Screening*

The MODIS AOD products produced by NASA have been screened for cloudiness as part of the data-processing algorithms (Levy et al., 2009). However, in tropical regions with highly variable cloud cover, correlation between AOD and PM may still be improved by performing additional or different cloud screening (Koelemeijer et al., 2006). The R^2 for the Bangkok stations with clear sky conditions of CF $\leq 2/10$ increased considerably as compared

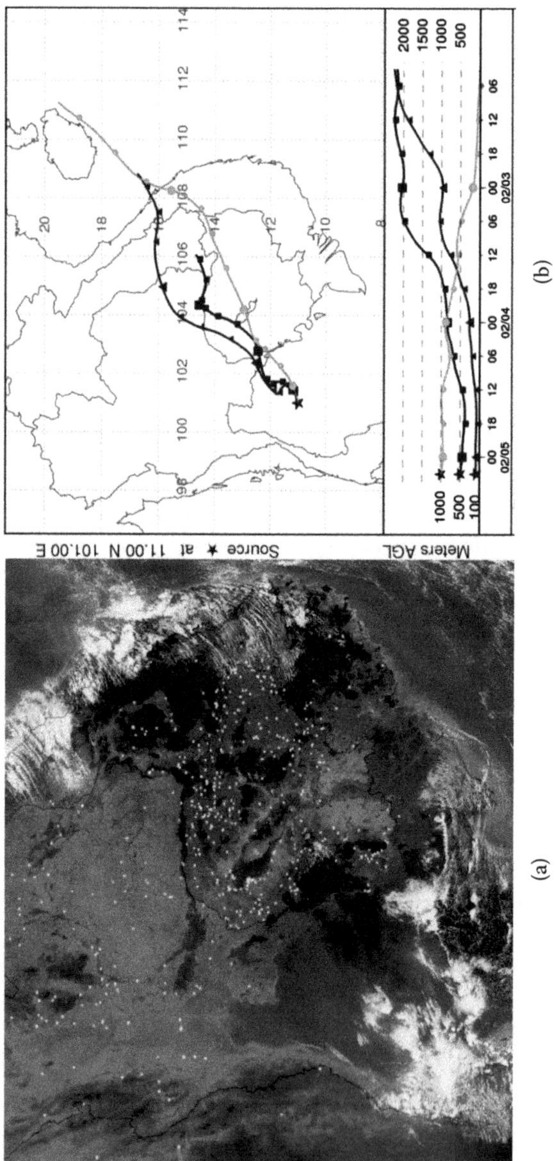

FIGURE 4.4
Satellite images with hot spots featuring the open burning (shown as white spots) over Indochina by MODIS Terra (a) and the 3-day HYSPLIT backward trajectories showing the potential pathways of air masses arriving at the Gulf of Thailand (b). Color image can be viewed at http://www.crcpress.com/product/isbn/9781439862254

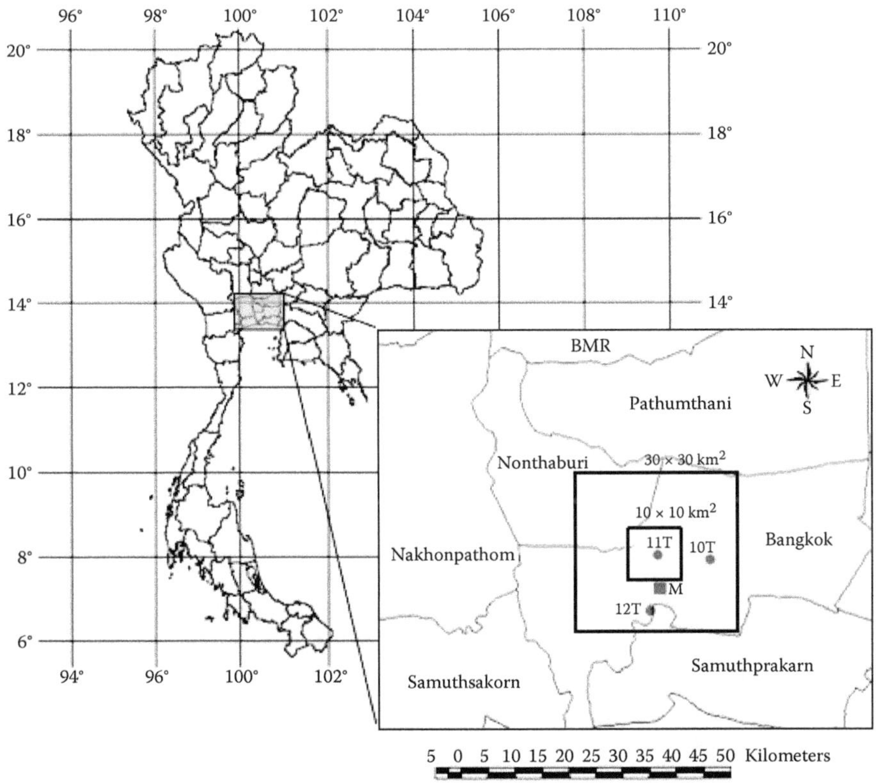

FIGURE 4.5
Locations of the PM_{10} monitoring stations (dark circles) and the meteorological station (M; dark square) illustrated with two grids of AOD retrieval for station 11T in BMR. Color image can be viewed at http://www.crcpress.com/product/isbn/9781439862254

to CF ≤ 3/10 (Table 4.1). Bangkok weather is quite cloudy all year; hence, the effects of cloud screening are worth further investigation. Table 4.2 presents the regression equations and R^2 values at three different cloudiness levels for station 11T, which had the most complete PM_{10} data series among the three stations. Overall, cloud screening improved the linear AOD-PM_{10} correlation significantly. For AOD retrieved at 10 × 10 km², the R^2 for cloudy sky conditions with CF ≤ 5/10 were negligible (0.04) but were increased to 0.25 for CF ≤ 3/10 and reached the highest of 0.43 for CF ≤ 2/10. A similar trend was also obtained for the larger grid size (30 × 30 km²).

4.3.1.2.2 Effects of the Horizontal Grid Size for AOD Data

The PM_{10} measurement data are point based, but PM itself in principle is subject to large spatial variations in relation to the emission sources and dispersion conditions (downwind vs. upwind of major sources). The AOD are a spatial average, 10 × 10 km² as default, but in cases of frequent missing data, the use of a larger horizontal grid size may increase the number of data

TABLE 4.1

Linear Regression between AOD550 MODIS and Hourly PM_{10} in BMR and Other Locations

Country	Location	PM_{10}-AOD Linear Regression	R^2
Thailand	CF ≤ 3/10		
(10 × 10 km²)	BMR, 10T	Hourly PM_{10} = 62 AOD + 14	0.23
	BMR, 11T	Hourly PM_{10} = 95 AOD + 13	0.25
	BMR, 12T	Hourly PM_{10} = 114 AOD + 3	0.34
	BMR, average	Hourly PM_{10} = 74 AOD + 17	0.23
	CF ≤ 2/10		
	BMR, 10T	Hourly PM_{10} = 91 AOD + 6	0.40
	BMR, 11T	Hourly PM_{10} = 127 AOD – 5	0.43
	BMR, 12T	Hourly PM_{10} = 118 AOD – 2	0.42
	BMR, average	Hourly PM_{10} = 115 AOD – 1	0.46
United States	Baltimore[a] (10 × 10 km²)	Daily average $PM_{2.5}$ = 25 AOD + 11	0.32
		Hourly $PM_{2.5}$ = 31 AOD + 5	0.42
	Southeastern[b] (50 × 50 km²)	Daily average $PM_{2.5}$ = 28 AOD + 16	0.27
		Hourly $PM_{2.5}$ = 29 AOD + 9	0.38
China	Beijing[c] (1 × 1 km²)	Hourly PM_{10} = 100 AOD + 23	0.52
		Hourly $PM_{2.5}$ = 20 AOD + 9	0.48

Note: PM_{10} and $PM_{2.5}$ are in micrograms per cubic meter (μg/m³).

[a] Engel-Cox J.A., Hoff R.M., Rogers R., Dimmick F., Rush A.C., Szykman J.J., Al-Saadi J., Chu D.A., and Zell E.R. 2006. Integrating lidar and satellite optical depth with ambient monitoring for 3-dimensional particulate characterization, *Atmospheric Environment*, 40: 8056–8067: Baltimore, Maryland, during summer (July 1 to August 31) 2004.

[b] Gupta P., and Christopher S.A. 2008. Seven year particulate matter air quality assessment from surface and satellite measurements, *Atmospheric Chemistry and Physics*, 8: 3311–3324: Southeastern United States. during the years 2000–2006.

[c] Wang Z., Chen L., Tao J., Zhang Y., and Su L. 2010. Satellite-based estimation of regional particulate matter (PM) in Beijing using vertical-and-RH correcting method, *Remote Sensing of Environment*, 114(1): 50–63: Beijing during July 2007–October 2008.

points for the AOD-PM_{10} relationship analysis. A larger grid size, such as 30 or 50 km, could also be considered (Engel-Cox et al., 2004).

This case study compared the relationship between PM_{10} collected at station 11T and AOD integrated over the 10 × 10 km² grid with that for AOD integrated over the 30 × 30 km² grid. The results showed that the larger grid size (30 × 30 km²) increased the number of the data points available for the regression analysis by about 50–60% but not R^2 (Table 4.2). The replacement of the missing AOD data (retrieved at 10 × 10 km²) by the average AOD value calculated for the large grid size may be a reason for the reduction in R^2 values.

4.3.1.2.3 *Effects of Other Meteorological Variables*

The effects of RH and WS were studied by incorporating these variables directly into the linear regression analysis. Thus, the relationships between

TABLE 4.2

Relationships between MODIS Terra AOD and Hourly PM_{10} Mass at Station 11T (2002–2007)

Parameters		CF ≤ 5/10			CF ≤ 3/10			CF ≤ 2/10		
		Equation	R^2	N	Equation	R^2	N	Equation	R^2	N
10 × 10 km²	AOD-PM_{10}	PM_{10} = 29 AOD + 39	0.04	171	PM_{10} = 95 AOD + 13	0.25	91	PM_{10} = 127 AOD - 5	0.43	29
	AOD-PM_{10}(RH)	PM_{10} = 89 AOD + 61	0.09	170	PM_{10} = 222 AOD + 16	0.24	91	PM_{10} = 290 AOD - 23	0.34	29
	AOD/WS-PM_{10}	PM_{10} = 17 AOD + 45	0.07	171	PM_{10} = 67 AOD + 26	0.34	91	PM_{10} = 116 AOD + 7	0.58	29
	AOD/WS-PM_{10}(RH)	PM_{10} = 48 AOD + 83	0.13	170	PM_{10} = 176 AOD + 40	0.42	91	PM_{10} = 310 AOD - 11	0.61	29
30 × 30 km²	AOD-PM_{10}	PM_{10} = 38 AOD + 34	0.07	289	PM_{10} = 71 AOD + 23	0.18	145	PM_{10} = 95 AOD + 8	0.30	56
	AOD-PM_{10}(RH)	PM_{10} = 115 AOD + 49	0.14	288	PM_{10} = 174 AOD + 29	0.24	144	PM_{10} = 231 AOD - 2	0.34	55
	AOD/WS-PM_{10}	PM_{10} = 22 AOD + 42	0.11	289	PM_{10} = 28 AOD + 41	0.16	145	PM_{10} = 35 AOD + 34	0.30	56
	AOD/WS-PM_{10}(RH)	PM_{10} = 63 AOD + 74	0.19	288	PM_{10} = 81 AOD + 69	0.28	144	PM_{10} = 106 AOD + 51	0.51	55

Note: $PM_{10}(RH) = PM_{10}/(1 - RH/100)$.

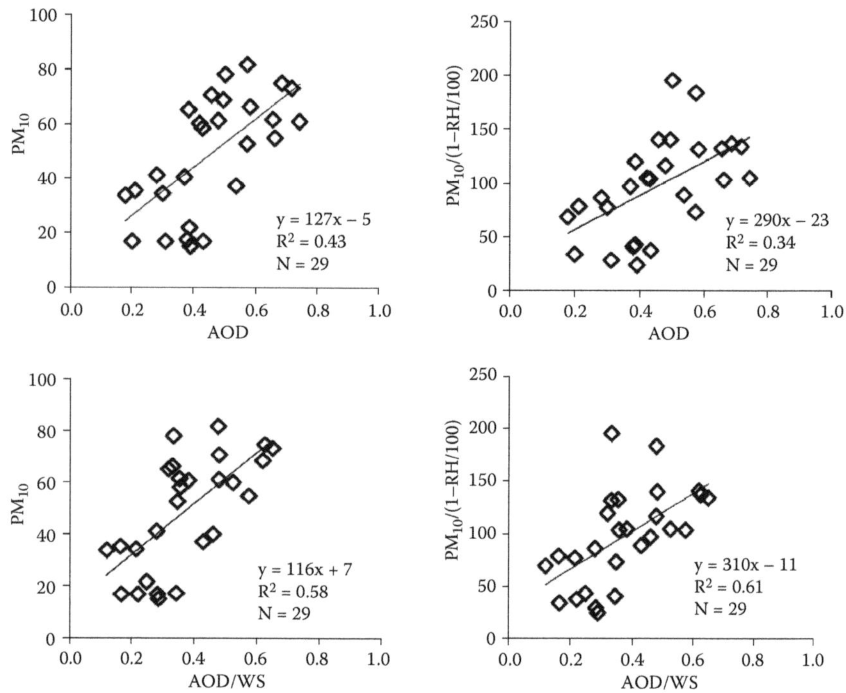

FIGURE 4.6
Scatter plots of PM_{10} and AOD in various combinations for CF ≤ 2/10 (PM_{10} in μg/m³, WS in m/s, and RH in percents).

AOD and a simple PM_{10} function of RH, namely, $PM_{10}/(1 - RH/100)$, was studied following the method applied in the work of Li et al. (2005) and Tsai et al. (2009). The WS was simply included as AOD/WS in the regression analysis with PM_{10}. The results for station 11T are presented in Table 4.2, which shows that the consideration of these meteorological parameters seemed to bring in a consistent improvement for higher cloudiness (CF ≤ 5/10), although R^2 remained quite small. For the clearer sky conditions (CF ≤ 3/10 and CF ≤ 2/10), the R^2 increased considerably with the inclusion of both meteorological variables, but the effects were variable when only RH or WS was included in the analysis. For illustration, scatter plots of AOD-PM_{10} relationships are presented in Figure 4.6.

At station 11T during the period 2002–2007, the mean ambient RH was approximately 67%, and the surface wind measured at 10 m (above the ground) was generally light, with a mean of 1.1 m/s. In principle, the ambient RH can significantly affect the AOD-PM relationship by changing the light-scattering properties of the aerosol due to its hygroscopic property. The higher the RH is, the larger the portion of the light that is scattered and the higher AOD value that is measured (Zhang et al., 2009). The effects may be insignificant for low RH, for instance below 70% (Wang and Martin, 2007) or

50–65% (Tsai et al., 2009). Accordingly, the average RH of 67% in this study may not introduce much effect on the R^2, although it is worth noting that the RH in Thailand is strongly seasonal; hence, the AOD-PM_{10} relationships may change considerably between the dry and wet season. The surface WS has an important role in aerosol generation and dispersion, which was also shown in the decrease of PM_{10} levels measured at station 11T when WS increased.

The results shown in Table 4.2 indicate that the inclusion of RH and WS had more effect on the AOD-PM_{10} relationship for the small grid size than the large grid size. This is may be due to the strong spatial variations of the RH and WS, especially for the observed low wind. Thus, colocated meteorology measurements would be most useful to improve the AOD-PM_{10} relationships. In addition, there are many meteorological variables that may affect the AOD and PM relationship, so a multivariate analysis approach may reveal these complex interrelations.

4.3.2 Semiquantitative Event Monitoring: Biomass Burning in Indonesia and Malaysia

Agricultural burning and wildfires are a common occurrence in Southeast Asia, especially during the dry season. These fires are typically agricultural, related to clearing and maintaining plantations and forests. A few times per year, the number of fires and the meteorological conditions produce intense regional haze, with transported smoke reaching major cities and populated areas. One such event was in late July 2009. Conditions were clear on July 28, 2009, but a review of the satellite images on July 29 showed the development of thermal hot spots and long smoke plumes on the island of Sumatra, Indonesia (Figure 4.7). By July 31, 2009, the smoke covered much of Sumatra,

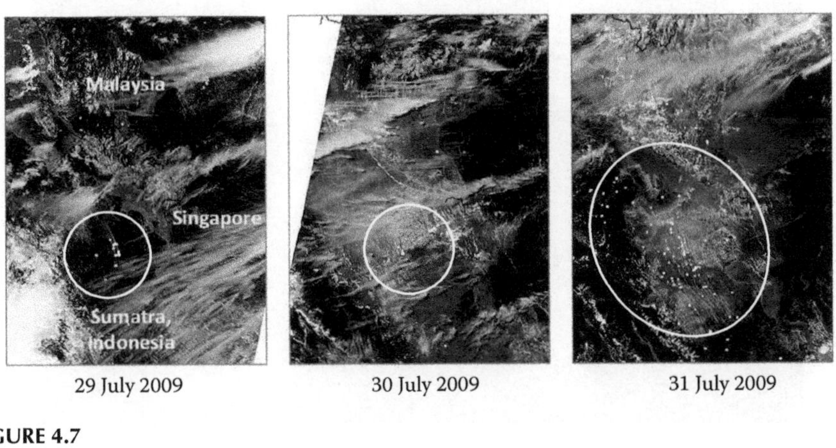

FIGURE 4.7
MODIS Terra images of smoke and haze in Indonesia and Malaysia. The small white dots are outlines of thermal hot spots, indicating the location of fires. Areas with dense hotspots are circled. (*Source:* NASA/GSFC, MODIS Rapid Response.) Color image can be viewed at http://www.crcpress.com/product/isbn/9781439862254

peninsula Malaysia, and Singapore. Dense gray smoke can be seen covering most of the middle of the island of Sumatra (Indonesia), with a plume crossing the Strait of Malacca into the central and southern portion of peninsular Malaysia and off the east coast over the South China Sea. AOD values exceeded 1.0 in several regions of Malaysia and Indonesia, corresponding to the dense smoke (Figure 4.8).

Kuala Lumpur was visibly hazy, and the Malaysian Department of Environment indicated that over 75% of the air quality monitors were reading "moderate" or "unhealthy" for PM_{10} according to their air pollution index (concentrations were around 70–80 μg/m^3). The Singapore Weather Information Portal has two products that provided information on this event (Figure 4.9). These indicated wind direction from Sumatra to Malaysia and Singapore, low visibility and haze, and thermal hot spots from fires. The data confirmed that the smoke from Sumatra was being transported throughout the region, as well as smoke from fires in the Malaysian and Indonesian parts of the island of Borneo.

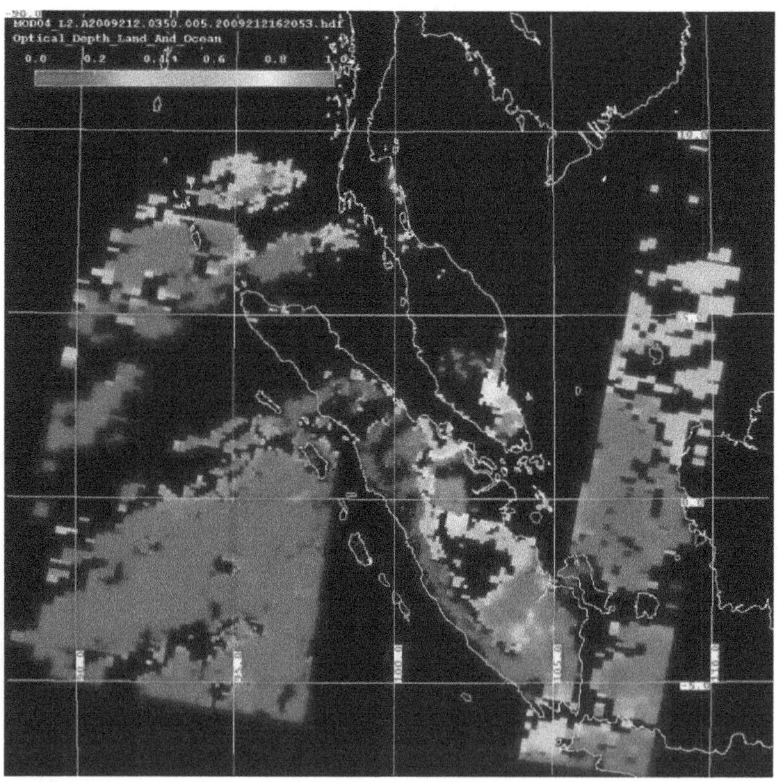

FIGURE 4.8
MODIS Terra aerosol optical depth in Indonesia and Malaysia, July 31, 2009. The black regions are areas without data mostly due to clouds. (*Source:* NASA/GSFC, LAADS.) Color image can be viewed at http://www.crcpress.com/product/isbn/9781439862254

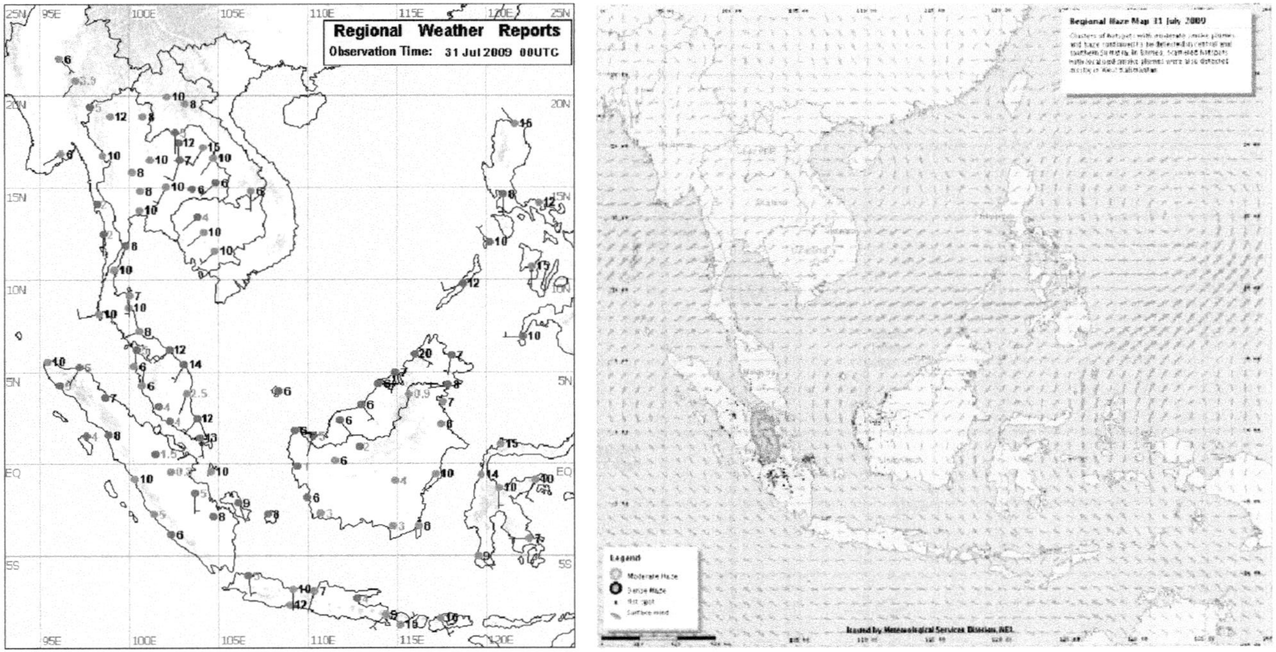

(a) (b)

FIGURE 4.9
Meteorological, visibility, and smoke data, July 31, 2009. On the left is the regional weather report. The dark dots indicate haze at the weather stations, the grey numbers are low visibility, and the wind barbs are the prevailing winds; all confirmed the smoke and haze. On the right is the regional haze map, with satellite-derived hot spots identified by the many black dots on Sumatra and Borneo (including East Malaysia on Borneo). The shaded area is moderate haze. (*Source:* Singapore Weather Information Portal.) Color image can be viewed at http://www.crcpress.com/product/isbn/9781439862254

The integration of the satellite hot spots, true color, and AOD data with the ground-based meteorological and PM_{10} concentration data enabled a more thorough understanding of the causes of the event than would be possible using ground-based data alone. The satellite data indicated the source of pollution was smoke from fires in Sumatra and in parts of Malaysia. Wind trajectories helped track the path of the pollutants through the event. The ground-based data were able to indicate that the effect on human health was moderate to unhealthy in southern Peninsular Malaysia and Singapore.

This combination of datasets is used in the region during the dry season to make policy decisions, such as closing schools and advising people to avoid exercise in times of increased concentrations of PM_{10}. These data have also been used to influence agricultural practices of plantations over the long term, which has reduced the number and intensity of events over the last decade.

4.3.3 Qualitative Daily Event Analysis: Regional Haze in India and Bangladesh

Thick regional haze is a common occurrence in India and Bangladesh from approximately October through February each year. During these cooler months, temperature inversions are common and are a contributing factor to the formation and persistence of haze. A temperature inversion occurs when cold air is located at the surface and warm air is aloft. This is a very stable atmospheric situation, so any pollutants that are present tend to remain trapped at the surface. The Himalaya Mountains form a natural barrier for winter haze that prevents its transport north into Nepal and China.

A typical regional haze event occurred on December 14, 2009. Pollution, likely from industrial processes, built up under a temperature inversion and created widespread haze. Smoke from wildfires may have contributed to the haze as well. Figure 4.10 shows MODIS true-color (a) and AOD (b) images of the haze. The haze is an opaque gray feature in the true-color image; high PM concentrations associated with the haze appear as dark areas in the AOD image. Table 4.3 shows several hours of meteorological observations on December 14, 2009, for Lucknow, India, which is located in northern India and was in the midst of the haze event. The winds in Lucknow were calm to light, which helped permit the buildup and persistence of the haze. There were also many reports of visible haze on the ground, particularly in the afternoon and evening, with one report of visible smoke. Haze events such as this one can continue for days until temperature inversions break or winds pick up to disperse trapped pollutants.

4.4 Summary and Recommendations

Satellite remote-sensing data have become an essential part of air quality monitoring programs. The consistent global coverage of satellite imagery

TABLE 4.3

Meteorological Data for Lucknow, India (26.75°N, 80.88°E) on December 14, 2009

Local Time	Temperature (°C)	Dew Point Temperature (°C)	Visibility (km)	Wind Direction	Wind Speed (m/s)	Conditions
12:00 a.m.	14.0	11.0	3.1	Calm	Calm	Scattered clouds
1:00 a.m.	14.0	11.0	3.1	Calm	Calm	Scattered clouds
2:00 a.m.	14.0	11.0	3.1	Calm	Calm	Mist
2:30 a.m.	13.9	10.6	1.6	Calm	Calm	Mist
3:00 a.m.	14.0	11.0	3.1	Calm	Calm	Mist
4:00 a.m.	13.0	10.0	3.1	Calm	Calm	Mist
5:00 a.m.	13.0	11.0	3.1	Calm	Calm	Scattered clouds
5:30 a.m.	12.8	11.1	1.6	Calm	Calm	Mist
6:00 a.m.	13.0	11.0	2.6	Calm	Calm	Scattered clouds
8:00 a.m.	15.0	11.0	1.1	Variable	1.0	Mist
8:30 a.m.	15.6	13.3	1.6	WNW	1.6	Mist
9:00 a.m.	17.0	13.0	1.1	NNW	1.6	Mist
10:00 a.m.	20.0	15.0	1.8	NNW	2.1	Mist
11:00 a.m.	22.0	13.0	2.6	NNW	1.6	Haze
11:30 a.m.	23.3	12.8	1.6	NNW	1.6	Haze
12:00 p.m.	24.0	13.0	3.1	NNW	2.1	Haze
1:00 p.m.	25.0	12.0	3.5	NNW	2.1	Haze
2:00 p.m.	26.0	12.0	4.0	NNW	2.6	Haze
2:30 p.m.	25.6	11.7	3.2	NNW	2.1	Haze
3:00 p.m.	25.0	12.0	4.0	Variable	1.0	Haze
4:00 p.m.	25.0	12.0	4.0	NNW	2.1	Haze
5:00 p.m.	24.0	12.0	4.0	Variable	1.0	Haze
5:30 p.m.	22.2	13.3	3.2		1.0	Haze
6:00 p.m.	21.0	14.0	4.0	Variable	1.0	Smoke
7:00 p.m.	19.0	13.0	4.0	Variable	1.0	Haze
8:00 p.m.	18.0	13.0	4.0	Variable	1.0	Haze
8:30 p.m.		12.2	3.2	North		Haze
9:00 p.m.	16.0	13.0	3.5	Variable	1.0	Scattered clouds
10:00 p.m.	16.0	13.0	3.5	Variable	1.0	Scattered clouds
11:00 p.m.	15.0	13.0	3.5	Variable	1.0	Scattered clouds
11:30 p.m.	14.4	12.2	1.6		1.0	Mist

Source: Weather Underground, http://www.wunderground.com.
Note the light-to-calm winds and reported "haze" conditions.

and data products, such as AOD, complements point-based, highly accurate ground monitor data. Both datasets have limitations, such as satellite data gaps due to clouds and ground-based monitor data gaps due to sparse monitoring networks. Yet, the integration of these datasets enables air quality analysts, policy decision makers, and the public to understand air pollution in

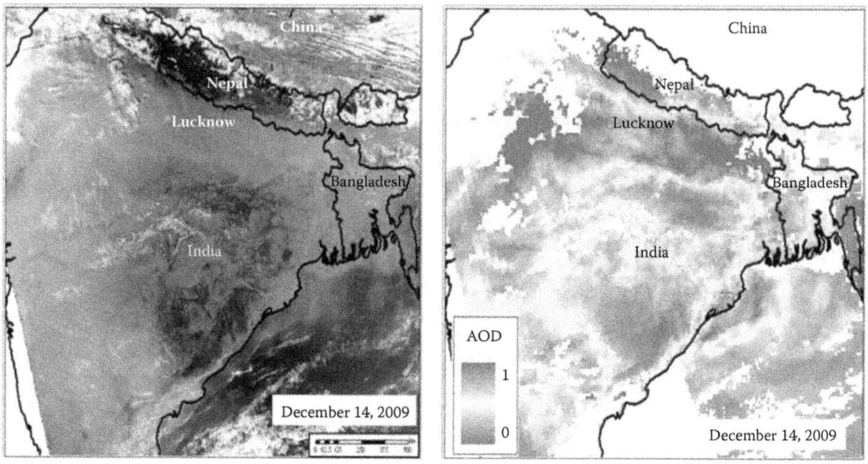

(a) (b)

FIGURE 4.10
MODIS (a) and AOD (b) images of south Asia showing widespread haze in India and Bangladesh (opaque gray feature in Figure 4.10a and dark colors in Figure 4.10b). (Data courtesy of NASA LAADS.) Color image can be viewed at http://www.crcpress.com/product/isbn/9781439862254

their region. In Asia, these integrated datasets are essential due to Asia's vast landmass, relatively few pollutant monitors, and multiple pollutant sources, from urban haze to desert dust to biomass burning, all of which can be transported hundreds or thousands of kilometers across oceans or remote regions The relationship between particulate ground monitor data and satellite AOD data can be established for sites in Asia where systematic PM data are available. The resultant AOD-PM relationships can provide useful information to assist air quality management in Asia. A multivariate analysis approach can be applied to improve the accuracy of the relationship between AOD and ground measurement data.

References

Combrink J., Diab R.D., Sokolic F., and Brunket E.G. 1995. Relationship between surface, free tropospheric and total column ozone in two contrasting areas in South Africa, *Atmospheric Environment*, 29: 685–691.

Deeter M.N., Emmons L.K., Francis G.L., Edwards D.P., Gille J.C., Warner J.X., Khattatov B., Ziskin D., Lamarque J.F., Ho S.P., Yudin V., Attie J.L., Packman D., Chen, J., Mao, D., and Drummond J.R. 2003. Operational carbon monoxide retrieval algorithm and selected results for the MOPITT instrument, *Journal of Geophysical Research*, 108(D14): 4399, doi: 10.1029/2002JD003186.

Dinoi A., Perrone M.R., and Burlizzi P. 2010. Application of MODIS products for air quality studies over Southeastern Italy, *Remote Sensing of Environment*, 2(7): 1767–1796, doi: 10.3390/rs2071767.

Engel-Cox J.A., Hoff R.M., Rogers R., Dimmick F., Rush A.C., Szykman J.J., Al-Saadi J., Chu D.A., and Zell E.R. 2006. Integrating lidar and satellite optical depth with ambient monitoring for 3-dimensional particulate characterization, *Atmospheric Environment*, 40: 8056–8067.

Engel-Cox J.A., Holloman C.H., Coutant B.W., and Hoff R.M. 2004. Qualitative and quantitative evaluation of MODIS satellite sensor data for regional and urban scale air quality, *Atmospheric Environment*, 38: 2495–2509.

Fairlie T.D., Szykman J., Gilliland A., Pierce R.B., Kittaka C., Weber S., Engel-Cox J.A., Rogers R.R., Tikvart J., Scheffe R., and Dimmick F. 2009. Lagrangian sampling of 3-D air quality model results for regional transport contributions to sulfate aerosol concentrations at Baltimore, MD, in summer 2004, *Atmospheric Environment*, 43: 3275–3288.

Falke S.R., Husar R.B., and Schictel B.A. 2001. Fusion of SeaWiFS and TOMS satellite data with surface observations and topographic data during extreme aerosol events, *Journal of Air and Waste Management Association*, 51: 1579–1585.

Fishman J.A., Wozniak A.E., and Creilson J.K. 2003. Global distribution of tropospheric ozone form satellite measurements using the empirically corrected tropospheric ozone residual technique: Identification of the regional aspects of air pollution, *Atmospheric Chemistry and Physics*, 3: 893–907.

Gupta P., and Christopher S.A. 2008. Seven year particulate matter air quality assessment from surface and satellite measurements, *Atmospheric Chemistry and Physics*, 8: 3311–3324.

Hoff R.M., and Christopher S.A. 2009. Remote sensing of particulate pollution from space: Have we reached the promised land? *Journal of Air and Waste Management Association*, 59: 645–675.

Kaufman Y.J., Gobron N., Pinty B., Widlowski J.L., and Verstraete M.M. 2002. Relationship between surface reflectance in the visible and mid-IR used in MODIS aerosol algorithm-theory, *Geophysical Research*, 29: 2116, doi: 10.1029/2001GL014492.

Koelemeijer R.B.A., Homan C.D., and Matthijsen J. 2006. Comparison of spatial and temporal variations of aerosol optical thickness and particulate matter over Europe, *Atmospheric Environment*, 40(27): 5304–5315.

Krotkov N.A., Carn S.A., Krueger A.J., Bhartia P.K., and Yang K. 2006. Band residual difference algorithm for retrieval of SO_2 from the Aura Ozone Monitoring Instrument (OMI), *IEEE Transactions on Geoscience and Remote Sensing*, 44(5): 1259–1266.

Levy R.C., Remer L.A., Mattoo S., Vermote E.F., and Kaufman Y.J. 2007. Second-generation operational algorithm: Retrieval of aerosol properties over land from inversion of moderate resolution imaging spectroradiometer spectral reflectance, *Journal of Geophysical Research*, 12(D13211): 1–21.

Levy R.C., Remer L.A., Tanré D., Mattoo S., and Kaufman Y.J. 2009. Algorithm for remote sensing of tropospheric aerosol over dark targets from MODIS: Collections 005 and 051. http://modis-atmos.gsfc.nasa.gov/, accessed June 14, 2009.

Li C., Mao J., Lau A.K.H., Zibing Y., Meihua W., and Xiaoyang L. 2005. Application of MODIS satellite products to the air pollution research in Beijng, *Science in China Series D (Earth Science)*, 209–219, doi: 10.1360/05yd0395.

Martin R.V. 2008. Satellite remote sensing of surface air quality, *Atmospheric Environment*, 42: 7823–7843.

McMillan W.W., Barnet C., Strow L., Chahine M.T., McCourt M.L., Warner J.X., Novelli P.C., Korontzi S., Maddy E.S., and Datta S. 2005. Daily global maps of carbon monoxide from NASA's Atmospheric Infrared Sounder, *Geophysical Research Letters*, 32(11), doi: 10.1029/2004GL021821.

Pollution Control Department (PCD). 2008. Regional area air quality data. http://pcd.go.th/AirQuality/Regional/Default.cfm, accessed 2008.

Pollution Control Department (PCD). 2010. Air quality monitoring stations in Thailand. http://aqnis.pcd.go.th/station/allstation.htm, accessed 2010.

Schmit T.A., Gunshor M.M., Menzel W.P., Gurka J.J., Li J., and Bachmeier A.S. 2005, August. Introducing the next generation advanced baseline imager on GOES-R, *Bulletin of American Meteorological Society*, 1079–1096.

Tsai T.C., Jeng Y.J., Chu D.A., Chen J.P., and Chang S.C. 2009. Analysis of the relationship between MODIS aerosol optical depth and particulate matter from 2006 to 2008, *Atmospheric Environment*, 107: 1–12.

Wang J., and Martin S.T. 2007. Satellite characterization of urban aerosols: Importance of including hygroscopicity and mixing state in the retrieval algorithms, *Journal of Geophysical Research*, 112: 1–18, doi: 10.1029/2006JD008078.

Wang Z., Chen L., Tao J., Zhang Y., and Su L. 2010. Satellite-based estimation of regional particulate matter (PM) in Beijing using vertical-and-RH correcting method, *Remote Sensing of Environment*, 114(1): 50–63.

Warner J., Comer M.M., Barnet C.D., McMillan W.W., Wolf W., Maddy E., and Sachse G. 2007. A comparison of satellite tropospheric carbon monoxide measurements from AIRS and MOPITT during INTEX-A, *Journal of Geophysical Research*, 112: 1–12.

Wenig M.O., Cede A.M., Bucsela E.J., Celarier E.A., Boersma K.F., Veefkind J.P., Brinksma E.J., Gleason J.F., and Herman J.R. 2008. Validation of OMI tropospheric NO_2 column densities using direct-Sun mode Brewer measurements at NASA Goddard Space Flight Center, *Journal of Geophysical Research*, D16S45: 113.

Zhang H., Hoff R.M., and Engel-Cox J.A. 2009. The relation between moderate resolution imaging spectroradiometer (MODIS) aerosol optical depth and $PM_{2.5}$ over the United States: A geographical comparison by U.S. environmental protection agency regions, *Journal of Air and Waste Management Association*, 59(11): 1358–1369.

Ziemke J.R., S. Chandra., B. N. Duncan., L. Froidevaux., P. K. Bhartia., P. F. Levelt., and J. W. Waters. 2006. Tropospheric ozone determined from Aura OMI and MLS: Evaluation of measurements and comparison with the Global Modeling Initiative's Chemical Transport Model, *Journal of Geophysical Research*, 111 (D19303).

Section II

Modeling Tool for Air Quality Management

5

Applications of Photochemical Smog Models for Assessment of Ozone, Particulate Matter Air Quality, and Acid Deposition in Asian Cities

Nguyen Thi Kim Oanh, Didin Agustian Permadi, Zhang Bao-Ning, Tran Nguyen Quang Huy, Nguyen Lu Phuong, Thongchai Kanabkaew, and Asif Iqbal

CONTENTS

5.1 Introduction

The air pollution levels in several big cities frequently exceed the National Ambient Air Quality Standards (NAAQS), especially for particulate matter (PM) and ground-level ozone, as highlighted in Chapter 1 of this book. Rapid population growth, industrialization, and urbanization result in high emissions of primary PM and major precursors for formation of secondary PM, ozone, and acid deposition in urban areas. Understanding the complex relationships between the source emission and the multiple air pollution issues is required to assess the present and project future effects of various development scenarios on the human health and environment. In particular, analysis of causal links between emissions and meteorology with the atmospheric concentrations and deposition of air pollutants is essential for development of the integrated air quality management (AQM) strategies.

Pollutants released from various sources are present and mixed in a giant reactor of the atmosphere, where multiple complex chemical and physical interactions occur. Through the atmospheric photochemical reactions, the precursors are transformed into secondary pollutants, such as ozone and secondary PM, including acidic particles. Nonlinear interactions between the precursor emission, meteorology, and surface ozone levels, for example, present a real challenge for formulation of appropriate ozone pollution abatement measures (Kim Oanh and Zhang, 2004; Marr et al., 2002). Comprehensive monitoring and modeling tools are commonly required to analyze these relationships. Air pollution monitoring alone may give important quantitative information about ambient concentrations and deposition of air pollution, but only at specific locations and time periods of measurement. Moreover, monitoring data do not directly provide clear guidance on the identification of the causes of the air quality problems. In many cases, comprehensive air quality models are considered the most suitable tool for quantification of the deterministic relationship between emissions and atmospheric concentrations/depositions.

Air quality models have evolved from the simple Gaussian plume models (Bosanquet and Pearson, 1936; Sutton, 1932) to the present new generation of models. The simple Gaussian plume model was applied first for calculating the maximum ground-level concentration and the distance (from the emission source) where it occurs, mainly for nonreactive pollutants. This type of model cannot handle the complex atmospheric processes that need to be incorporated to consider secondary pollutants such as ozone or secondary PM (sulfate, nitrate, and secondary organic particles). The latest generation of air quality models, the so-called one-atmosphere models, theoretically can handle all pollutants present in the atmosphere with their complex interactions in relation to the emission and atmospheric processes at multiple scales. Such models have been equipped with chemical mechanisms incorporating complex reactions between species involved to simulate, for example, the formation of ozone, secondary PM, and acidic compounds in the atmosphere. These models can treat multiple air quality issues and simulate the consequences of past and future emission scenarios to assess the effectiveness of abatement strategies (Daly and Zannetti, 2007). Models-3/Community Multiscale Air Quality (CMAQ) and the comprehensive air quality model with extensions (CAMx) are examples of one-atmosphere models.

Applications of air quality models for AQM are attracting increasing attention among researchers in Asian developing countries. The models in many cases are used to generate air pollution information, especially where the monitoring data are still scarcely available. However, air quality models require updated and detailed emission input data, which are rarely available in the developing Asian countries. In addition, lack of observation data at required spatial and temporal resolutions has been an obstacle for model performance evaluation. This chapter presents case studies of applications of CAMx/MM5 (mesoscale meteorological model), CMAQ/MM5, and variable-grid Urban Airshed Model (UAM-V)/Systems Applications International mesoscale model (SAIMM) modeling systems in the Asian cities conducted at the Asian Institute of Technology (AIT), Thailand. The model applications were mainly for ozone AQM (Bangkok metropolitan region [BMR], Jakarta, Hanoi, Ho Chi Minh City [HCMC]) with a case study for PM AQM in BMR and a case study for acid deposition in Dhaka. The content focuses on the steps in photochemical smog model applications for AQM purposes and highlights the approaches to preparation of the input data and model performance evaluation specifically when only limited emission inventory (EI) and measurement data are available.

5.2 Photochemical Smog Models

5.2.1 Description of Modeling Systems and Applications

The majority of the photochemical smog models are based on the Eulerian and a few on the Lagrangian reference approach. Kim Oanh and Permadi (2009) provided a more detailed description of available photochemical smog models and the common chemical mechanisms incorporated. The model systems presented in this chapter belong to the three-dimensional (3D) Eulerian group in operation at the air pollution modeling computer laboratory at the AIT and are summarized in Table 5.1. The earlier studies applied UAM-V/SAIMM and CHIMERE/ECMWF (European Center for Medium-Range Weather Forecasts), while the majority of the later case studies used CAMx/MM5; one case study used CMAQ/MM5 for the simulations. For the modeling systems with nesting capability, the two domains—outer (coarse) and inner (fine)—were simulated. Nesting was conducted by either "one-way" or "two-way" methods. In one-way nesting, the coarse resolution domain simulation is run independently of the nested domain, with no feedbacks between both domains (e.g., CMAQ/MM5 application for ozone forecast in BMR) (Nghiem, 2008). In this case, the simulations for the outer domain provided the boundary conditions for the inner domain simulation. Two-way nesting was applied through a simultaneous running of two domains to allow feedbacks between both domains during the simulations (e.g., CAMx/MM5 applications for BMR, HCMC, Jakarta, and Dhaka). When the nested simulation approach was not applied, the boundary conditions were prepared from the observed data (i.e., UAM-V/SAIMM for the BMR domain). Carbon bond mechanism IV (CB-IV), described by Atkinson et al. (1997), was used in all modeling attempts with updated reaction rates. Due to the lack of the volatile organic compound (VOC) profile for the region, the VOC emission profile developed for Europe presented by Kuhn et al. (1998) was used to prepare the input data. The modeling domains are shown in the regional map in Figure 5.1. The main features and modeling purposes for each case study are highlighted in the following section. Anthropogenic emission data for the outer domain were taken from the Center for Global and Regional Environmental Research (CGRER) Web site (http://www.cgrer.uiowa.edu/EMISSION_DATA/), while for the inner urban domains, commonly the input data were prepared through independent emission inventories. Biogenic VOC emission data were mainly downloaded from the Global Emission Inventory Activity (GEIA) Web site (http://www.geiacenter.org/) for the coarse domains. For the inner/fine domains, they were estimated either manually or using a software called Global Biosphere Emissions and Interaction Systems (GLOBEIS), available at http://www.globeis.com; both were based on the approach given by Guenther et al. (1995).

5.2.1.1 CMAQ and CAMx with MM5

CMAQ, the model 3 community multiscale air quality model (http://www.cmaq-model.org), and CAMx, the comprehensive air quality model with extensions (http://www.camx.com), are the "one-atmosphere" models developed to simulate tropospheric ozone and a range of other atmospheric phenomena. The models consider the complex interactions between atmospheric pollutants at the urban and regional scales.

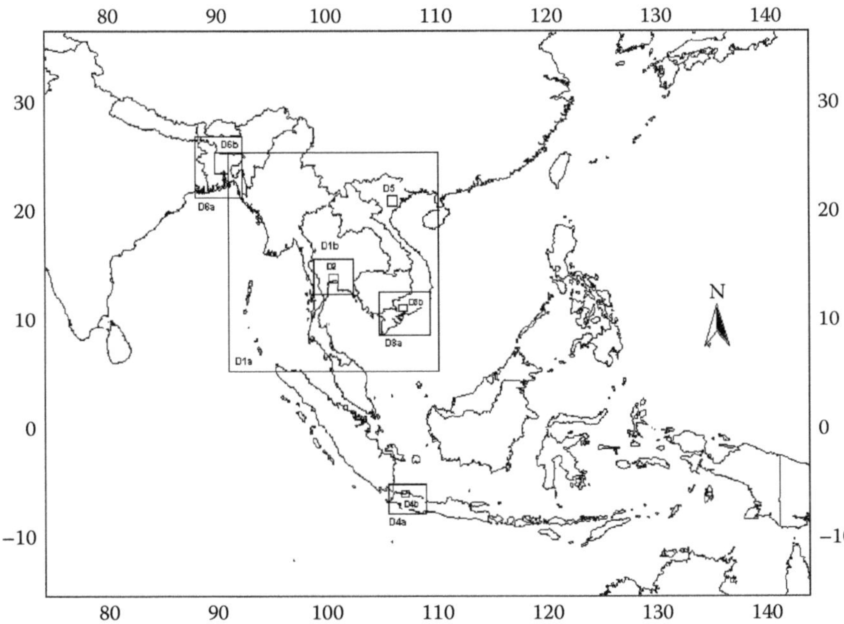

FIGURE 5.1
Modeling domains used in the case studies. D1a, CSEA outer domain; D1b, CENTHAI inner domain; D2, BMR domain; D3a, HCMC outer domain; D3b, HCMC inner domain; D4a, Jakarta outer domain; D4b, Jakarta inner domain; D5, Hanoi domain; D6a, Dhaka outer domain; and D6b, Dhaka inner domain.

The Fifth-Generation Pennsylvania State University/National Center for Atmospheric Research (PSU/NCAR) MM5 generates the meteorological field inputs for the CMAQ and CAMx simulations. The model input includes the global reanalysis data and global surface and upper air observation data, which are obtained from the National Center for Atmospheric Research Data Support Section (NCAR-DSS) (http://dss.ucar.edu/datasets/ds083.2/). In the case studies, MM5 was exercised in four-dimensional data assimilation (FDDA) mode using 3D grid and surface nudging over the entire modeling period. The MM5 output files were processed with the Meteorology Chemistry Interface Processor (MCIP) for the subsequent chemical transport simulations by CMAQ or CAMx. Meteorological schemes used in the studies are summarized in Table 5.2.

The CMAQ/MM5 modeling system was applied for the regional CSEA and the nested BMR domain (outer domain D1a and inner domain D1b) for ozone forecast experiments. The CAMx/MM5 modeling system was used to simulate surface ozone pollution in BMR (D2), Jakarta (D4a, D4b), and HCMC (D3a and D3b). The modeling system was also used for simulation of the PM pollution in BMR (D2, D1b) and the acid deposition assessment over Dhaka (D6a, D6b).

TABLE 5.1

Information on Modeling Systems Used in AIT Studies

Case Study	Photochemical Model	Meteorological Model	Emission Data	Domain[a]	Simulation Period	Grid Size	Simulation Purpose
1	CMAQ	MM5	2000 CGRER[b] GEIA[c]	CSEA (D1a)	Jan 26–29 & Mar 24–26, 2004	56 km	Historical ozone episode simulation and to produce boundary data for ozone prediction in BMR
2	CMAQ	MM5	1997 PCD projected to 2006 GEIA	CENTHAI (D1b)	Jan 1–30, 2006	11.2 km	Next-day ozone prediction
3	CAMx	MM5	2000 CGRER GEIA Independent EI[d]	JMA (D4a, D4b)	Oct 17–18, 2003	Outer: 10.8 km Inner: 3.6 km	Episode simulation of ozone air quality
4	CAMx	MM5	1997 PCD updated with rice straw field-burning EI[e]	BMR (D2)	Jan 26–29 & Mar 24–26, 2004	4 km	Impact of agroresidue field burning on ozone air quality
5	CAMx	MM5	2000 CGRER Independent EI[f] GEIA for outer & GLOBEIS for inner domain	HCMC (D3a, D3b)	Mar 1–13, 2005	Outer: 10.8 km Inner: 3.6 km	Episode simulation of ozone air quality and analysis of ozone formation sensitivity
6	CAMx	MM5	1997 PCD projected to 2007 2006 CGRER GEIA Crop residue field burning[g]	BMR (D2, D1b)	Jan 2007 Aug 2007	Outer: 12 km Inner: 4 km	Simulations for PM and its composition

7	CAMx	MM5	2006 CGRER Independent EI Updated UNEP Male EI[h]	Dhaka (D6a, D6b)	June 2006 Dec 2006	Outer: 16.67 km Inner: 5.55 km	Assessment of acid deposition (dry, wet) in wet and dry seasons
8	UAM-V	SAIMM	Independent EI[i] Manual calculation for biogenic VOC	Hanoi (D5)	Mar 3–4, 2003	4 km	Episode simulation of ozone air quality
9	UAM-V	SAIMM	1997 PCD Manual calculation for biogenic VOC	BMR (D2)	Jan 13–14, 1997	4 km	Episode simulation of ozone air quality and analysis of ozone formation sensitivity
10	CHIMERE	ECMWF	1997 PCD Manual calculation of biogenic VOC	BMR (D2)	Jan 13–14, 1997	4 km	Episode simulation of ozone to compare with UAM-V/SAIMM

[a] Domain abbreviations: CSEA, continental Southeast Asia; CENTHAI, central Thailand; JMA, Jakarta metropolitan area; BMR, Bangkok metropolitan region; HCMC, Ho Chi Minh City (see Figure 5.1).

[b] CGRER: emission data from http://www.cgrer.uiowa.edu/EMISSION_DATA/ for 2000 and 2006 years.

[c] GEIA: Global Emission Inventory Activity, downloaded from http://www.geiacenter.org/.

[d] Permadi (2007): gridded EI for anthropogenic sources of Jakarta.

[e] Phuong (2006) emission inventory for rice straw field burning in BMR.

[f] Huy (2008) emission inventory for anthropogenic sources in HCMC.

[g] Kanabkaew and Kim Oanh (2010) EI for crop residue field burning in 2007 for Thailand.

[h] Iqbal and Kim Oanh (2011) updated emission inventory for anthropogenic sources in Dhaka.

[i] An (2005) emission inventory for anthropogenic sources in Hanoi.

TABLE 5.2

Physical Schemes Used in MM5 Simulations

Case Study	Domain (see also Figure 5.1)	Cumulus Convection	Boundary Layer Parameterization	Cloud Microphysics	Radiation	Surface/Soil Temperature	Shallow Convection
1	CSEA (D1a)	Grell[a]	MRF[b]	Schultz[c]	RRTM[d]	Noah[e]	No[f]
2	CENTHAI (D1b)	Grell	MRF	Schultz	RRTM	Noah	No
3	JMA (D4a, D4b)	Grell	MRF	Schultz	RRTM	Noah	No
4	BMR (D2)	Grell	MRF	Schultz	RRTM	Noah	No
5	HCMC (D3a, D3b)	Scheme 1: Grell (D3a, D3b) Scheme 2: Kuo[g] (D3a, D3b) Scheme 3: Grell (D3a), No Cumulus[h] (D3b)	MRF	Schultz	RRTM	Noah	No
6	BMR (D2, D1b)	Grell	MRF	Schultz	Cloud[i]	Noah	No
7	Dhaka (D6a, D6b)	Grell	MRF	Schultz	RRTM	Noah	No

Source: PSU/NCAR mesoscale modeling system tutorial class notes and user's guide: MM5 Modeling System version 3. 2005, January. http://www.mmm.ucar.edu/mm5/documents/MM5_tut_Web_notes/MM5/mm5.htm.

[a] Grell: Simple single-cloud scheme with updraft and downdraft fluxes and compensating motion determining heating/moistening profile; it is suitable for small grid sizes, 10–30 km.

[b] MRF: Medium range forecast, a representation of countergradient term and K-profile in the well-mixed planetary boundary layer (PBL) that is suitable for high resolution in PBL.

[c] Schultz: This is a highly efficient and simplified scheme designed for running fast and easy tuning for real-time forecast systems.

[d] RRTM: The rapid radiative transfer model for representing the effects of the detailed absorption spectrum, taking into account water vapor, carbon dioxide, and ozone.

[e] Noah: This scheme is capable of predicting soil moisture and temperature in four layers (10-, 30-, 60-, and 100-cm thick), as well as canopy moisture and water-equivalent snow depth.

[f] No shallow convection: This scheme does not handle nonprecipitating clouds.

[g] Kuo: This scheme uses convective instability and moisture convergence as a measure to parameterize cumulus convection.

[h] No-cumulus: It is applicable to the grid length less than 5 km, and grid size is sufficient to resolve updrafts and downdrafts.

[i] Cloud radiation scheme: This scheme takes the cloud-radiation interactions into account (i.e., long-wave and short-wave interactions with explicit cloud and clear air).

5.2.1.2 UAM-V/SAIMM

The UAM-V model, developed by Systems Applications International (SAI), is a significant update to the UAM system (http://uamv.saintl.com). The prognostic meteorological model of SAIMM is used to drive the UAM-V model. SAIMM utilizes a terrain-following vertical coordinate system with variable vertical grid spacing and FDDA (SAI, 1995). Although UAM-V has been reported to produce some additional uncertainty that may be propagated in complicated terrains (http://www.camx.com/reports/stimpe.pdf) both the BMR (D2) and Hanoi (D5) domains are quite flat to justify applications of the model.

5.2.1.3 CHIMERE/ECMWF

The CHIMERE (http://lmd.polytechnique.fr/chimere) is a 3D chemistry transport model that is mainly intended for real-time forecasts and long-term simulations. CHIMERE runs over a range of spatial scales, from the regional scale (several thousand kilometers) to the urban scale (100–200 km), with resolutions from 1–2 to 100 km. CHIMERE uses hourly meteorological data provided by several meteorological models, such as weather research and forecasting (WRF), MM5, or the ECMWF. The case study presented in this chapter used the CHIMERE and UAM-V/SAIMM simulations for selected ozone episodes in BMR (D2) for comparison (Zhang et al., 2005).

5.2.2 Model Performance Evaluation

Model performance evaluation in the presented case studies was conducted by both qualitative and quantitative approaches. The qualitative approach for meteorological models mainly focused on the consistency between the wind field and pressure field as well as the temporal variations of temperature and wind during the simulation periods. The evaluation for the photochemical smog models focused on the consistency in the spatial distribution of meteorology fields and ozone over the domains and between ozone and emission sources of precursors and comparison between the temporal variations of simulated ozone and the available monitoring data.

Photochemical models are normally evaluated based on the performance in simulation of a few historical high-ozone episodes before they are used for emission scenario simulations to analyze various abatement strategies or for forecasting. With the present availability of high computational facilities, many models can be evaluated based on a long (a few months) simulation, which will certainly help to examine the performance more thoroughly. In the case studies presented in this chapter, the performance of different modeling systems was evaluated based on simulations of selected ozone and PM historical episodes. The selection of representative historical episodes for model evaluation is an important step that should be done properly. The following criteria were used in the ozone episode selection in the case studies:

- The area in which hourly ozone concentrations exceeded a threshold level (normally 100 ppb, which is a common hourly ambient ozone standard in many countries) across the selected domain should be as large as possible.
- High hourly ozone concentrations at a few stations in the domain should be observed at least 2 hours during 1 day and for at least 2 consecutive days.
- Meteorological conditions of episodes should be representative and occur with a high frequency in the domain.

The quantitative evaluation of photochemical smog models (for ozone) was done following the USEPA (U.S. Environmental Protection Agency) guidelines (USEPA, 1991) and for meteorological parameters following the values proposed by Emery et al. (2001), which are summarized in Table 5.3. In the model performance evaluation, the simulation results of the first (lowest) layer were compared with available measurement data collected from air quality monitoring stations. The performance statistics used to evaluate O_3 simulation included the mean normalized bias error (MNBE), the mean normalized gross error (MNGE), and unpaired peak prediction accuracy (UPA). For meteorological parameters, the statistical measures used were mean bias (MB), mean absolute gross error (MAGE), root mean-squared error (RMSE), and index of agreement (IOA). MNGE and MNBE were used for O_3 model performance evaluation, considering a cutoff (minimum) value of the observation threshold (for example, 60 ppb) (Boylan and Russell, 2006). Excluding the lower O_3 concentration from performance statistics is acceptable since the NAAQS for O_3 are directed at peak concentrations.

Unlike O_3 model performance criteria, which were well established for many decades (USEPA, 1991), the criteria for PM have not been readily available. PM simulation consists of many PM components, and the PM regulations (NAAQS) are concerned with 24-h or annual average rather than the hourly concentrations. It is therefore difficult to set a minimum threshold for the model performance evaluation. Normally, very large values of MNGE and MNBE can be obtained when the observed levels are close to zero. The mean fractional bias (MFB) and mean fractional error (MFE) have been suggested for the evaluation of the simulated PM results (Boylan and Russell, 2006). Table 5.3 presents the PM performance "goals" (the level of accuracy that is considered to be close to the best a model can be expected to achieve in that application) and the PM performance "criteria" (the level of accuracy that is considered to be acceptable for standard modeling applications).

TABLE 5.3

Statistical Parameters and Criteria for Model Performance Evaluation

Parameters	Formula	Suggested Criteria
Mean bias (MB)[a]	$MB = \frac{1}{N}\sum_{i=1}^{N}(M_i - O_i)$	WS: ≤ ±0.5 m/s WD: ≤ ±10 deg T: ≤ ±0.5°C RH: ≤ ±1 g/kg
Mean normalized gross error (MNGE)[a]	$MNGE = \frac{1}{N}\sum_{i=1}^{N}\frac{\lvert M_i - O_i\rvert}{O_i}\ 100$	Ozone: ≤ +35%
Unpaired peak accuracy (UPA)[b]	$UPA = \frac{M_{max} - O_{max}}{O_{max}}\ 100$	Ozone: ≤ ±20%
Mean absolute gross error (MAGE)[a]	$MAGE = \frac{1}{N}\sum_{i=1}^{N}\lvert M_i - O_i\rvert$	WD: ≤ +30° T: ≤ +2°C RH: ≤ +2 g/kg
Root mean-squared error (RMSE)[a]	$RMSE = \sqrt{\frac{1}{N}\sum_{i=1}^{N}(M_i - O_i)^2}$	WS: ≤ +2 m/s
Index of agreement[a]	$IOA = 1\ - \left[\frac{N \times RMSE^2}{\sum_{i=1}^{N}\left(\lvert M_i - \overline{O}\rvert + \lvert O_i - \overline{O}\rvert\right)^2}\right]$	WS: ≥ +0.6 T: ≥ +0.8 RH: ≥ +0.6
Mean normalized bias error (MNBE)[b]	$MNBE = \frac{1}{N}\sum_{i=1}^{N}\frac{M_i - O_i}{O_i}$	Ozone: ≤ ±15%
Mean fractional bias (MFB)[c]	$MFB = \frac{2}{N}\sum_{1}^{N}\left[\frac{(M_i - O_i)}{(M_i + O_i)}\right].100$	PM: ≤ ±30% (goal): ≤ ±60% (criteria)
Mean fractional error (MFE)[c]	$MFE = \frac{2}{N}\sum_{1}^{N}\left[\frac{\lvert M_i - O_i\rvert}{(M_i + O_i)}\right].100$	PM: ≤+50% (goal): ≤+75% (criteria)

Notes: RH, relative humidity; *M*, modeled value (model outputs for the first layer in different grids); *N*, number of observations; *O*, observations; $\overline{O}$, mean observation; T, temperature; WD, wind direction; WS, wind speed.

[a] Emery et al. (2001)

[b] USEPA (1991)

[c] Boylan and Russell (2006).

5.3 Simulation Results

5.3.1 CMAQ/MM5 Application for Next-Day Ozone Forecast in BMR

5.3.1.1 Synoptic Meteorology of Selected Episodes Used for Model Performance Evaluation

Two ozone episodes in 2004 (January 26–29 and March 24–26) were selected for model performance evaluation. Typical regional synoptic patterns during the two ozone episodes are shown in Figure 5.2. The January episode was characterized by the presence of a high-pressure system over China with a ridge that reached deep into Southeast Asia (SEA). This synoptic pattern occurs mainly in December and January, featuring the influence of the northeast monsoon during winter in the study area (Kim Oanh et al., 2005). Low winds due to the weak-pressure gradient over the southern part of SEA would result in stagnant air conditions horizontally, while subsidence and radiative inversions typical for high-pressure systems may limit the vertical dispersion. These conditions would lead to the accumulation of air pollutants, including ozone precursors and ozone after formation. The March episode was characterized by several thermal lows over India and northern Thailand. In this synoptic pattern, hot weather conditions with mostly calm wind are normally observed over BMR, HCMC, and other southern parts of the SEA region. The Bangkok meteorological station, for example, recorded high temperature and solar radiation, low pressure, and dry weather during this period, which are favorable conditions for ozone formation (Nghiem, 2008).

5.3.1.2 CMAQ Model Setup and Input Data Preparation

5.3.1.2.1 Model Setup

A coarse domain of CSEA and a nested inner domain of CENTHAI were used for both MM5 and CMAQ simulation (see also Figure 5.1) to produce the next-day O_3 forecast in the inner domain. One-way nesting was applied for both MM5 and CMAQ simulations. The same horizontal grid size was used for both model simulations, 56 km (~0.5°) for CSEA and 11.2 km (~0.1°) for the CENTHAI domain. Vertically, in the sigma coordinate system, both domains extended from mean sea level to approximately 16 km, segregated into 30 layers for MM5 and 15 layers for CMAQ simulations (Nghiem and Kim Oanh, 2008).

5.3.1.2.2 Anthropogenic Emissions

The available database of emission sources for the CSEA model domain was not sufficient to apply the Sparse Matrix Operator Kernel Emissions (SMOKE) model that is known to be compatible with CMAQ. Therefore, detailed hourly emissions for major source types in CSEA were prepared. The annual emissions of nitrogen oxides (NO_x), CO, and VOCs for the year 2000 were obtained from the anthropogenic emission database provided by CGRER (http://www.

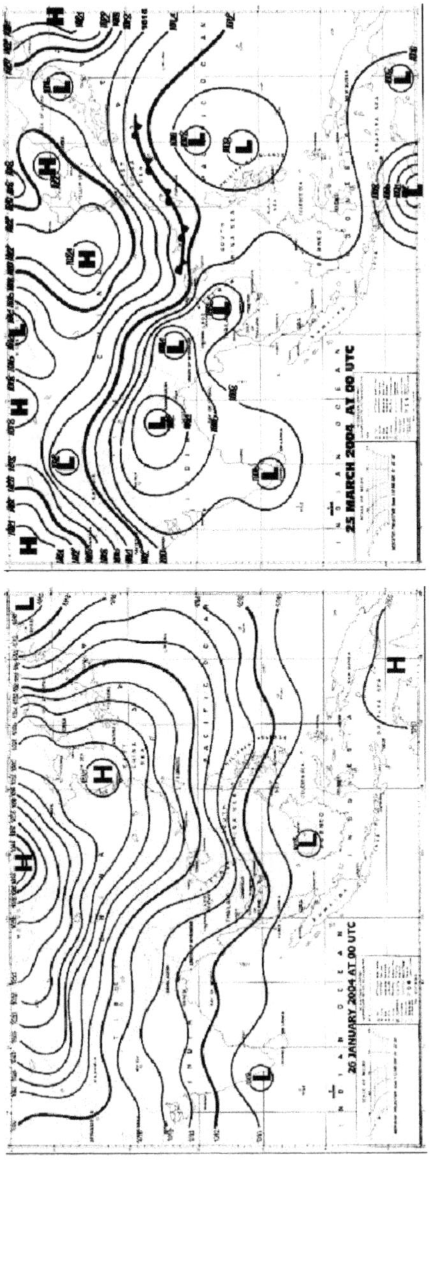

(a) (b)

FIGURE 5.2
Surface synoptic meteorological charts at 0700 LST on January 26, 2004 (a), and on March 25, 2004 (b), when high surface ozone was observed in Southeast Asia. (*Source:* Adapted from the Thai Meteorological Department [TMD] http://www.tmd.go.th/en/.)

cgrer.uiowa.edu/EMISSION_DATA_new/). The gridded emission data at 0.5° × 0.5° resolution were used for the CSEA domain, while that at the 0.1° × 0.1° resolution were used for the inner (CENTHAI) domain. Assuming the emission increased at the same pace as the economic growth, the emission data were projected to the year 2004 using the average gross domestic product (GDP) growth of the CSEA region, which was averaged at 7% per year during this period (Asian Development Bank [ADB], 2006). The monthly emission data were estimated using the fractions of the combustion-related species provided by Streets et al. (2003). To compute the hourly emission, the diurnal emission patterns were required; however, these were not available in the study domain. The measured hourly ambient CO concentrations in BMR, a relatively conservative pollutant in the ambient air, were used as the basis for estimation of hourly emissions (Nghiem and Kim Oanh, 2008).

5.3.1.2.3 *Biogenic Emissions*

The monthly global inventory of biogenic emissions of NO_x and VOCs from GEIA of 1° × 1° resolution were obtained from the Web site (http://www.geiacenter.org) for the selected months. Further, the emissions were interpolated to the 0.5° × 0.5° grid size for the CSEA domain simulation. The hourly biogenic emission rate was estimated based on the diurnal temperature and solar radiation profiles, which were generated by averaging 4-year (2000–2004) recorded hourly data of these variables at 20 air quality monitoring stations in the domain.

5.3.1.2.4 *Initial and Boundary Conditions*

For the CSEA domain, clean initial conditions were applied, and a spin-up period of 10 days was used to minimize the effects of the initial concentration fields on the simulation results. The initial conditions for the inner domain were taken from both monitoring data, whenever available, and the outputs of the outer domain simulation (Nghiem, 2008). The boundary conditions of the outer domain were developed based on monthly mean concentrations of O_3, NO_2, CO, peroxyacetyl nitrate (PAN), CH_4, C_2H_6, HCHO, and HNO_3 taken from the global model of ozone and related tracers (MOZART), version 2 (Horowitz et al., 2003). The boundary conditions for other species in the CB-IV mechanism were set to zero and allowed to evolve during the simulation. The boundary conditions for the inner CENTHAI domain were obtained from the simulations of the outer domain.

5.3.1.3 Model Performance Evaluation

5.3.1.3.1 *MM5 Performance*

The MM5 performance (with schemes presented in Table 5.2) was evaluated by comparing the simulated hourly meteorological fields with the observed at six selected surface weather stations in the domain (four in Thailand and two in Vietnam). MM5 generally satisfactorily reproduced wind and temperature fields, but the performance appeared to be better at some stations than others. The model performance statistics (Table 5.3) for two airport stations, Don Muang in Thailand and Tan Son Hoa (TSN) in Vietnam, during the selected episodes were satisfactory in terms of MAGE and IOA for temperature and MB, IOA and RMSE for wind speed. During the two historical episodes, the daily average MB for simulated wind speeds at Don Muang ranged from -0.02 to -0.25 m/s, that is, within the criteria of ±0.5 m/s (Figure 5.3), while the corresponding range for TSN was -0.09 to -1.11 m/s (Nghiem and Kim Oanh, 2008). The diurnal variations of temperature were satisfactorily reproduced, but the MB was from +0.93°C to +1.5°C for Don Muang and from +0.75°C to +1.45°C for TSN, which were well out off the criteria range of ±0.5°C.

5.3.1.3.2 CMAQ/MM5 System Performance

The CMAQ/MM5 performance for CSEA was analyzed for three subregions in the inner domain, western BMR (WBMR), the eastern region of BMR (EBMR), and HCMC, and three individual stations in northern Thailand where sufficient monitoring data were available. The average observed ozone levels from three to five monitoring stations available over each subregion were compared with subregion simulated results, which would bring in better agreement than for individual stations. The performance statistics calculated for the observed O_3 greater than 40 ppb (the cutoff value of 40 ppb; i.e., only values above 40 ppb were considered) were satisfactory. The calculated statistics for MNBE, MNGE, and UPA were within the USEPA guideline ranges for most stations and subregions. Ozone levels were generally overpredicted at one station, which was located near a local road and had more missing observation data points during the simulation period (Nghiem and Kim Oanh, 2008).

The temporal ozone distributions during the episodes were well captured by the modeling system (Figure 5.4). Generally, the performance was better for daytime than nighttime, especially for smaller cities outside BMR. Observed nighttime hourly ozone levels at some stations in those cities were higher than the modeled values. There was a second (lower) peak at night, more noticeable in Chiangmai, for example, that was not reproduced by the model. The complex mountainous terrain and the higher altitude of Chiangmai may be among the causes for the discrepancy, which needs further research. Finer grids for both the EI data and simulations should be done to better capture local-scale processes in this situation (Nghiem and Kim Oanh, 2008).

The spatial distributions of O_3 concentrations in the lowest layer (top at 38 m) over the CSEA domain at 1400 LST (local standard time) of episode days showed consistency with the wind fields (Figure 5.5). During the January episode, when the northeast monsoon prevailed, the ozone plume extended southwestward, while during the March episode the opposite plume direction was observed. Large urban areas in the domain, such as BMR, HCMC, and Chiangmai, generally showed high O_3 concentrations, with the ozone plume extending downwind from the urban centers. This regional transport of ozone may affect agricultural crops in areas surrounding large cities.

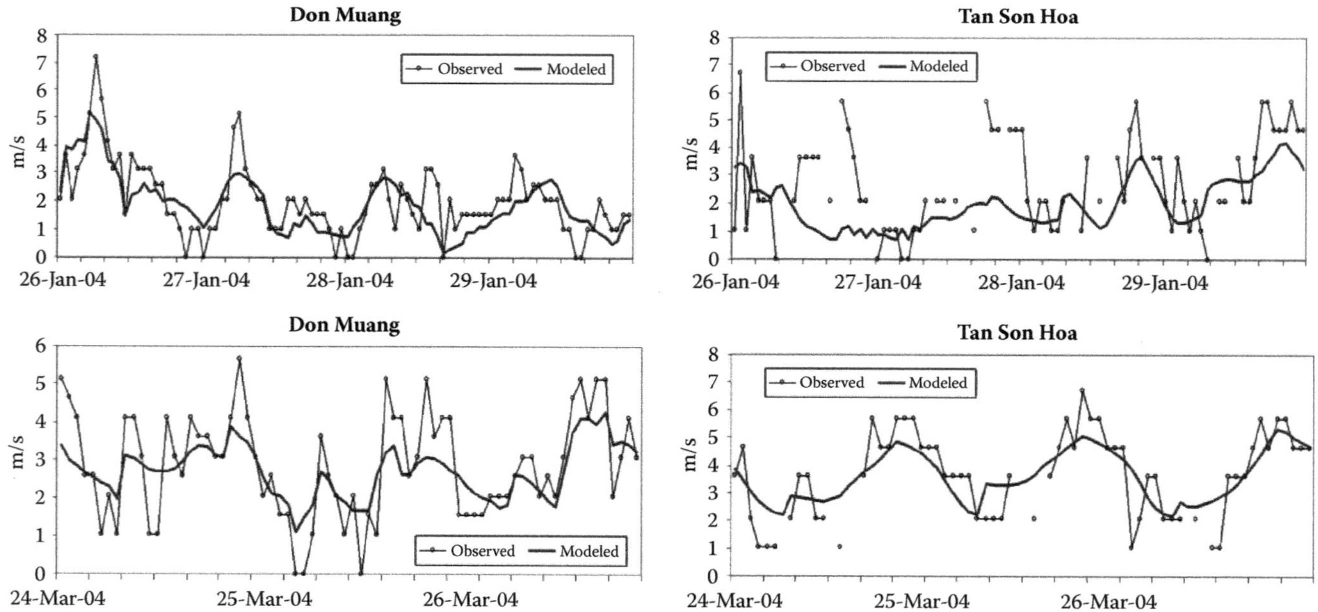

FIGURE 5.3
Gridded simulated hourly wind speed compared to observed at Don Muang and Tan Son Hoa meteorology stations, January and March 2004 episodes. (*Source:* Extracted from Nghiem L.H. 2008. Photochemical modeling for simulation of ground-level ozone over the continental Southeast Asian region to assess potential impacts on rice crop, doctoral dissertation, Asian Institute of Technology, Pathumthani, Thailand.)

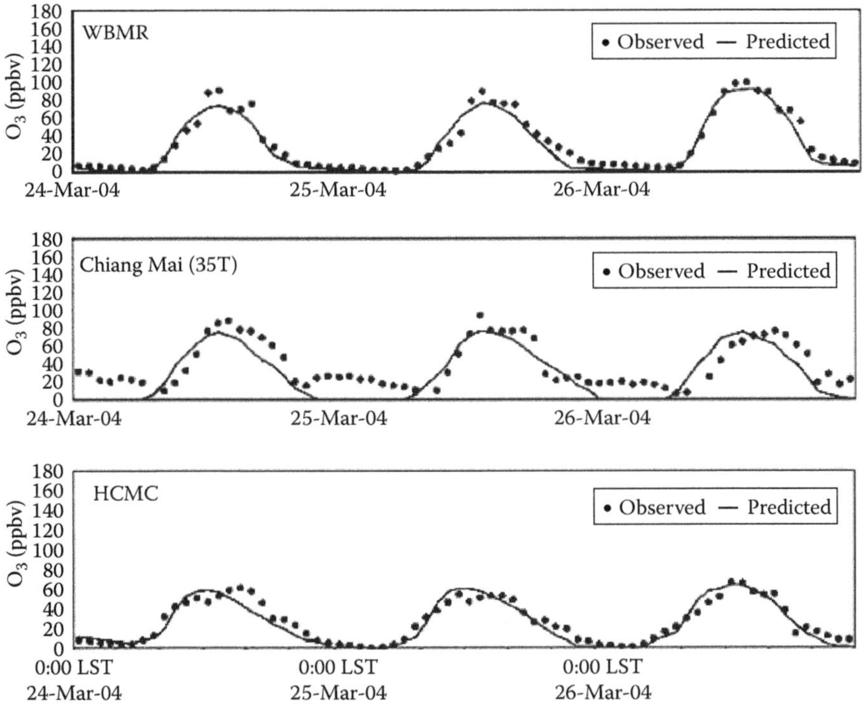

FIGURE 5.4
Gridded simulated hourly ozone levels compared to monitoring data at stations located inside the grid in the March episode. (*Source:* Nghiem L.H., and Kim Oanh N.T. 2008. Evaluation of the mesoscale meteorological model (MM5)-community multi-scale air quality model (CMAQ) performance in hindcast and forecast of ground-level ozone, *Journal of Air and Waste Management Association*, 58: 1341–1350. With permission.)

5.3.1.4 Forecast of Next-Day Hourly Ozone for Central Thailand

The CSEA domain (D1a, Figure 5.1) simulation produced the boundary conditions for the inner domain of CENTHAI (D1b). The experiment forecast of the next-day hourly ozone over the CENTHAI was done for 1 month, January 1–30, 2006, using the emission data projected for 2006. The forecast results were evaluated by discrete type (point-by-point observed vs. modeled values) using MB, MNBE, RMSE, and normalized mean error (NME). In addition, the two-category evaluation method (USEPA, 2003) was used to evaluate the model forecast skill in terms of correctly or incorrectly predicting the next-day peak 1-h ozone above or below the threshold of 100 ppb.

The observed data showed high 1-h O_3 levels over the CENTHAI domain during January 2006, with a maximum of 170 ppb, and the maximum 1-h ozone concentrations above 100 ppb were recorded on 18 days of the month (Figure 5.6). The scatterplot of the domain's highest ozone values observed versus predicted (gridwise average) showed a reasonable high R^2 of 0.82.

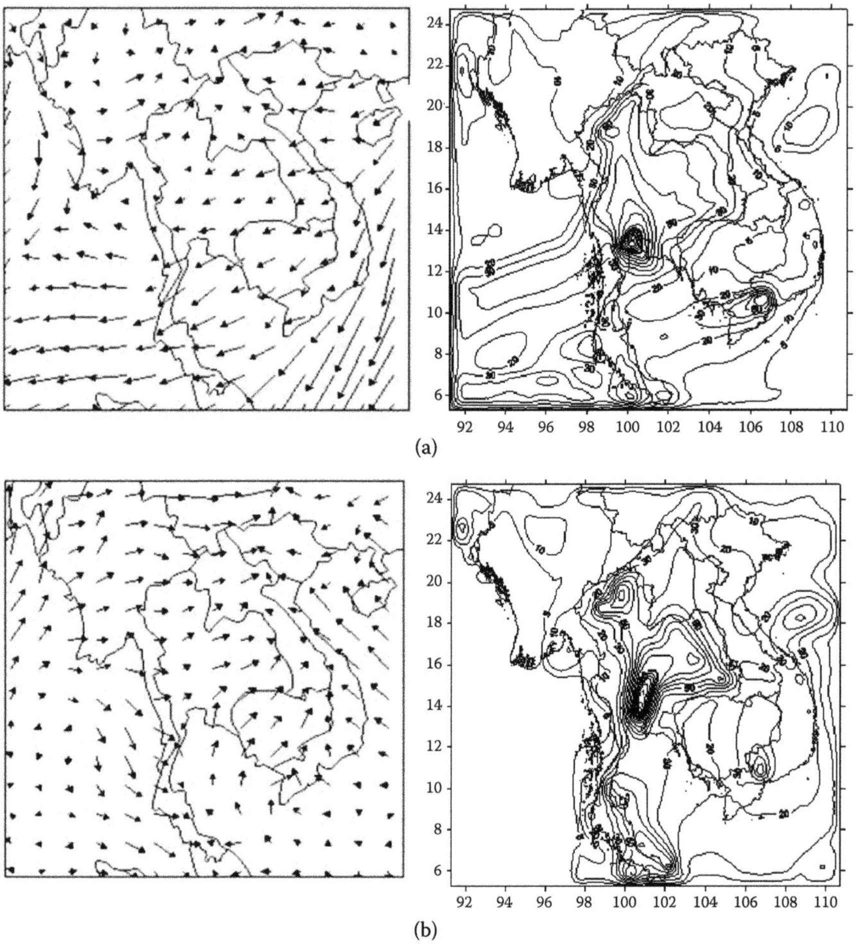

FIGURE 5.5
Typical wind fields and ozone plumes (values in ppb) over CSEA at 1400 LST during January episode (a) and March episode (b). (*Source:* Nghiem L.H., and Kim Oanh N.T. 2008. Evaluation of the mesoscale meteorological model (MM5)-community multi-scale air quality model (CMAQ) performance in hindcast and forecast of ground-level ozone, *Journal of Air and Waste Management Association*, 58: 1341–1350. With permission.)

The gridwise average values produced by the model were lower than the observed values, as expected. In general, underprediction was observed, especially for high measured ozone values; hence, MB was mostly negative (-6.4 ppb). The two-category forecast evaluation showed satisfactory performance, especially in predicting the single highest value (one over the domain) with an accuracy (A) of 94% (Nghiem and Kim Oanh, 2008), which was higher than the minimum requirement of 90% for the initial implementation of the air quality forecasting capacity (Davidson et al., 2004).

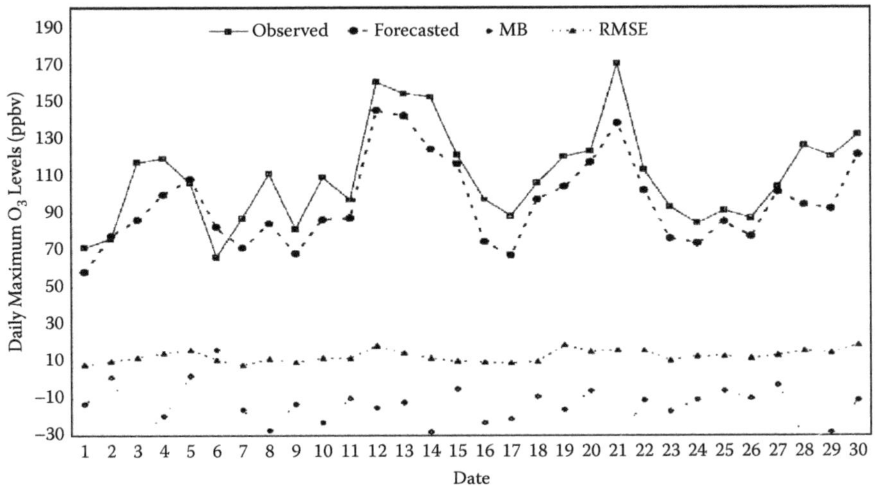

FIGURE 5.6
Daily observed (solid line) and forecasted (dashed line) peak ozone concentrations in CENTHAI during January 2006. The observed peak concentration is the single highest hourly ozone level observed in the domain in each day of the month. (*Source:* Nghiem L.H., and Kim Oanh N.T. 2008. Evaluation of the mesoscale meteorological model (MM5)-community multi-scale air quality model (CMAQ) performance in hindcast and forecast of ground-level ozone, *Journal of Air and Waste Management Association,* 58: 1341–1350. With permission.)

The results of this experimental forecast are considered promising, although a longer study period is required. Further improvements should focus, for example, on refining the grid size. However, the emission data remain the weakest part of this forecast attempt. In particular, crop residue field-burning emission, which may be significant over the domain (Kim Oanh et al., 2011), was not accounted for in the input data.

5.3.2 CAMx/MM5 Applications for Urban-Scale Simulations

The CAMx/MM5 model system was used for urban areas in Asia for ozone air quality analysis in BMR, Thailand; HCMC, Vietnam; and Jakarta, Indonesia and for PM air quality analysis in BMR and acid deposition assessment in Dhaka, Bangladesh. Details of each case study are discussed in this section.

5.3.2.1 Bangkok Metropolitan Region, Thailand

5.3.2.1.1 Model Setup

A single domain for CAMx simulation covering the Universal Transverse Mercator (UTM) coordinate from the southwestern corner (615 km, 1,488 km) to the northeastern corner (703 km, 1,560 km) was selected with a grid size of 4 km × 4 km. The MM5 simulation was done using two-way nesting, with the outer domain being the same CSEA domain presented in Section 5.3.1 and the inner domain of BMR.

5.3.2.1.2 Emission Input Data

This case study aimed to assess the impact of rice straw (RS) field-burning activity, which commonly occurs around BMR, on the ozone air quality. The original anthropogenic emission database prepared by the Pollution Control Department (PCD) of Thailand for 1997 does not cover the crop residue field burning. Therefore, the first EI for this source type was conducted for 2004 using the cultivation areas derived from the BMR land use map (Phuong, 2006). According to the survey, typically in the BMR provinces there are two rice crop cycles per year. The major crop is cultivated during June–September and harvested during November–February. The second crop growth period is October–February, with harvesting during the March–May period. The survey also showed most RS of these two crop cycles is burned in the field. In this area, there is also a third crop cycle in some places but the RS from this crop is not normally field burned. Further, the emission was segregated into daily and then hourly resolutions using the MODIS (Moderate Resolution Imaging Spectroradiometer) hotspot counts as a proxy after cross-checking with the BMR land use map. Biogenic emission (isoprene and monoterpenes) was estimated using the algorithms developed by Guenther et al. (1995) for the identified land use patterns of BMR, with detailed steps given in the work of Kim Oanh and Zhang (2004). The boundary and initial conditions for CAMx simulations were taken from available monitoring data as two-way nesting was not applied.

5.3.2.1.3 Scenario Simulation and Results

The simulation was done for two cases: case 1, without RS field burning (original PCD database), and case 2, PCD emission plus 100% RS field-burning emission. For the model evaluation, the same episodes (January and March 2004) as presented in Section 5.3.1 for the CSEA domain were used. The MM5 simulation was done with the same schemes as for the CSEA domain (Table 5.2), which, as discussed, showed satisfactory performance in the central part of Thailand. The simulation results were compared between these two cases for March episode (case 1M and case 2M) and January episode (case 1J and case 2J).

5.3.2.1.3.1 March Episode In case 1M, the simulations were done without RS field burning and with the meteorological conditions of the March episode. The gridded simulated hourly ozone levels were compared with the measurements at 10 PCD monitoring stations in BMR. Generally, the model system could capture the temporal variations of O_3 (Figure 5.7), although the peak ozone levels were underestimated, largely related to the model's grid-averaged outputs. The model performance statistics, calculated using the observed ozone cutoff value of 60 ppb, were mostly satisfactory, with an

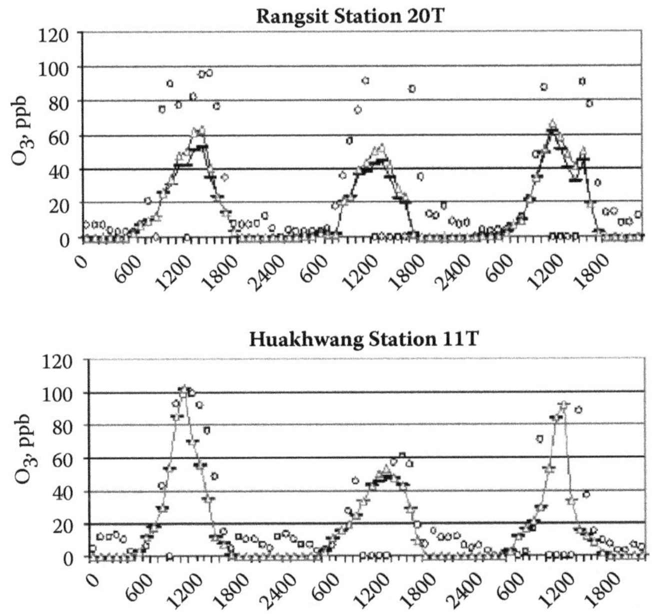

FIGURE 5.7
Gridded hourly simulated ozone concentrations for case 1M and case 2M scenarios during March episode days compared to monitoring data at two stations (o, observed; –, case 1M; Δ, case 2M).

MNGE of 26% and a UPA of -6.5%, which were within the USEPA guideline values (Table 5.3), but the MNBE of -17% was out of the guideline range (±15%).

At the Rangsit station, located in Pathumthani Province, the underestimation was more significant than at the Huakhwang station, which is downtown of Bangkok City (Figure 5.7). Pathumthani Province was the largest rice-growing province in BMR, which consisted of about 50% of the total rice cultivation area of BMR. The Rangsit station was surrounded by rice paddies; hence, the contribution of RS burning emission may have been more significant. Also, this station is located within 500 m of a large highway, so the fresh traffic emission (NO) would efficiently titrate ozone. In fact, the observed and simulated wind directions were mainly from south to west during the March episode following the synoptic pattern, which means that Pathumthani (and Rangsit station) was downwind of Bangkok City.

Typical spatial distributions in this case showed high simulated O_3 spots in the east and northeast of the domain (Phuong, 2006), which agrees with the results of CMAQ/MM5 simulation for CSEA presented previously in Figure 5.5(b).

In case 2M, the gridded hourly emission of RS field burning was added to the PCD emission database in the BMR domain to simulate the impact of RS field burning on ozone air quality. Overall, the RS burning emission did not bring about significant changes in the O_3 levels over BMR except for the Rangsit station. A maximum increase of 10 ppb in hourly O_3 levels was shown for this station at around 1500–1600 on March 24. On average, the hourly ozone levels at this station increased by 4 ppb over the period, which was attributed to RS burning emission. The low combustion temperature and incomplete combustion nature of this emission source would release more VOCs than NO_x. The EI showed that RS burning over BMR contributed 16 kt VOCs annually compared to 4 kt NO_x (Phuong, 2006). The ozone formation in BMR was shown more VOC sensitive (Kim Oanh and Zhang, 2004); hence, an increase in the VOC/NO_x ratio would enhance ozone formation.

5.3.2.1.3.2 January Episode Similar to the March episode, for the January episode the simulation results of case 1J and case 2J were compared (Figure 5.8). In the January episode, the NE wind was prevalent, and the ozone plume was extending southwestward; the maximum ozone appeared at the SW corner of the domain. Pathumthani Province, where intensive RS burning took place, was upwind. A slight increase in ozone levels was seen over the BMR domain, with the maximum increase of 0.5 ppb near the city center (Figure 5.8). As compared to the March episode, when more noticeable effects were seen at the Rangsit station, in this January episode the effects were spread more over the BMR domain owing to the prevalent wind direction.

5.3.2.2 Ho Chi Minh City, Vietnam

5.3.2.2.1 Model Setup

The simulation was done for a historical ozone episode in HCMC, March 1–13, 2005, for model performance evaluation. A sensitivity study was also conducted to analyze the ozone formation response to the emission reductions. Two-way nesting was applied for the CAMx simulation with the inner domain (18 × 24 cells each of 3.6 km × 3.6 km) and the outer domain (50 × 45 cells of 10.8 km × 10.8 km) shown in Figure 5.1. The boundary conditions for the inner domain were generated by the outer domain simulation. Each CAMx domain was matched with an MM5 domain with a similar grid size but of larger coverage (i.e., 46 × 48 cells for the inner MM5 domain and 90 × 66 cells for the MM5 outer domain). Simulations with MM5 were performed by three different sets of cumulus parameterizations—Grell, Kuo, and Grell and no cumulus for scheme 1, scheme 2, and scheme 3, respectively—to evaluate their effects on the model performance. Other parameterizations were identical among the tested schemes as presented in Table 5.2.

5.3.2.2.2 Emission Data

The anthropogenic emission data for the outer domain were extracted from the annual EI dataset obtained from CGRER for 2000; the biogenic emission data were prepared using data from GEIA. For the inner

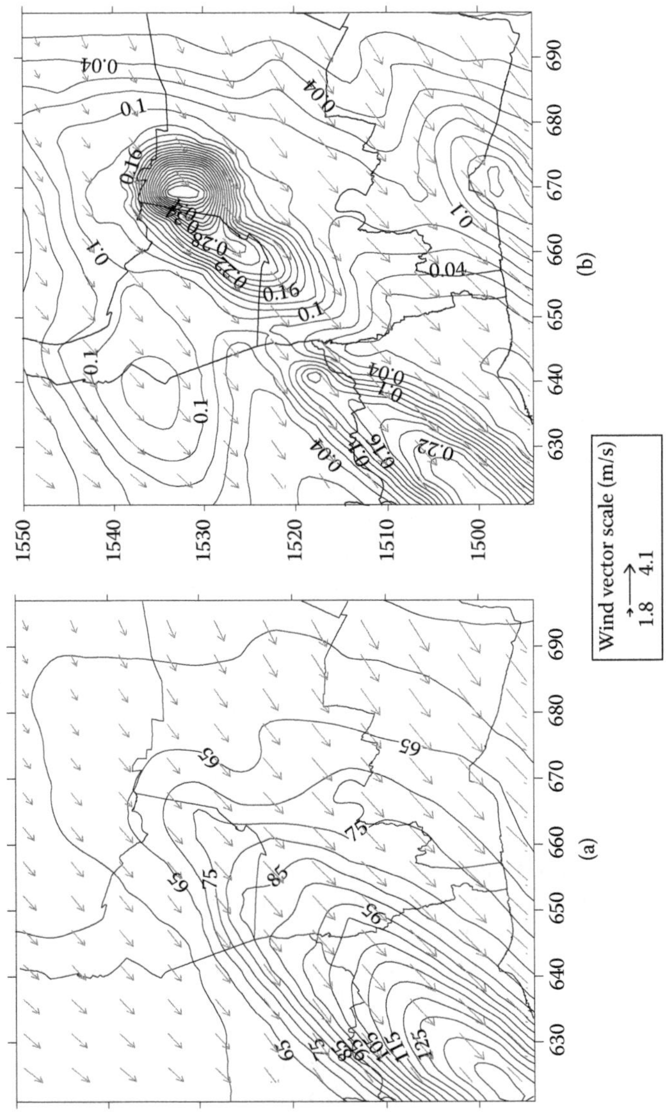

FIGURE 5.8
Spatial distribution of simulated O_3 (ppb) and wind field at BMR at 1300 LST, January 27, 2004: (a) simulated O_3 for case 2J (with RS burning emission) and (b) the differences between simulated O_3 of case 2J and case 1J (without RS burning emission) .

domain, an EI was conducted that produced the hourly gridded anthropogenic emission for major sources in HCMC, including elevated point sources; area sources (industrial and residential combustion, gasoline service station); and on-road mobile sources. The biogenic VOC emission was calculated using the GLOBEIS model, which uses the algorithms developed by Guenther et al. (1993) (available at http://www.globeis.com).

5.3.2.2.3 MM5 Performance

The MM5 performance was evaluated by comparing the simulated meteorological variables with those observed at the Tan Son Nhat airport station, including wind speed and directions and temperature profiles during the episode. All three simulated schemes produced similar outputs, which were in reasonably good agreement with the measured temperatures. The model appeared to reproduce the wind directions during the period reasonably but strongly overestimated the wind speeds for 4 days (March 4–7), when mostly calm-to-light winds were observed, as seen in Figure 5.9.

5.3.2.2.4 CAMx/MM5 Performance

The simulation results of the CAMx/MM5 modeling system were compared with the available ozone monitoring data in the domain. The overestimation of wind speeds during these days in the selected period obviously affected the performance of the CAMx model as it would lead to higher dispersion of pollutants. Figure 5.10 presents the temporal variation of observed versus simulated O_3 level at three monitoring stations in HCMC, which shows the underestimation of ozone levels especially on three days (March 5–7) at all the stations. Nevertheless, the temporal variation of O_3 was reasonably reproduced by the model, although the simulated nighttime levels at a station (QT) were significantly lower compared to the measured values of around 20 ppb. Performance of the model was better at the other two stations (D2 and ZOO), with simulated and monitored value matching well for several days. Several reasons may explain the discrepancies such as the inaccuracy of the emission data. In addition, the measured values at a monitoring station represented the concentration at a specific point, while the simulated value was the average over a grid; hence, in principle, there is always a difference. Nevertheless, the direction of the ozone plumes during all episode days appeared to be consistent with the wind directions in the domain (Huy, 2008).

Using the 40-ppb cutoff point of the observed ozone levels, the modeling performance statistics were estimated, which were satisfactory during the period excluding the 3 days of low MM5 performance (March 5–7). All the performance statistics averaged for three stations were within the suggested criteria ranges (Table 5.3) for MNBE (-13%), MNGE (19%), and UPA (-0.6%) when these 3 days were not included. However, for the entire period (including these 3 days), the average performance statistics were only within the criteria of MNGE (31%) and UPA (-14%) and not for MNBE (–27%).

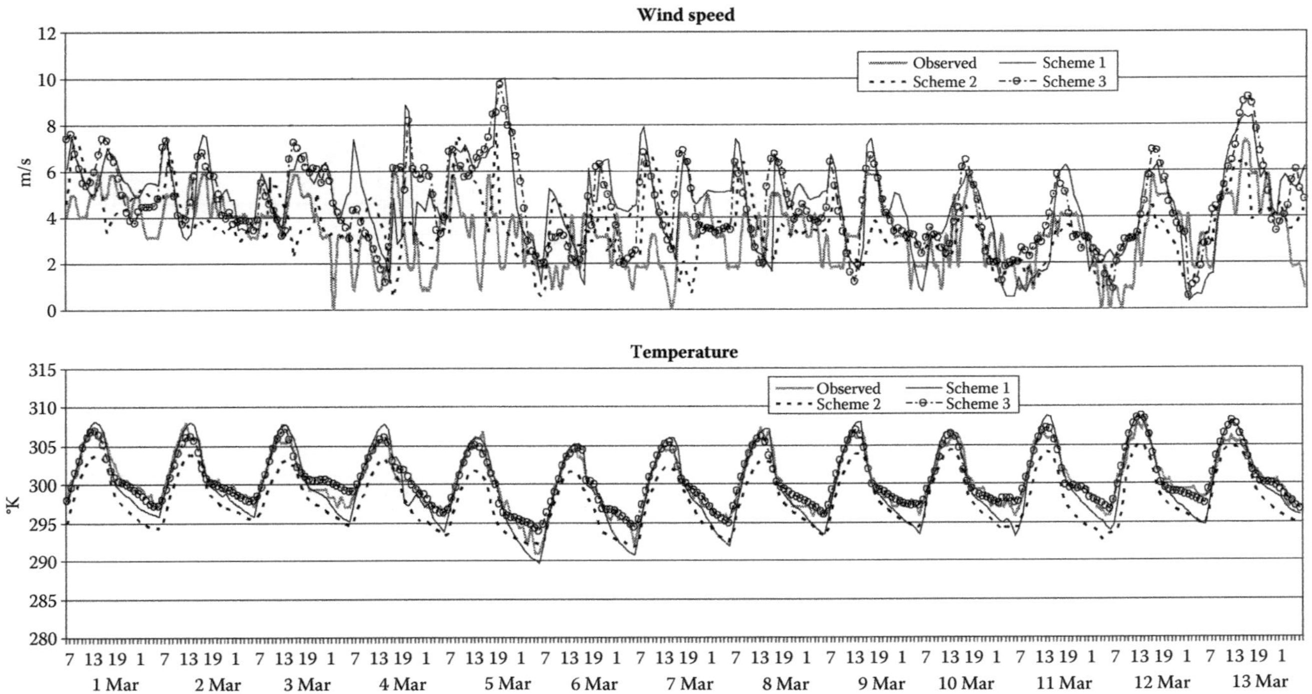

FIGURE 5.9
Gridded simulated hourly wind speed compared to observation data at the Tan Son Nhat airport by different meteorology schemes for MM5 (March 1–13, 2005). (*Source:* Adapted from Huy T.N.Q. 2008. Photochemical smog modeling for ozone air quality management in Ho Chi Minh City, Vietnam, master's thesis, Asian Institute of Technology, Pathumthani, Thailand.)

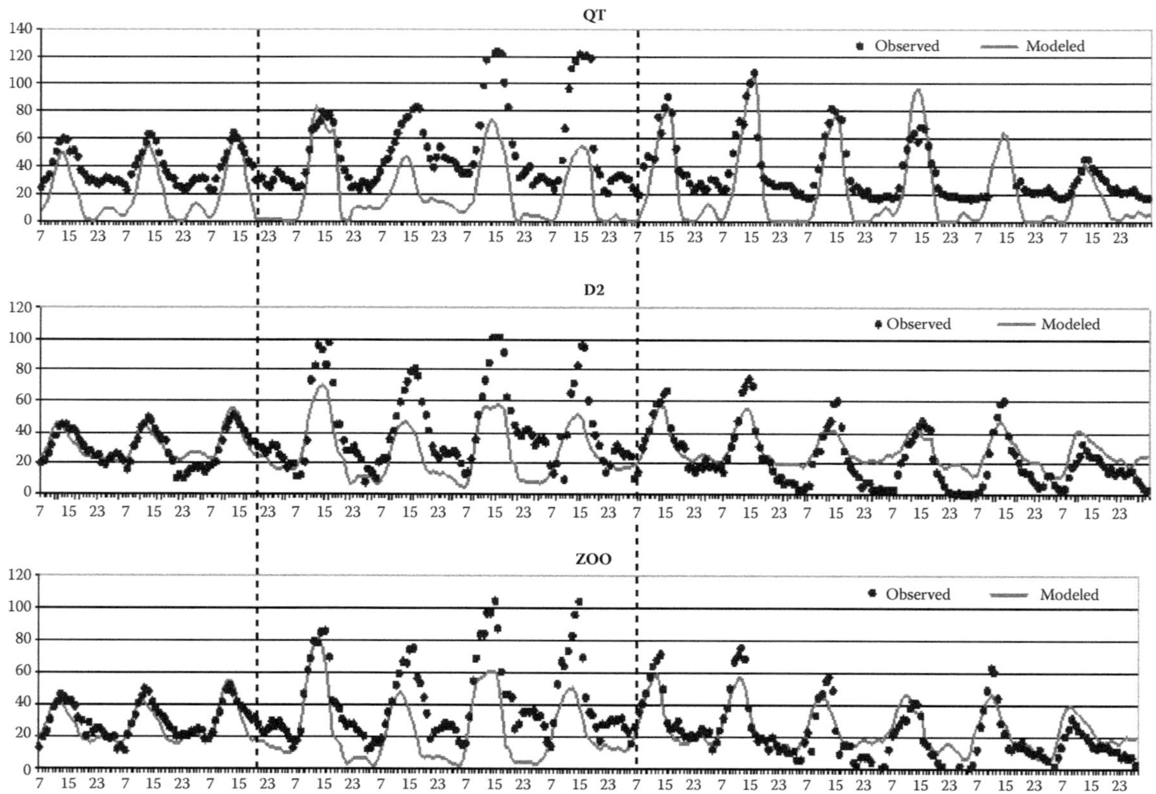

FIGURE 5.10
Gridded simulated hourly ozone compared to monitored ozone (ppb) at three stations located inside the respective grid, March 1–13, 2005, HCMC. (*Source:* Adapted from Huy T.N.Q. 2008. Photochemical smog modeling for ozone air quality management in Ho Chi Minh City, Vietnam, master's thesis, Asian Institute of Technology, Pathumthani, Thailand.)

5.3.2.2.5 Ozone Formation Sensitivity

Sensitivity analysis of NO_x and VOC reduction on ozone air quality in HCMC was conducted. The maximum modeled ozone concentration at 1300 LST on March 9, 2005, was selected for this sensitivity study. Isopleths of ozone formation sensitivity at the grid having the highest ozone level (No. K18) are presented in Figure 5.11. It indicates that initially ozone levels were sensitive to both NO_x and VOC emission reductions. However, when NO_x and VOCs were reduced by over 40%, the isopleths are almost parallel to the *x*-axis, meaning that further reduction in VOCs would not affect ozone concentrations, while further NO_x reduction would reduce ozone significantly. Large urban areas are reported to have a more VOC-sensitive ozone formation regime, such as Bangkok (Zhang and Kim Oanh, 2002); Los Angeles (Milford et al., 1989; Sillman et al., 1997); New York (Sistla et al., 1996); and Milan (Prevot et al., 1997). The NO_x-sensitive regime is more likely observed in less-polluted areas (Sillman, 1999). Overall, the ozone formation in HCMC was more sensitive to NO_x than to VOC emission reduction.

Thus, based on the modeling results, to attain the ozone hourly standard of 90 ppb in HCMC, the emission control strategies should be designed to

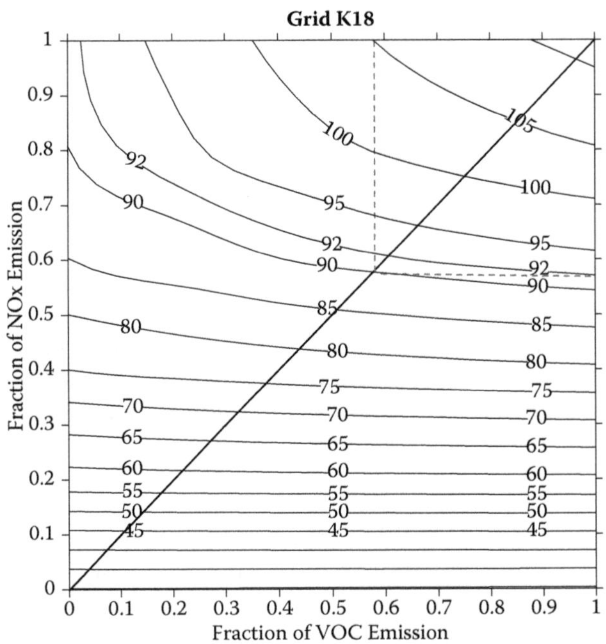

FIGURE 5.11
Response of maximum ozone concentration to NO_x and VOC reductions in HCMC. (*Source:* Adapted from Huy T.N.Q. 2008. Photochemical smog modeling for ozone air quality management in Ho Chi Minh City, Vietnam, master's thesis, Asian Institute of Technology, Pathumthani, Thailand.)

reduce either NO_x by over 45% or collectively to reduce both NO_x and VOC emission by over 40%. VOC emission reduction alone would reduce ozone but not enough to reach the 90-ppb level.

5.3.2.3 Jakarta, Indonesia

5.3.2.3.1 Model Setup

The simulation was done for a historical ozone episode on October 17–18, 2003, with high ozone levels monitored over Jakarta. Similar to the case study for HCMC, the two-way nesting technique was applied to produce the boundary and initial conditions for the inner domain. Two CAMx domains and three MM5 domains are presented in Figure 5.1. The finest domain of CAMx was 20 × 20 cells (3.6 km × 3.6 km), while the innermost domain of MM5 was 36 × 36 cells (3.6 km × 3.6 km). The schemes used were similar to those in the CSEA and BMR case studies, as presented in Table 5.2.

5.3.2.3.2 Emission Data

The anthropogenic emission data for the CAMx coarse domain was prepared using an annual emission database obtained from CGRER for 2000, while biogenic emission was based on GEIA. For the inner domain, the hourly gridded anthropogenic emission from major sources, including elevated point sources, area sources (industrial and residential combustions and gasoline service stations), and on-road mobile sources were prepared (Permadi, 2007). The biogenic VOC emission was prepared using the algorithms developed by Guenther et al. (1995), similar to the case study of BMR presented in Section 5.3.2.

5.3.2.3.3 MM5 Performance

The MM5 simulation results for Jakarta, including air temperature and wind during the episode, were compared with the observations at Soekarno Hatta Airport (SKH) meteorological station (Figure 5.12). The model did not satisfactorily reproduce both temperature and wind speed, with the MB significantly over the criteria range (Table 5.3).

5.3.2.3.4 CAMx/MM5 Performance

The simulated ozone levels by CAMx/MM5 modeling system were compared with observations collected from three ambient air monitoring stations in Jakarta (Figure 5.13). In general, the model captured the temporal variations of ozone but overestimated the peak ozone values at one station (JAF 1) while underestimating ozone at two other stations. The model also produced low ozone levels at nighttime, which is quite similar to the situation at a station (QT) in HCMC presented in Figure 5.10. Overestimation of NO emission and the uncertainty in the simulated wind speeds (overestimation at nighttime) may be some reasons for the low simulated ozone levels

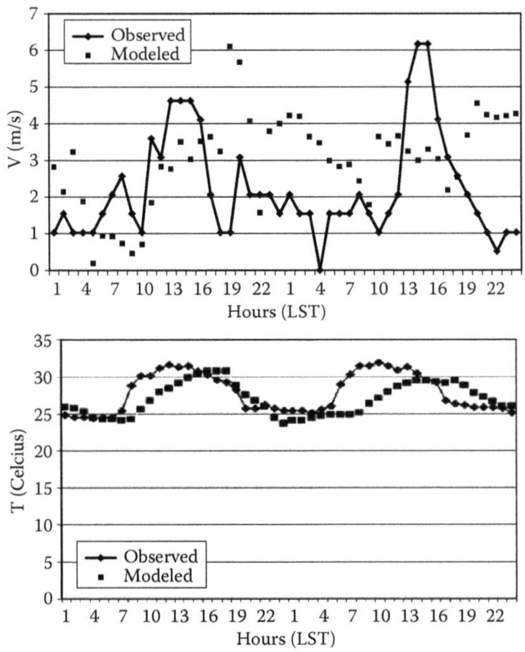

FIGURE 5.12
Gridded simulated hourly wind speed and temperature compared to observation data at the Soekarno Hatta Airport (SKH) during October 17–18, 2003. (*Source:* Adapted from Permadi D.A. 2007. Surface O_3 air quality in Jakarta: current status, precursor emission inventory, and modeling efforts, master's thesis, Asian Institute of Technology, Pathumthani, Thailand.)

at night. Accordingly, the model performance statistics for ozone were not satisfactory for all three criteria given in Table 5.3. Typical spatial distributions of simulated O_3 together with wind field at 1300 LST for October 17 and October 18 were analyzed by Permadi (2007). On October 17, the spatial distribution of simulated O_3 appeared to be consistent with the wind field, which showed high O_3 in the southern part of the domain (downwind of the city center). However, on October 18, high O_3 occurred in the eastern part of the domain (upwind of the city center).

The performance of the CAMx/MM5 system cannot be better than the MM5 performance. Therefore, the simulated ozone levels were not expected to be satisfactory once MM5 performance was not good. In fact, the city of Jakarta has a complex topography with the Salak mountain range (with an average height of 1,500 m and a maximum height of more than 3,000 m) located to the south of the city. This mountain range has an almost V shape, which is directed NE on one side and NW on the other. Sofyan et al. (2007) also noted that due to the surface terrain and sea breeze, the lower atmosphere of Jakarta is not largely affected by the synoptic-scale wind. Jakarta, with its near-equatorial location and specific topography, may be a challenge

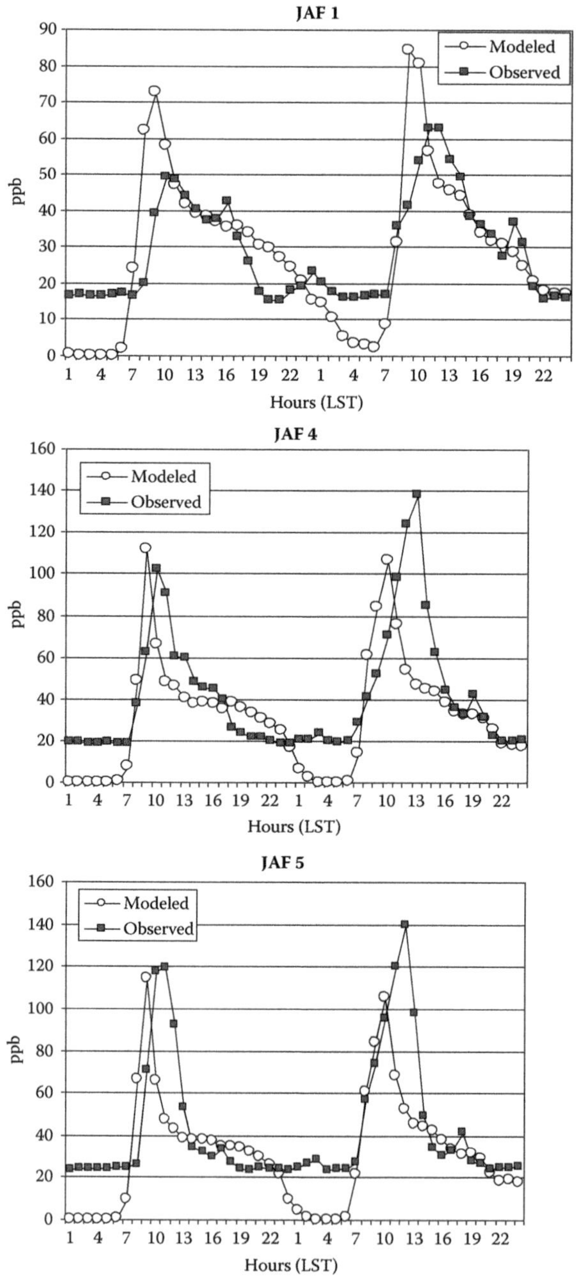

FIGURE 5.13
Gridded simulated hourly versus observed ozone levels at three stations located inside the respective grid in Jakarta. (*Source:* Adapted from Permadi D.A. 2007. Surface O_3 air quality in Jakarta: current status, precursor emission inventory, and modeling efforts, master's thesis, Asian Institute of Technology, Pathumthani, Thailand.)

for meteorology modeling (Permadi and Kim Oanh, 2008). Further, MM5 modeling should be investigated with different schemes to find the most suitable scheme for the domain. The EI should also be updated with more sources and detailed spatial and temporal distributions.

5.3.2.4 Acid Deposition in Dhaka

The case study in Dhaka was set up to assess the acid deposition in the central part of Bangladesh caused by the anthropogenic emission. The Dhaka division is the most populated portion of the country and is the center of all activities. A number of emission sources, including vehicles, brick kilns, industries, residential combustion, and agroresidue open burning, are present in the domain and produce a large amount of acid precursor emissions. Long-range transport of air pollution may also potentially contribute to the acid deposition in the area.

5.3.2.4.1 Model Setup

The simulation for the acid deposition assessment over the Dhaka domain was done using the nesting approach with three domains for MM5 (grid sizes of 0.5°, 0.167°, and 0.055°) and two domains for CAMx (grid sizes of 0.167° and 0.055°). The outer CAMx domain simulations were done with clean air initial conditions for a 10-day spin-up period. The boundary values from Environ (2008) were used in the outer domain simulations to produce the boundary initial conditions for the inner domain. The CAMx model considered both dry and wet depositions and provided the respective output separately. The simulation was done for December (dry) and June (wet) 2006. The simulation results for these 2 months were used to estimate the respective seasonal deposition rates, which were roughly six times the respective monthly deposition, assuming similar meteorological conditions and emission over a season. Further, the sum of depositions in the dry and wet season provided the annual deposition rate.

5.3.2.4.2 Emission Data

The anthropogenic emission data taken from CGRER for the year 2006 were used for the outer CAMx domain simulation. For the inner domain of the Dhaka division, the emission data were prepared using the updated EI results conducted for the year 2006 from an existing emission database. A detailed EI for 2006 was conducted for two major emission source categories: brick kilns and on-road vehicles. For other sources, a previous EI conducted under the Malé Declaration project of the United Nations Environment Program (UNEP) for 2000 was upgraded considering the sectoral growth statistics for the country from 2000 to 2006 (Bangladesh Bureau of Statistics [BBS], 2008). The emission obtained was segregated into gridded emission following the Dhaka land use map. Emissions from natural sources were not considered in this case study (Iqbal and Kim Oanh, 2011).

5.3.2.4.3 Simulation Results and Evaluation

The annual average nitrogen deposition was quite evenly distributed over the domain, with some higher values appeared in the NE part, which was consistent with the distribution of the dominant NO_x sources (Iqbal and Kim Oanh, 2011). The sulfur deposition was mainly concentrated in the SW corner, where the brick kilns were located; hence, there was high sulfur oxide (SO_x) emission during the dry season. Overall, the maximum gridded average sulfur deposition rate in 2006 over the Dhaka division was lower than that of nitrogen. The annual wet deposition rates of both sulfur and nitrogen were higher than the dry deposition rates. Domainwise, the highest annual sulfur plus nitrogen deposition was observed in the southwest part, while several locations with high depositions were also simulated in the northeastern and central parts of the domain (Figure 5.14).

The acid deposition monitoring in Bangladesh has not been studied intensively; hence, there were no published data to compare with the model results. The model results were thus only qualitatively evaluated considering the consistency in the spatial distribution of acid deposition and meteorology fields as well as with the gridded emission strength. The spatial distribution of the simulated acid deposition showed reasonable agreement with the simulated meteorological fields (wind direction and precipitation). In both seasons, the simulated deposition was higher in the downwind region (Figure 5.14). In the dry season, when the northerly wind prevails, higher deposition was observed in the southern to southwestern corners of the domain, which was also the area of maximum sulfur emission. In contrast, during the wet season, the wind direction is mainly from the south; hence, the deposition was higher in the northern parts of the domain. Also, during the wet season, the precipitation rate appeared to be a dominant factor for acid deposition. Consequently, the simulated acid deposition was higher in the center of the domain where higher precipitation occurred (Figure 5.14). Iqbal and Kim Oanh (2011) presented a more detailed evaluation of the simulated ambient concentrations (NO_x and SO_x) and meteorology (wind and temperature) and suggested that more measurement data are required to evaluate model performance adequately.

5.3.2.5 Particulate Matter Air Quality in BMR

5.3.2.5.1 Model Setup

The CAMx/MM5 modeling system was applied to simulate a month of high (February) and a month of low (August) PM in 2007 for the BMR domain. Two-way nesting was used with two CAMx domains, D1b and D2, as shown in Figure 5.1. The clean air conditions were used as the initial conditions for the outer domain (D1b) simulations, which produced the boundary conditions for the inner domain.

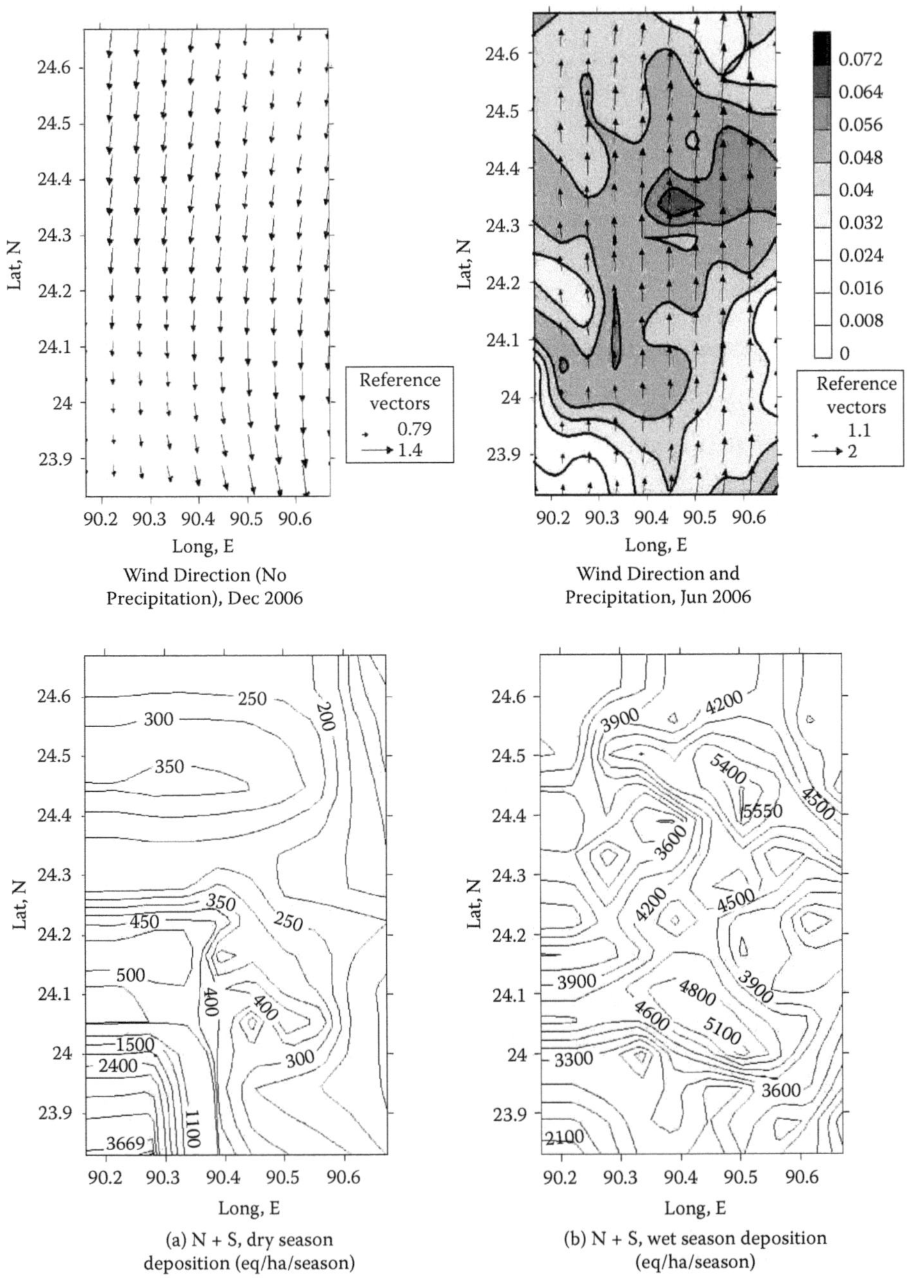

FIGURE 5.14
Simulated meteorological conditions (wind and precipitation) and seasonal acid deposition rates (eq $ha^{-1}season^{-1}$) in Dhaka division. (*Source:* Iqbal A., and Kim Oanh N.T. 2011. Assessment of acid deposition over Dhaka division using CAMx-MM5 modeling system, *Atmospheric Pollution Research,* doi: 10.5094/APR.2011.05.)

5.3.2.5.2 Emission Input Data

The anthropogenic emission data of 2006 were extracted from CGRER for the outer domain simulations. To prepare the input data for the inner domain simulation, the anthropogenic emission data by PCD for 1997 were projected to 2007 and were updated with the crop residue field-burning emission information prepared by Kanabkaew and Kim Oanh (2010). Further, the inventory was also conducted for the annual NH_3 emission, which is an important species for secondary PM simulation. The biogenic VOC emission extracted from GEIA was used, which was distributed into various grids of the BMR domain based on the land use map. The annual emission data were segregated into hourly emission using the same approach presented in Section 5.3.1.

5.3.2.5.3 Simulation Results

Hourly average concentrations of PM_{10} simulated for the months of February 2007 and August 2007 are presented together with the observation data (by the beta gauge method) at five monitoring stations in BMR, and the results for the grid containing station 13T, an ambient station located at the city's center, are presented in Figure 5.15 for illustration. Overall, the model results appeared to underestimate the observed PM_{10} levels in February but overestimated those in August. The model could not capture the hourly observed PM peaks. The performance statistics of CAMx/MM5 in these 2 months did not meet the goals (Table 5.3) but still were within the suggested criteria proposed by Boylan and Russell (2006) for MFB (±60%) and MFE (<75%).

Modeling for PM is in general more complex than other pollutants because it involves several PM compositions. The simulated PM mass is the sum of the primary PM (elemental carbon [EC], organic carbon [OC], and others); secondary inorganic aerosol (SIA) (PSO_4, PNH_4, and PNO_3); and secondary organic aerosol (SOA). In this case study, the average contribution of the primary PM components over BMR was up to 70–80% of the total simulated PM_{10} mass. Measurement data for EC and SIA (sum of ammonium, sulfate, and nitrate) are available for BMR (Kim Oanh et al., 2006, and Chapter 3 of this book), which show higher seasonal average values (dry season: EC of 9.4 $\mu g/m^3$, SIA of 10.8 $\mu g/m^3$; wet season: EC of 6.3 $\mu g/m^3$ and SIA of 5 $\mu g/m^3$) but still in compatible ranges compared to the simulated monthly averages for February and August, such as for station 13T as shown in Figure 5.16. Kim Oanh et al. (2006) reported 27 $\mu g/m^3$ organic matter (OM) in PM_{10} in BMR for the dry season; but the value was not separated into SOA and primary organic aerosol (POA) to be compared with the simulation results.

The spatial distribution of simulated SOA showed the highest values in the northeast area of BMR, which coincided with the locations of the highest emission strength for biogenic VOCs (isoprene and monoterpenes). The gridded maximum simulated SOA was higher in the dry season (5.0 $\mu g/m^3$) and lower (1.5 $\mu g/m^3$) in the wet season, which agrees with the seasonal variation in surface ozone levels as

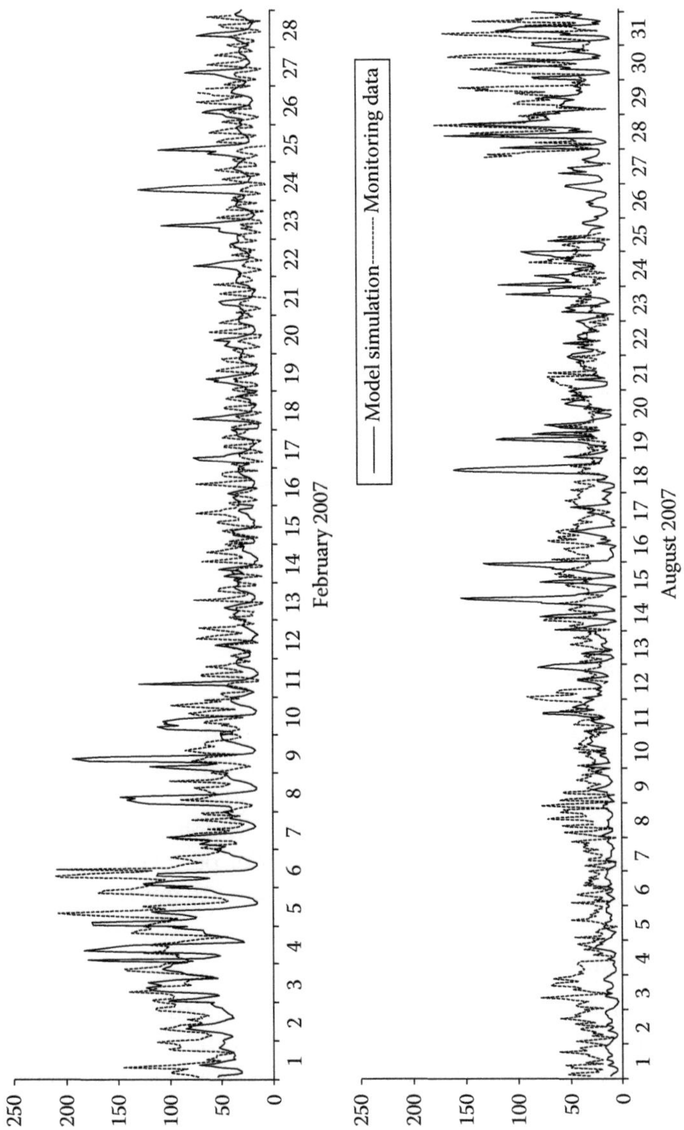

FIGURE 5.15
Hourly gridded simulated versus monitored PM_{10} at station 13T located inside the grid in February and August 2007.

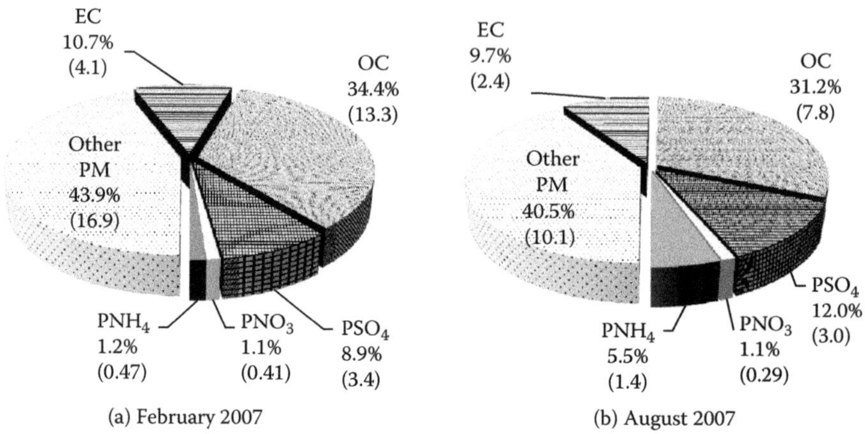

FIGURE 5.16
Simulated monthly average PM compositions (%) for the grid containing station 13T (in parentheses is concentration in μg/m³), Bangkok.

expected as they both are products of photochemical reactions. Preliminary SOA monitoring results, based on the EC/OC tracer analysis technique conducted for AIT site (Dahal, 2008), showed the same range as the simulated SOA but with higher peak values in both dry (0–14 μg/m³, average at 4.1 μg/m³) and wet seasons (0–10 μg/m³, average at 0.9 μg/m³) in 2007–2008.

Some important reasons for the discrepancies between the simulated and observed PM composition levels include (1) grid averaging effects of the simulation results as compared to the point-based measurement results, (2) emission input data were not detailed enough in terms of the spatial and temporal distribution, and (3) composition measurement data, especially for SOA, were limited and contained variable uncertainty. Updated emission input data with detailed spatial and temporal distributions are required to reduce the uncertainty of the PM modeling results. Lack of the PM composition monitoring data is a great challenge for comprehensive model performance evaluation, especially for SOA, for which the measurement data are not readily available.

5.3.3 UAM-V/SAIMM Modeling

The UAM-V/SAIMM model system was used for simulation of the surface ozone air quality in BMR and Hanoi (Table 5.1); both have flat terrains. For BMR, the UAM-V results are presented in comparison with the CHIMERE/ECMWF results.

5.3.3.1 Hanoi, Vietnam

5.3.3.1.1 Model Setup

The UAM-V/SAIMM system was used to simulate the formation and accumulation of surface ozone in the Hanoi metropolitan region domain (as of 2005) during an ozone episode March 3–4, 2003 (An, 2005). The model domain contained 10 × 10 cells of 4 km resolution. Hourly monitoring data available at continuous monitoring stations in the domain were used to prepare the initial and surface boundary conditions, while the default values of UAM-V (SAI, 1995) were used for the upper lateral boundary conditions. A 2-day period spin-up was used to minimize the effects of the initial conditions on the episode simulation.

5.3.3.1.2 *Emission Input Data*

An EI was conducted to produce the annual gridded emission (4 km × 4 km) for major anthropogenic source categories (industry, traffic, residential combustion, and gasoline stations). Biogenic VOC (isoprene and monoterpenes) emission was manually calculated using the algorithm developed by Guenther et al. (1995).

5.3.3.1.3 *Simulation Results*

In general, SAIMM output wind fields were consistent with the observed winds that appeared on the synoptic charts. The model output was also consistent with the gridded emission data (An, 2005). The spatial distribution showed the highest simulated O_3 concentrations in the central part of the domain, with the hourly peak of around 90 ppb on March 3, 2003. The available ozone-monitoring data in the city were screened, and the data recorded at two automatic ambient air quality monitoring stations in Hanoi were used to compare with the simulated results (Figure 5.17). The UAM-V/SAIMM model system captured the temporal variations of O_3 as compared to the observed pattern. The model performance statistics for ozone during the episode were in the suggested ranges (Table 5.3) for all three criteria: MNBE (+12.5%), MNGE (21%), and UPA (+18.3%). However, the model underestimated the peak O_3 for both days at one station (Lac Long Quan), which is located in the suburb upwind area of the city center, while it overestimated the peak values at the other station (Lang) located inside the city. The simulated O_3 levels at night were also lower than the observed values. Several factors may have caused the discrepancies, specifically the lack of updated and detailed emission input data. In addition, insufficiency of ozone monitoring data in the city also prevented a more comprehensive model evaluation.

5.3.3.2 UAM-V/SAIMM versus CHIMERE/ECMWF Performance for BMR

5.3.3.2.1 *Model Setup*

Two photochemical smog models, UAM-V/SAIMM and CHIMERE/ECMWF were used to simulate photochemical smog pollution over BMR. The CB-IV mechanism was used in UAM-V, while the MELCHIOR (Modèl Lagrangien de la Chimie de l'Ozone à l'échelle Régionale) mechanism, which is a lumped species mechanism, was used in CHIMERE modeling (Zhang et al., 2005). The

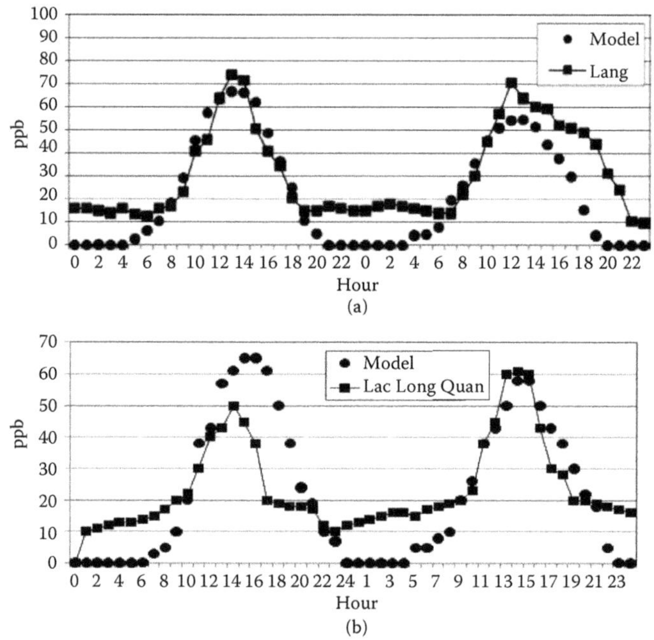

FIGURE 5.17
Simulated gridded ozone compared to monitoring data at the station located inside the respective grid during the episode (March 3–4, 2003): (a) Lang station and (b) Lac Long Quan station. (*Source:* Adapted from An D.D. 2005. Photochemical smog modeling for air quality management in the Hanoi metropolitan region, Vietnam, master's thesis, Asian Institute of Technology, Pathumthani, Thailand.)

BMR domain, 88 km × 72 km with a grid size of 4 km × 4 km, was selected for the simulations (Figure 5.1). The model performance was tested on a historical episode, January 11–14, 1997, and the model was applied to assess the impacts of several emission control measures on the ozone air quality in the domain. The model was also used for the analysis of sensitivity of ozone formation to precursor emission reductions (Kim Oanh and Zhang, 2004). Similar to the case of Hanoi, the initial and surface boundary conditions were prepared from the monitoring data in the domain, while the upper lateral and top boundary conditions were the default values provided by SAI (1995). A 2-day spin-up period was used to minimize the effects of initial conditions.

5.3.3.2.2 Emission Data

The anthropogenic emission data prepared by PCD, Thailand, for 1997 were used in the simulation. The biogenic emission (isoprene and monoterpenes) data were prepared using the algorithms developed by Guenther et al. (1995).

5.3.3.2.3 Simulation Results

Both SAIMM and ECMWF reproduced the temporal variation of temperature satisfactorily, but SAIMM simulation results appeared to be more

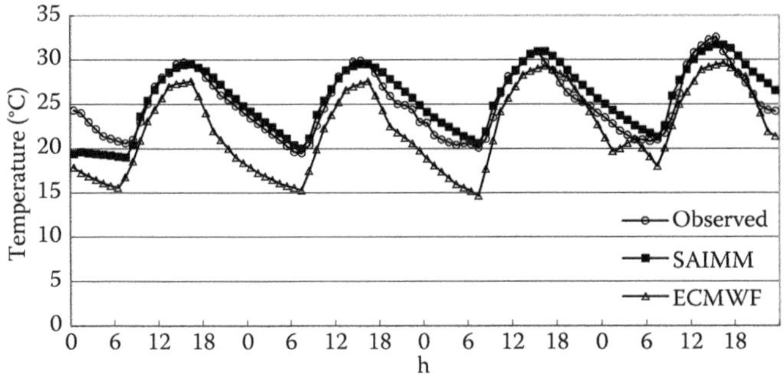

FIGURE 5.18
Simulated gridded temperature compared to monitoring data at Bangkok metropolis station located inside the grid, January 11–14, 1997. (*Source:* Zhang B.N., Vautard R., and Kim Oanh N.T. 2005. Comparison of two photochemical modeling systems in a tropical urban area, *Journal of Air and Waste Management Association*, 55: 903–918. With permission.)

consistent with the observations (Figure 5.18). The SAIMM simulation relied more on the surface meteorological measurements as the input and hence may improve the model performance at the urban scale as compared to the large-scale simulation results by ECMWF.

The simulated temporal variations in ozone levels as compared with the observations for the selected monitoring sites are presented in Figure 5.19. Better agreements between the simulated and the observed peak O_3 values were obtained on January 14 than on January 13, for which both models underestimated the peak values. Accordingly, on January 14 all performance statistics (MNBE, MNGE, and UPA), estimated for the cutoff observed value of 60 ppb, were within the suggested criteria ranges, while on January 13 only the UPA criterion was satisfactorily met. UAM-V simulated O_3 data were higher than for CHIMERE, which may be related to the difference in the simulated meteorological variables. Particularly, lower temperatures simulated by ECMWF as compared to SAIMM (also to the observations) would reduce the simulated O_3 formation rates. Higher temperature (e.g., by 40°F or 4.4°C) would significantly increase the overall rates of photochemical reactions (i.e., correspondingly by a factor of 2–4) (Wark et al., 1998).

Typical examples of simulated ozone spatial distribution by both models are presented in Figure 5.20 and show consistent ozone plume directions with higher ozone values in the downwind areas of Bangkok's city center, where the maximum values were above 190 ppb.

5.3.3.2.4 Sensitivity Analysis

The two modeling systems produced similar results when used to assess the contributions of different source emission categories to O_3 formation in Bangkok. The photochemical pollution in the domain appeared to be mainly caused by the anthropogenic emission from the sources inside the city (70%).

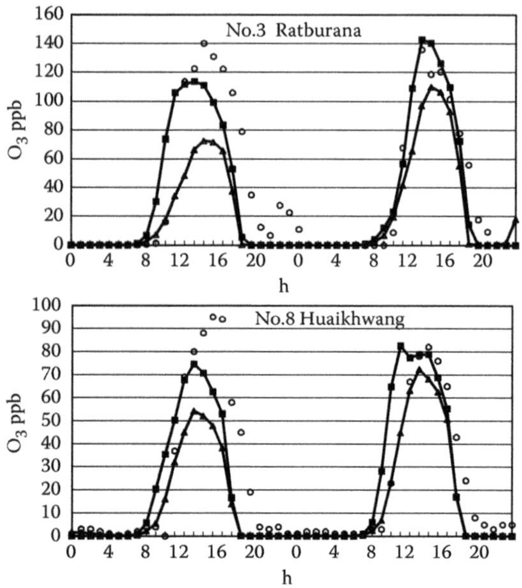

FIGURE 5.19
Simulated gridded hourly O_3 versus observed values at stations located inside the grid, January 13–14, 1997. ○, observed; ■, simulated by UAM-V; ▲, simulated by CHIMERE. (*Source:* Zhang B.N., Vautard R., and Kim Oanh N.T. 2005. Comparison of two photochemical modeling systems in a tropical urban area, *Journal of Air and Waste Management Association*, 55: 903–918. With permission.)

The biogenic emission and anthropogenic emission from other surrounding provinces in BMR had only relatively small contributions. Sensitivity analysis of the ozone formation in BMR by both models showed that ozone in the domain was more sensitive to VOC than to NO_x emission reductions. To attain the hourly ozone NAAQS of 100 ppb in BMR, the VOC emission should be at least reduced by about 40–50% (Figure 5.21).

Overall, the agreement between the two models for all considered parameters in spatial and temporal distributions was better than the agreement with the observations. The results of performance statistics for both models were similar.

5.3.3.2.5 Ozone Abatement Strategies

Effects of various emission reduction strategies on the ozone air quality in Bangkok were analyzed by UAM-V/SAIMM. Overall, the strategies that brought VOC emission reductions would also lower ozone levels in Bangkok, while those that reduced NO_x or increased VOC emission would increase ozone levels in the domain (Table 5.4). The modeling tool thus can help identify the suitable options to improve ozone air quality in the domain.

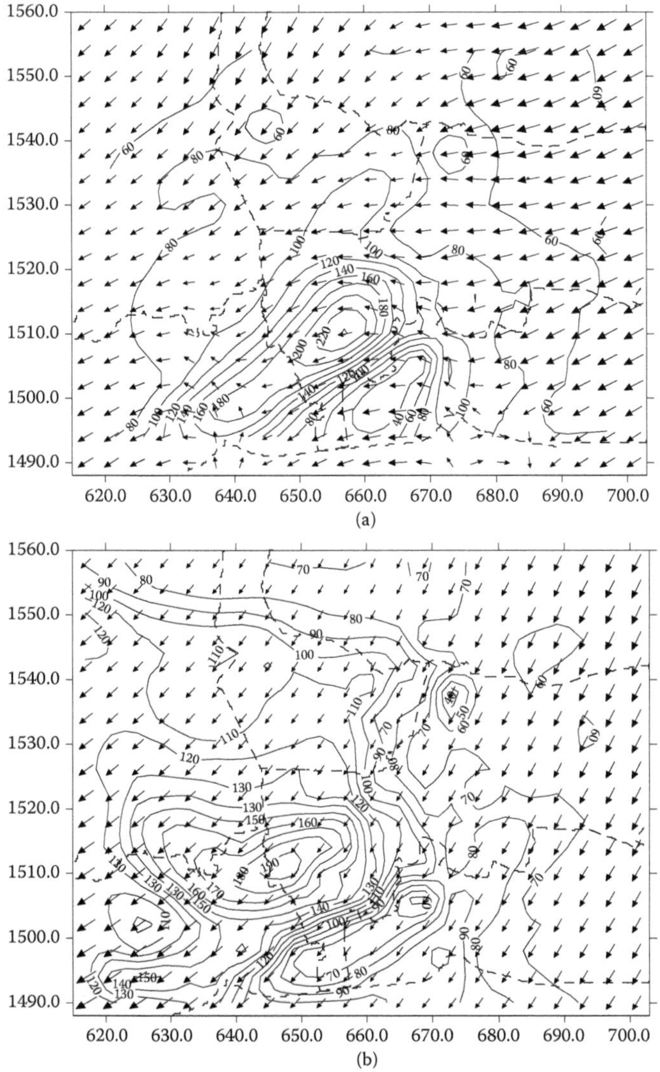

FIGURE 5.20
Simulated O_3 (ppb) at 1300 January 14, 1997, by SAIMM/UAM-V (a) and by ECMWF/CHIMERE (b). (Adapted from Zhang B.N. 2002. Photochemical smog modeling for air quality management of Bangkok metropolitan region, doctoral dissertation, Asian Institute of Technology, Pathumthani, Thailand.)

The information can be useful in the development of strategies to improve air quality in the area, taking into account multiple pollutants and a given priority.

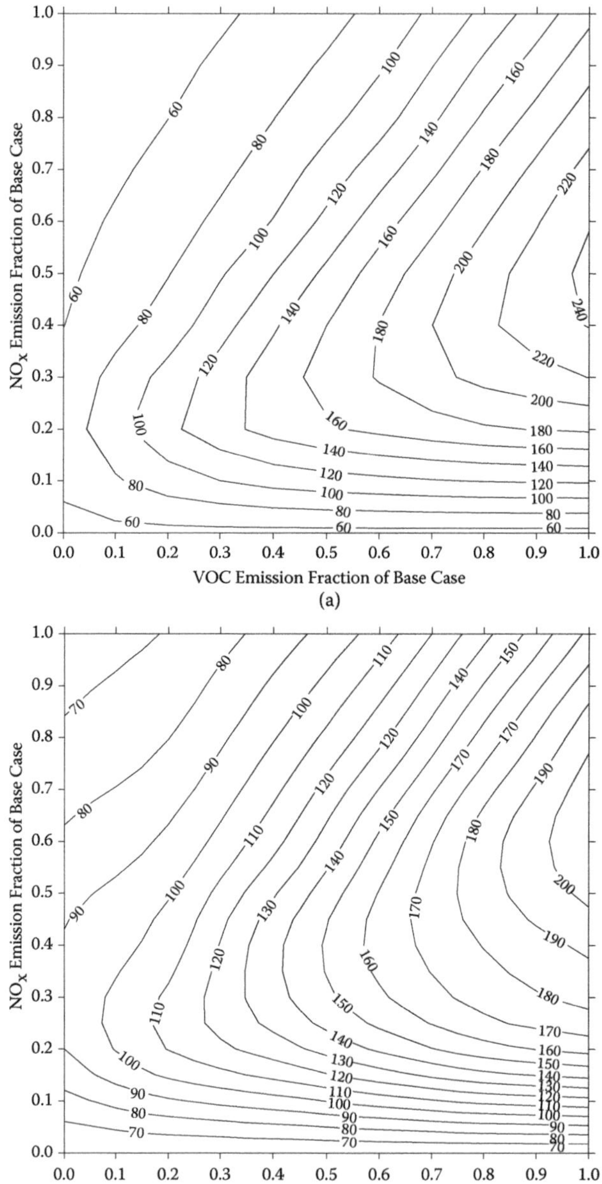

FIGURE 5.21
Response of peak ozone concentrations (ppb) for spatially uniform VOC and NO_x emission reductions at selected grids by UAM-V (a) and CHIMERE (b). (Adapted from (a) Zhang B.N. 2002. Photochemical smog modeling for air quality management of Bangkok metropolitan region, doctoral dissertation, Asian Institute of Technology, Pathumthani, Thailand; (b) Zhang B.N., Vautard R., and Kim Oanh N.T. 2005. Comparison of two photochemical modeling systems in a tropical urban area, *Journal of Air and Waste Management Association*, 55: 903–918. With permission.)

TABLE 5.4

Changes (%) in Peak Ozone Associated with Different Emission Control Strategies as Compared to the Base Case (UAM-V/SAIMM Results)

Emission Control Measures	Change, %	Remark
Implementation of stage I vapor control at fuel distribution and service stations	−2.6	Reduce VOCs
Implementation of stages I and II vapor control at fuel distribution and service stations	−5.0	Reduce VOCs
Switch all motorcycles to 4 stroke: 50% Honda + 50% Yamaha branch	−25.3	Reduce VOCs significantly, slightly increase NO_x emission
100% natural gas (NG) fired power plants in BMR with (SCR)[a]	+8.5	Reduce NO_x
100% compressed NG powered buses	+6.6	Significantly reduce VOCs, CO, and NO_x
Replacement of MTBE with ethanol[b]	+17.2	Reduce CO, slightly increase NO_x, significantly increase evaporative VOCs, increase highly reactive acetaldehyde in exhaust emission

Source: Data from Zhang B.N., and Kim Oanh N.T. 2002. Photochemical smog pollution in the Bangkok metropolitan region of Thailand in relation to O_3 precursor concentrations and meteorological conditions., *Atmospheric Environment*, 36(26): 4211–4222.

[a] SCR (selective catalytic reduction) was assumed to remove 90% NO_x.

[b] Conservative estimate with high-VOC emission.

5.4 Summary and Conclusions

Lack of emission data appeared to be a most challenging obstacle for applications of 3D photochemical model systems in Asian developing countries. The case studies presented in this chapter provided approaches to prepare such input data, both from the available global emission database and from the independent emission inventories for the selected urban areas. The modeling systems (CMAQ/MM5, CAMx/MM5, UAM-V/SAIMM, and CHIMERE/ECMWF) mostly showed satisfactory performance judging from the consistency between the simulated spatial distribution (ozone, PM, acid deposition) with the meteorology fields and source emission strength distribution.

The simulated ozone levels satisfactorily met the performance statistical criteria in most of the case studies. The models also reasonably reproduced the temporal variations in the observed ozone, although the insufficiency of the ozone monitoring data in some selected domains remains an obstacle for a more comprehensive model performance evaluation. Overall, the selected modeling systems performed more satisfactorily for ozone air quality simulation in BMR and Hanoi, which may be due to better MM5 performance, and less satisfactorily for HCMC and Jakarta. Further investigation should be conducted to improve the MM5 performance, with different schemes tested to select the best options for the tropical and near-equatorial areas. The ozone formation in BMR appeared to be more VOC sensitive, while that in HCMC would be more NO_x sensitive. A modeling tool thus helped identify suitable policy options to formulate effective management strategies to improve air quality in selected domains.

The one-atmosphere CAMx/MM5 modeling system also showed reasonable performance in simulation of PM and acid deposition at urban scales. However, the serious lack of measurement data, including the PM components, and the dry and wet deposition in the study areas presented an even greater challenge for model performance evaluation.

Modeling is a powerful tool for analyzing multiple air quality issues and developing AQM strategies. However, to use this modeling tool effectively for integrated AQM for Asian developing countries, serious efforts are required to prepare updated regional and urban-scale emission databases with necessary spatial and temporal distributions. The regional monitoring networks should be developed to provide sufficient data for model performance evaluation. Satellite monitoring data with the large spatial coverage, as highlighted in Chapter 4, may be used for the model performance evaluation, especially in large-scale simulations.

References

An D.D. 2005. Photochemical smog modeling for air quality management in the Hanoi metropolitan region, Vietnam, master's thesis, Asian Institute of Technology, Pathumthani, Thailand.

Asian Development Bank (ADB). 2006. Key indicators 2006: measuring policy effectiveness in health and education. Asian Development Bank. http://www.adb.org/Documents/Books/Key_indicators/2006/.

Atkinson R., Baulsch D.L., Cox R.A., Hampton R.F., Kerr J.A., Rossi M.J., and Troe J. 1997. Evaluated kinetics, photochemical and heterogeneous data, *Journal of Physical and Chemical Reference Data*, 26: 521–1012.

Bangladesh Bureau of Statistics (BBS). 2008. *Statistical Yearbook of Bangladesh 2007*, Bangladesh Bureau of Statistics, Dhaka, Bangladesh.

Bosanquet C.H., and Pearson J.L. 1936. The spread of smoke and gas from chimneys, *Transactions of Faraday Society*, 32: 1249–1263.

Boylan J.W., and Russell A.G. 2006. PM and light extinction model performance metrics, goals, and criteria for three-dimensional air quality models, *Atmospheric Environment*, 40: 4946–4959.

Dahal B. 2008. Determination of primary and secondary organic aerosols in the Bangkok metropolitan region with the focus on water soluble organic carbon, master's thesis, Asian Institute of Technology, Pathumthani, Thailand.

Daly A., and Zannetti P. 2007. Ambient air pollution. In *Air Pollution Modeling—An Overview*, Chapter 2, edited by Zannetti P., Al-Ajmi D., and Al-Rashied S., The EnviroComp Institute, Fremont, CA, and Arab School for Science and Technology, Kuwait.

Davidson P.M., Seaman N., Schere K., Wayland R.A., Hayes J.L., and Carey K.F. 2004. National air quality forecasting capability: first steps toward implementation, paper presented at the Sixth Conference on Atmospheric Chemistry: Air Quality in Megacities, Seattle, January 2004.

Emery C.A., Tai E., and Yarwood G. 2001. Enhanced meteorological modeling and performance evaluation for two Texas ozone episodes, Environ International, Novato, CA. http://www.tceq.state.tx.us/assets/public/implementation/air/am/contracts/reports/mm/EnhancedMetModelingAndPerformanceEvaluation.pdf.

Environ. 2008. *User's Guide: Comprehensive Air Quality Model with Extensions (CAMx)*, version 4.31, Environ International. http://www.camx.com.

Guenther A., Hewitt C., Erickson D., Fall R., Geron C., Graedel T., Harley P., Klinger L., Lerdau M., McKay W., Pierce T., Scholes B., and Steinbrecher R. 1995. A Globe model of natural volatile organic compound emissions, *Journal of Geophysical Research*, 100: 8873–8892.

Guenther A.B., Zimmerman P.R., and Harley P.C. 1993. Isoprene and monoterpene emission rate variability: model evaluations and sensitivity analysis, *Journal of Geophysical Research*, 98: 12609–12617.

Horowitz L.W., Walters S., Mauzerall D., Emmons L., Rasch P., Granier C., Tie X., Lamarque J., Schultz M., and Brasseur G. 2003. A global simulation of tropospheric ozone and related tracers: description and evaluation of MOZART, *Journal of Geophysical Research*, 108: D24.

Huy T.N.Q. 2008. Photochemical smog modeling for ozone air quality management in Ho Chi Minh City, Vietnam, master's thesis, Asian Institute of Technology, Pathumthani, Thailand.

Iqbal A., and Kim Oanh N.T. 2011. Assessment of acid deposition over Dhaka division using CAMx-MM5 modeling system, *Atmospheric Pollution Research*, doi: 10.5094/APR.2011.05.

Kanabkaew T., and Kim Oanh N.T. 2010. Development of spatial and temporal emission inventory for crop residue field burning, *Environmental Modeling and Assessment*, doi: 10.1007/s10666-010-9244-0.

Kim Oanh N.T., Chutimon P., Ekbordin W., and Supat W. 2005. Meteorological pattern classification and application for forecasting air pollution episode potential in a mountain-valley area, *Atmospheric Environment*, 39(7): 1211–1225.

Kim Oanh N.T., and Permadi D.A. 2009. Photochemical smog modeling for ozone air quality management. In *Modelling of Pollutants in Complex Environmental Systems* (Vol. 1), edited by Hanrahan G., ILM, Hertfordshire, UK, 291–319.

Kim Oanh N.T., Thuy L.B., Tipayarom D., Manadhar D.R., Pongkiatkul P., Simpson C.D., and Liu S.L.J. 2011. Characterization of particulate matter emission from open burning of rice straw, *Atmospheric Environment*, 45(2): 493–502.

Kim Oanh N.T., Upadhyay N., Zhuang Y.H., Hao Z.P., Murthy D.V.S., Lestari P., Villarine J.T., Chengchua K., Co H.X., Dung N.T., and Lindgren E.S. 2006. Particulate air pollution in six Asian cities: Spatial and temporal distributions, and associated sources, *Atmospheric Environment*, 40: 3367–3380.

Kim Oanh N.T., and Zhang B. 2004. Photochemical smog modeling for assessment of potential impacts of different management strategies on air quality of the Bangkok metropolitan region, Thailand, *Journal of Air and Waste Management Association*, 54(10): 1321–1338.

Kuhn M., Builtjes P.J.H., Poppe D., Simpson D., Stockwell W.R., Andersson-Sköld Y., Barrt A., Das M., Fiedler F., Hov, Ø., Kirchner F., Makar P.A., Milford J.B., Roemer M.G.M., Ruhnke R., Strand A., Vogel B., and Vogel H. 1998. Intercomparison of the gas-phase chemistry in several chemistry and transport models, *Atmospheric Environment*, 32: 693–709.

Marr L.C., Black D.R., and Harley R.A. 2002. Formation of photochemical air pollution in central California 1. Development of a revised motor vehicle emission inventory, *Journal of Geophysical Research*, 107(D6): 4047, doi: 10.1029/2001JD000689.

Milford J.B., Russell A.G., and McRae G.J. 1989. A new approach to photochemical pollution control: implications of spatial patterns in pollutant responses to reductions in nitrogen oxides and reactive organic gas emissions, *Environmental Science and Technology*, 23: 1290–1301.

Nghiem L.H. 2008. Photochemical modeling for simulation of ground-level ozone over the continental Southeast Asian region to assess potential impacts on rice crop, doctoral dissertation, Asian Institute of Technology, Pathumthani, Thailand.

Nghiem L.H., and Kim Oanh N.T. 2008. Evaluation of the mesoscale meteorological model (MM5)-community multi-scale air quality model (CMAQ) performance in hindcast and forecast of ground-level ozone, *Journal of Air and Waste Management Association*, 58: 1341–1350.

Permadi D.A. 2007. Surface O_3 air quality in Jakarta: current status, precursor emission inventory, and modeling efforts, master's thesis, Asian Institute of Technology, Pathumthani, Thailand.

Permadi D.A., and Kim Oanh N.T. 2008. Episodic ozone air quality in Jakarta in relation to meteorological conditions, *Atmospheric Environment*, 42(28): 6806–6815.

Phuong N.L. 2006. Modelling impact of open rice-straw burning emission on ozone air quality in the Bangkok metropolitan region, master's thesis, Asian Institute of Technology, Pathumthani, Thailand.

Prevot A.S.H., Staehelin J., Kok G.L., Schillawski R.D., Neininger B., Staffelbach T., Neftel A., Wernli H., and Dommen J. 1997. The Milan photooxidant plume, *Journal of Geophysical Research*, 102: 23375–23388.

Sillman S. 1999. The relation between ozone, NO_x and hydrocarbons in urban and polluted rural environments, *Atmospheric Environment*, 33: 1821–1845.

Sillman S., He D., Cardelino C., and Imhoff R.E. 1997. The use of photochemical indicators to evaluate ozone-NO_x-hydrocarbon sensitivity: case studies from Atlanta, New York and Los Angeles, *Journal of Air and Waste Management Association*, 47: 1030–1040.

Sistla G., Zhou N., Hou W., Ku J.Y., Rao S.T., Bornstein R., Freedman F., and Thuns P. 1996. Effects of uncertainties in meteorological inputs on urban airshed model predictions and ozone control strategies, *Atmospheric Environment*, 30: 2011–2025.

Sofyan A., Toshihiro K., and Gakuji K. 2007. Difference of sea breeze in Jakarta between dry and wet seasons: implication in NO_2 and SO_2 distributions in Jakarta, *Journal of Global Environment Engineering*, 12: 63–85.

Streets D.G., Bond T.C., Carmichael G.R., Fernandes S.D., Fu Q., He D., Klimont Z., Nelson S.M., Tsai N.Y., Wang M.Q., Woo J.H., and Yarber K.F. 2003. An inventory of gaseous and primary aerosol emissions in Asia in the year 2000, *Journal of Geophysical Research*, 108 (D21), doi: 10.1029/2002JD003093.

Sutton O.G. 1932. A theory of eddy diffusion in the atmosphere, *Proceedings of the Royal Society of London, Series A*, 135–143.

Systems Applications International (SAI). 1995. *User's Guide to the Systems Applications International Mesoscale Model; SYSAPP-95/070*, Systems Applications International, San Rafael, CA.

U.S. Environmental Protection Agency (USEPA). 1991. *Guideline for the Regulatory Application of the Urban Airshed Model. USEPA EPA-450/4-19-013*, Office of Air Quality Planning and Standards, Research Triangle Park, NC.

U.S. Environmental Protection Agency (USEPA). 2003. *Guidelines for Developing an Air Quality (Ozone and $PM_{2.5}$) Forecasting Program. EPA-456/R-03-002, U.S*, United States Environmental Protection Agency, Research Triangle Park, NC.

Wark K., Warner C.F., and Davis W.T. 1998. *Air Pollution: Its Origin and Control.* (3rd ed.), Addison-Wesley, Boston, MA.

Zhang B.N. 2002. Photochemical smog modeling for air quality management of Bangkok metropolitan region, doctoral dissertation, Asian Institute of Technology, Pathumthani, Thailand.

Zhang B.N., and Kim Oanh N.T. 2002. Photochemical smog pollution in the Bangkok metropolitan region of Thailand in relation to O_3 precursor concentrations and meteorological conditions., *Atmospheric Environment*, 36(26): 4211–4222.

Zhang B.N., Vautard R., and Kim Oanh N.T. 2005. Comparison of two photochemical modeling systems in a tropical urban area, *Journal of Air and Waste Management Association*, 55: 903–918.

6

Regional-Scale Modeling for Projection of Ozone Air Quality and Potential Effects on Agricultural Crops in Southeast Asia

Didin Agustian Permadi, Frank Murray, and Nguyen Thi Kim Oanh

CONTENTS

6.1 Introduction

Photochemical smog pollution is presently a well-recognized air pollution problem worldwide. High levels of ozone pollution observed near the Earth's surface (ground-level ozone) in photochemical smog episodes can cause adverse effects on human health (Burnett et al., 2000; Lippmann, 1991; Weisel et al., 1995), materials, and ecosystems (Royal-Society, 2008). Ozone also causes damage to vegetation and consequently reduces growth and yield of agricultural crops and forests (Emberson et al., 2003; Hogsett et al., 1997; Novak et al., 2003; Orendovici et al., 2003).

In Asia, problems of high ground-level O_3 concentrations are encountered in many areas, as highlighted in Chapter 1 of this book. Since around the year 2000, emissions of nitrogen oxides (NO_x) and nonmethane volatile organic compounds (NMVOCs), the major precursors of O_3, have rapidly increased in the region (Kato and Akimoto, 2007). As one of the world's fastest-growing subregions, Southeast Asia (SEA) is expected to experience growing air pollution emissions in the future. Hot and with plenty of sunlight all year, this tropical region has favorable natural conditions for tropospheric ozone formation. Daily maximum hourly ozone levels are reported often to exceed the National Ambient Air Quality Standards (NAAQS) in large urban areas (Lee et al., 2002; Shiu et al., 2007; Tu et al., 2007; Zhang and Kim Oanh, 2002). The influence of major monsoon regimes brings two distinct seasons (dry and wet) to SEA, with the observed ozone levels significantly higher in the dry season. For example, in the Bangkok metropolitan region (BMR), Thailand and Ho Chi Minh City (HCMC), Vietnam high-ozone episodes are commonly observed during the dry and hot months of January to March, while in the rainy season, from June to September, there is a low frequency of exceeding NAAQS (Nghiem and Kim Oanh, 2008; Zhang and Kim Oanh, 2002).

6.1.1 Emission of Ozone Precursors in Continental Southeast Asia

Table 6.1 presents ozone precursor emission rates extracted from the emission inventory database provided by the Center for Global and Regional Environmental Research (CGRER), for 2006 (http://www.cgrer.uiowa.edu/EMISSION_DATA). In Cambodia, Laos, and Myanmar, biomass open burning has a substantial share in national emissions (i.e., over 50% of the national total for NO_x, NMVOCs, and CO). Compared to other countries, Thailand and Vietnam have lower shares of biomass open-burning emission (generally below 40%) but higher shares of transportation emission (30–53%) in the total emissions of NO_x, NMVOCs, and CO. For SO_2, industry has a higher share in the total national emissions in these two countries (above 50%) than other countries (below 30%).

Emission of methane, an important ozone precursor, from the continental SEA (CSEA) is generally dominated by the rice paddy cultivation and livestock sectors, as presented in Figure 6.1. This is expected as the region has an agriculture-based economy. In general, Thailand and Vietnam are among the top major contributors to air pollution emissions in this region.

6.1.2 Ground-Level Ozone Monitoring

Ground-level ozone has been monitored in several Asian countries for many years, but the monitoring network is still sparse in SEA. Of the five CSEA countries presented in Table 6.2, only Thailand and Vietnam have reported

TABLE 6.1

Sectoral Pollution Emission Shares in Selected SEA Countries, 2006

	Sectoral Emission Shares (%)					
Pollutants	**Power**	**Industry**	**Residential**	**Transportation**	**Biomass Open Burning[a]**	**Total Emission, Tg/year**
Cambodia						
NO_x	4.0	6.5	3.6	16.0	69.8	0.09
NMVOC	0.0	2.0	18.6	16.5	62.9	0.31
CO	0.0	0.6	28.4	4.3	66.6	1.71
SO_2	17.3	33.8	18.8	15.1	15.0	0.04
Laos						
NO_x	0.1	0.3	2.4	16.0	81.1	0.10
NMVOC	0.0	0.9	4.6	7.1	87.5	0.50
CO	0.0	0.0	6.4	1.2	92.4	2.55
SO_2	0.0	12.6	11.6	14.0	61.9	0.02
Myanmar						
NO_x	5.6	4.0	7.9	16.8	65.7	0.25
NMVOC	0.0	6.2	14.7	14.8	64.3	1.80
CO	0.0	0.3	21.8	7.0	70.9	8.33
SO_2	0.7	22.5	34.7	2.2	39.9	0.09
Thailand						
NO_x	18.2	13.3	2.7	53.0	12.9	1.50
NMVOC	0.1	12.9	23.4	37.5	26.1	3.60
CO	0.0	4.1	15.3	38.5	42.1	12.40
SO_2	35.1	58.4	2.0	2.3	2.1	1.33
Vietnam						
NO_x	10.1	19.6	11.2	30.4	28.8	0.46
NMVOC	0.1	5.2	30.3	38.5	25.9	1.95
CO	0.0	1.0	37.5	38.6	22.9	12.70
SO_2	14.5	52.4	23.7	5.6	3.8	0.40

Source: Extracted from CGRER Web site: http://www.cgrer.uiowa.edu/EMISSION_DATA.
Note: Only inland emission is included; ship emission is not included in the presented values.
[a] Biomass open burning includes savanna/grassland, forest, and crop residue open burning.

the ozone measurements generated by the urban air quality monitoring networks operated by local environmental protection agencies. In addition, the available air quality monitoring stations are mostly located inside large cities.

In Thailand, of over 50 stations operated by the Pollution Control Department (PCD), 41 stations have concentrations of ground-level ozone reported with an established quality assurance/quality control (QA/QC) procedure, and the monitoring data are available online (PCD, 2010). In Vietnam, ozone is mainly monitored in Hanoi and HCMC (An, 2005; Huy,

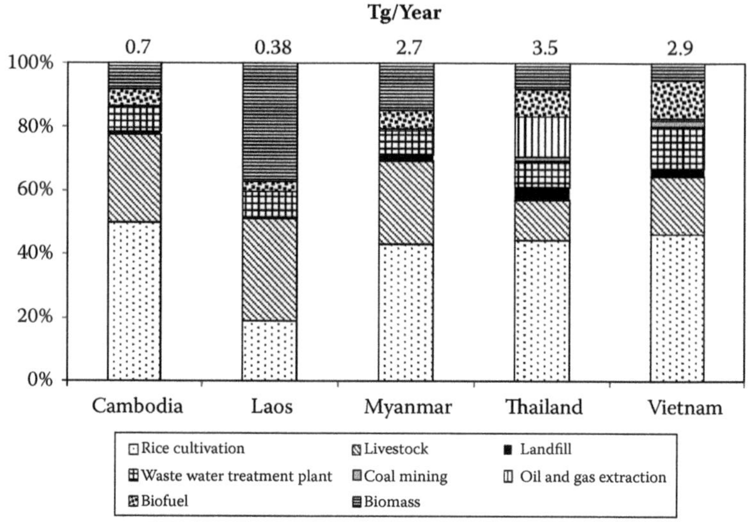

FIGURE 6.1
Emission shares of CH_4 (%) by different source sectors in 2006. (*Source:* Extracted from CGRER Web site, http://www.cgrer.uiowa.edu/EMISSION_DATA.)

TABLE 6.2
Ground-Level Ozone Monitoring Network in Selected SEA Countries

		EANET[c]		
SEA Countries	**National Urban Air Monitoring Network with Ozone Records**	**Total No. of Stations**	**Number of Stations with O_3 Monitoring**	**Characteristics**
Thailand	41 (56)[a]	5	3	Urban and rural
Laos	na	1	na	Urban
Cambodia	na	1	na	Urban
Myanmar	na	1	na	Urban
Vietnam	9 (20)[b]	2	na	Urban and rural

Sources: [a] Pollution Control Department (PCD). 2010. Air quality monitoring stations in Thailand, http://www.aqnis.pcd.go.th/station/allstation.htm; [b]Long D.H. 2006. Quality assurance and quality control for ambient air pollution data produced by automatic monitoring network in Hanoi, Vietnam, master's thesis, Asian Institute of Technology, Pathumthani, Thailand; [c]http://www.eanet.cc/product.html.

Note: na, not available; in parentheses are the total number of stations in the countries.

2008). In addition to the national monitoring networks, there are air quality stations provided by the Acid Deposition Monitoring Network in East Asia (EANET) operated in several countries of the region, but ozone measurements are available only for Thailand (Table 6.2).

The limitations on spatial and temporal distribution of observed data significantly prevent a comprehensive study of ozone pollution and its potential impacts in SEA. In particular, the current urban air quality monitoring networks do not capture potentially high ozone levels that occur at downwind locations outside city centers. For example, the lack of ozone measurement data in agricultural areas surrounding cities is an obstacle to a comprehensive assessment of ozone impacts on agricultural crops. Modeling tools capable of simulating these complex photochemical air pollution problems are normally used to fill in the data gaps and to assess impacts of present and future ozone levels on human health and ecosystems (Kim Oanh and Permadi, 2009; Russell and Dennis, 2000).

6.1.3 Applications of Photochemical Smog Models for Ozone Impact Study in Asia

Models have been used to investigate ozone impacts on various crops in China and some other East Asian countries (Takigawa et al., 2009; Wang and Mauzerall, 2004). Wang and Mauzerall (2004) projected that in 2020 ground-level ozone concentrations would increase substantially, by 5–10 ppb in April and by 5–15 ppb in October, in China, Japan, and Korea due to the increased emissions. This would cause an increase in grain loss by 2–16% for wheat, rice, and corn and 28–35% for soybeans as compared to the base year of 1990. Takigawa et al. (2009) projected that in East Asia, between 2000 and 2020, ozone concentrations will rise by 5–20 ppb, and the crop yield loss may be increased by 7–15%. As yet, there is no published assessment of future ozone and potential impacts on agricultural crops for the SEA region.

This chapter presents a case study of using the Community Multiscale Air Quality and mesoscale meteorological (CMAQ/MM5) modeling system, which has been set up and operated over the CSEA domain (Nghiem and Kim Oanh, 2008) to simulate ozone for the base year 2006 and project the levels in 2020. Future ozone concentrations in CSEA were simulated under two emission scenarios extracted from the Intergovernmental Panel on Climate Change (IPCC) Special Report on Emission Scenarios (SRES): SRES-A2 and SRES-B2. The accumulated ozone exposure over a threshold of 40 ppb (AOT-40 exposure metric) was used to assess the potential impacts on crops. The chapter is designed to provide an easy-to-follow, step-by-step procedure to set up the modeling system and apply it for scenario simulation and assessment of the potential impacts of ozone on agricultural crops at regional scales.

6.2 Methodology

6.2.1 Domain Description

The model domains used in this study, shown in Figure 6.2, covered the CSEA region. The MM5 domain covered a total of 56 × 56 grid cells extended from

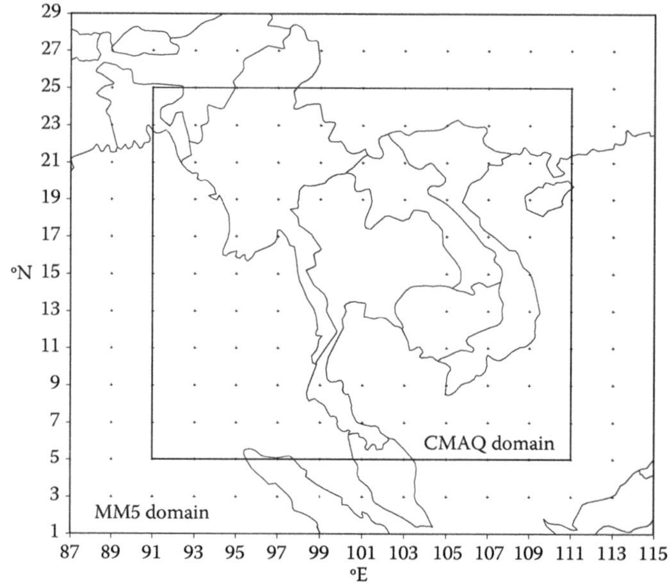

FIGURE 6.2
Model domains for MM5 and CMAQ simulations.

87°E to 115°E and from 1°N to 29°N, whereas the CMAQ domain of 40 × 40 grid cells extended from 91°E to 111°E and from 5°N to 25°N. The horizontal grid size of 56 km (0.5°) is similar for both CMAQ and MM5. Vertically, the domains extended from mean sea level (0 m) to approximately 15,797 m for the MM5 simulations and to 13,681 m for the CMAQ simulations. MM5 has 31 vertical layers, whereas CMAQ has 15 layers in the sigma coordinate system as presented in Figure 6.3, which is similar to the setup used in Nghiem (2008).

6.2.2 CMAQ/MM5 Modeling Setup

CMAQ is an Eulerian-type model developed by the U.S. Environmental Protection Agency (EPA) to simulate tropospheric O_3 and a range of other atmospheric phenomena (Byun and Ching, 1999). Model-3 CMAQ is a community model that can be freely downloaded from the Web site (http://www.cmascenter.org/). The simulations in this study were conducted using CMAQ version 4.3. The Fifth-Generation Pennsylvania State University/National Center for Atmospheric Research (PSU/NCAR) MM5, version 3.6.2, was used to generate meteorological field input for CMAQ.

The boundary conditions of the domain were developed based on monthly mean concentrations of O_3, nitrogen dioxide (NO_2), carbon monoxide (CO), peroxyacetyl nitrate, methane (CH_4), ethane (C_2H_6), formaldehyde (HCHO), and nitric acid (HNO_3) taken from the global model of ozone and related tracers, version 2 (MOZART-2) (Horowitz et al., 2003). The boundary conditions

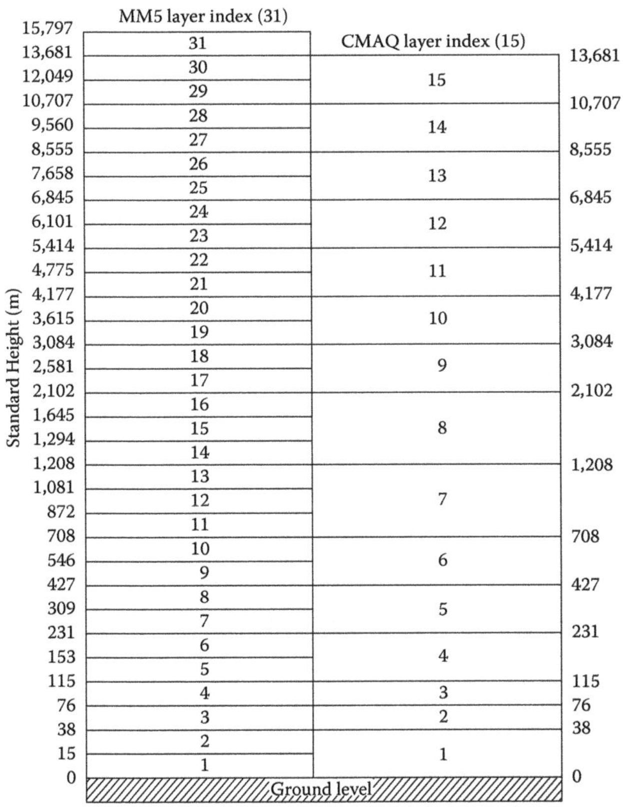

FIGURE 6.3
Harmonization of the vertical structure between MM5 and CMAQ.

for other species in the carbon bond mechanism IV (CB-IV) were set to zero and allowed to evolve during the simulations. The clean air over the domain was used as the initial condition.

MM5 was run with Lambert conformal projection (LCP) centered at 15°N and 101°E (in the central region of Thailand). MM5 used the terrain databases derived from the U.S. Geological Survey (USGS) and the global 25-category data of vegetation type and land use data, with a resolution of 5 min (~9 km). MM5 was exercised in four-dimensional data assimilation (FDDA) mode using a three-dimensional grid and surface nudging over the entire modeling period. The input meteorological data used for the MM5 initialization included the global reanalysis data, the global surface and upper air observation data obtained from the Data Support Section at the National Center for Atmospheric Research (NCAR-DSS) (available at http://dss.ucar.edu/datasets/ds083.2/). The physical schemes used for MM5 simulation were Schultz (moisture scheme); rapid radiative transfer model (RRTM; radiation scheme); Grell (cumulus parameterizations); and MRF (medium-range forecast, with representation of counter gradient term and K-profile in the well-mixed planetary boundary layer [PBL]), and there was no shallow convection option (no treatment of nonprecipitating cloud). The MM5 output files were processed with the Meteorology Chemistry Interface Processor Version 2.2 (MCIP2.2) to produce the meteorological fields for the subsequent chemical transport simulations by CMAQ.

6.2.3 Simulation Period

The simulation was conducted for the month of January (January 1–31, 2006, and January 1–31, 2020) as this is a month of high ozone reported in many places in CSEA, such as BMR, Hanoi, and HCMC (An, 2005; Huy, 2008; Nghiem and Kim Oanh, 2008; Zhang and Kim Oanh, 2002). January is also the month with a strong influence of the northeast (NE) monsoon, which would feature the long-range transport of air pollution emitted from the upwind regions to the CSEA. Meteorology of January 2006 was used for both base year and future ozone simulations.

6.2.4 Emission Input Data Preparation and Projection

6.2.4.1 Base Year Emission (January 2006)

Data on the annual anthropogenic emissions of NO_x, CO, volatile organic compounds (VOCs), SO_2, PM_{10} (particles with an aerodynamic diameter less than 10 μm), $PM_{2.5}$ (particles with an aerodynamic diameter less than 2.5 μm), black carbon (BC), and organic carbon (OC) for 2006 were obtained from the latest updated version of the regional anthropogenic annual emission inventory with a resolution of 0.5° × 0.5° for Asia, prepared by the CGRER (Zhang et al., 2009). Data on the monthly global biogenic emissions of VOCs were obtained from the Global Emissions Inventory Activity (GEIA) with a spatial resolution of 1° × 1° (available at http://www.geiacenter.org). The biogenic emission data for the domain were then extracted for the target simulation month of January 2006. Further, the emissions data were interpolated to the same grid size as the anthropogenic emissions (0.5° × 0.5°). Summaries of emissions data of selected pollutants for the CSEA countries are presented in Table 6.1 and Figure 6.1. Note that no updated emission data for biomass open burning were available, so the emission data for 2000 of CGRER were used. Also, only inland emission was considered, thus no oceanic emissions, including ship emission, were accounted for in the model input.

Hourly emission profiles of the input species are required in the CMAQ simulation, which is always a challenge to prepare. In this study, the hourly variation of anthropogenic emissions was prepared following the methods proposed by Nghiem and Kim Oanh (2008). Briefly, for the combustion-related species, monthly anthropogenic emission of modeled species (SO_2, CO, NO, NO_2, and VOCs) for January 2006 was estimated based on the monthly fraction of the annual emission suggested by Streets et al. (2003). Hourly anthropogenic emission profiles were made based on the ambient CO concentration profile. The observed diurnal solar radiation was used to segregate the monthly biogenic emission data extracted from the GEIA Web site to produce the hourly emission profile. Finally, the hourly anthropogenic and biogenic VOC emissions were aggregated among the lumped hydrocarbons of CB-IV using the VOC profiles presented by Fu et al. (2004).

6.2.4.2 Projected Emission (January 2020)

This study used the SRES scenarios proposed by IPCC SRES (2000) for the future ozone projection (January 2020). There are four general narrative storylines of SRES, labeled A1, A2, B1, and B2, that are characterized by different key driving forces, such as demographic, social, economic, technological, and environmental developments, to present an evolution of emissions during the period 1990–2100. A1 assumes very rapid economic growth with global population that peaks in midcentury and declines thereafter and a rapid introduction of new and more efficient technologies. A2 has a continuously regional population increase throughout the period (1990–2100) and economic growth that is more fragmented and slower than in other storylines. In B1, the population growth is similar to A1 but with more rapid changes in economic structures that are accompanied by the introduction of clean and resource-efficient technologies. B2 assumes a lower global population growth than A2 and an intermediate economic growth with an emphasis on environmental sustainability.

This study projected the future ozone concentrations in CSEA in January 2020 under two anthropogenic emission scenarios: (a) IPCC SRES-A2 and (b) IPCC SRES-B2. For Asia, the A2 emission scenario is based on the large-scale energy-environment model of the Atmospheric Stabilization Framework (ASF) for the Asian region, while B2 is based on the Model for Energy Supply Strategy Alternatives and their General Environmental Impact (MESSAGE). There is no country-specific emission scenario yet produced by these global energy-environment models. Thus, the estimation of the future emission over the domain was done based on the emission growth rate over all of Asia during 2000–2100 provided by Sankovski et al. (2000) and Nakicenovic et al. (1998) (http://sres.ciesin.org/final_data.html). Accordingly, the emission increase from 2006 to 2020 (14 years) was linearly interpolated from that over the total period of 2000–2100 to find the emission change (ratio) between 2020 and 2006 under the selected scenarios. The ratio obtained for a particular species was then applied to its emission in the base year of 2006 to estimate the emission in the target year (2020). Emissions from the biogenic and biomass open-burning sources in 2020 were assumed to be the same as those in the base year of 2006.

A summary of the emission change between 2020 and 2006 in the selected scenarios is presented in Figure 6.4. The A2 scenario projected slightly higher emission for NO_x, CO, and SO_2 as compared to B2. In 2020, the emission of

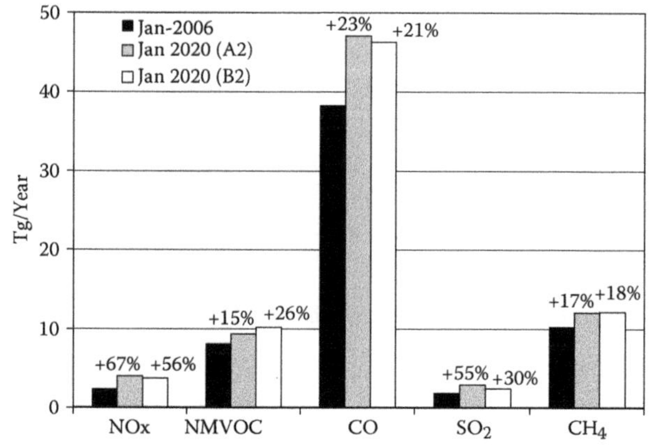

FIGURE 6.4
Projected 2020 emissions under A2 and B2 as compared to 2006 (increase in percentage is shown in labels).

NO_x would increase by around 67% in A2 and 56% in B2 as compared to 2006. CO and SO_2 emission increases would be around 23% and 55%, respectively, for A2, which are higher than those for B2 (21% and 30%, respectively). NMVOC emissions under A2 would be lower than in B2 (i.e., corresponding to an increase by 15% in A2 vs. 26% in B2).

6.2.5 Calculation of AOT-40

Potential impacts of ozone on crops and vegetation were analyzed using the AOT-40 exposure metric developed in Europe (EEA, 2003). In principle, AOT-40 over a crop growth season is calculated using Equation 6.1:

$$AOT-40=\sum_{i\in M}\sum_{j=1}^{n}\sum_{k\in H}\left(C_{ijk}-C_t\right) \tag{6.1}$$

where C_{ijk} is the hourly ozone concentration observed/simulated for the ith month, j^{th} day, and k^{th} hour; C_t is threshold concentration (i.e., 40 ppb); M is a set of months of a growth season of the crop; H is a set of daylight hours (hours with a globe radiation ≥ 50 W/m²) (Fuhrer et al., 1997); and n is the number of days in a month.

The risk to the crops and seminatural vegetation loss was assessed by comparing the obtained AOT-40 values with the World Health Organization's (WHO's) recommended critical level of 3,000 ppb.h over a crop's growing season (WHO, 2000).

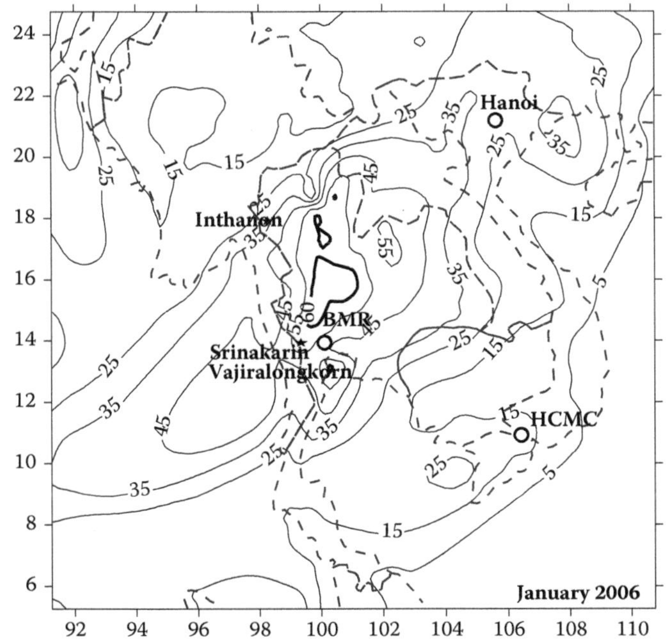

FIGURE 6.5
Spatial distribution of monthly average ozone in January 2006.

6.3 Projection of Ozone and Potential Impacts

6.3.1 Spatial Distribution of Ozone over CSEA

6.3.1.1 Monthly Ozone

The January monthly average ozone concentrations in 2006 and 2020 were calculated from the hourly simulated ozone in the lowest modeling layer (38 m) for each grid in the domain. The spatial distribution of monthly average O_3 levels in 2006 (Figure 6.5) showed high O_3 concentrations in the central and northern parts of Thailand. The highest monthly average ozone levels in the domain (60–65 ppb) appeared at the northwestern (NW) part of Thailand. The simulated monthly average ozone concentrations at remote sites (marked with stars in Figure 6.5) of around 35 ppb (Inthanon), 55 ppb (Srinakarin), and 50 ppb (Vajiralongkorn dam) are in the same range but slightly higher than the observed January monthly levels in 1996–1997, that is 30 and 40 ppb for the first two sites (Pochanart et al., 2001) and 34 ppb during January–December 2009 for the last site (Asian Development Bank

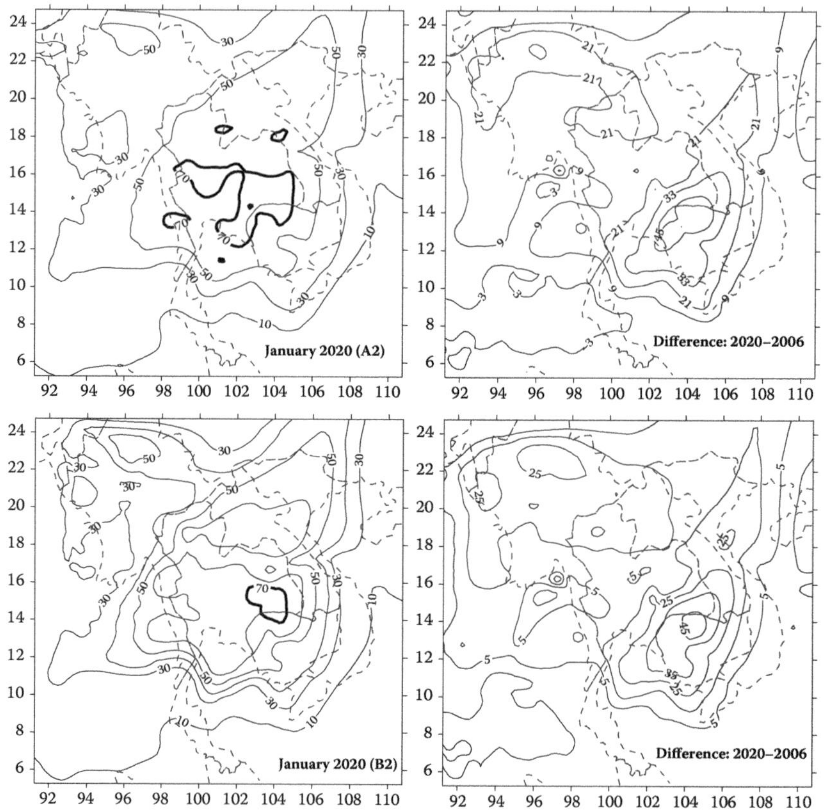

FIGURE 6.6
Spatial distribution of projected ozone in January 2020 and the change in ozone over 14 years (2006–2020) under A2 and B2 scenarios.

and Murdoch University [ADB/MU], 2010). In the urban areas such as BMR, Thailand, Hanoi and HCMC, Vietnam (marked with circles in Figure 6.5), simulated monthly average ozone concentration ranges were 38–45, 25–30, and 15–20 ppb, respectively. These values are similar to the observed values of around 25–30 ppb for BMR (Nghiem, 2008), around 18–27 ppb for Hanoi (An, 2005), and 28 ppb for HCMC (Huy, 2008).

The simulated increase of ozone concentration over the 14-year period in both scenarios is presented in Figure 6.6. Under scenario A2, the highest January monthly average of ozone concentration in the CSEA in 2020 would be about 80 ppb, and high ozone concentrations would be observed over a large area of central Thailand, extending from NW to NE parts of the country. The highest levels of ozone concentrations simulated for B2 would be about 70 ppb (i.e., lower than A2). In particular, the spatial variations of simulated future ozone concentrations under B2 showed the highest concentrations over northeastern Thailand (i.e., a shift from NW Thailand as compared to the January 2006 pattern). In general, the ozone plume would move southwestward following the NE monsoon in most parts of the domain, which reflects the prevalent monsoon direction in January.

The changes in January monthly ozone levels in 2020 as compared to 2006 under scenario A2 would be in the range of –15 to +51 ppb. Smaller changes were simulated for scenario B2 (i.e., in the range of –5 to +45 ppb). The maximum increase would occur over northern Cambodia in both scenarios (Figure 6.6). The simulated increases of monthly ozone concentrations in CSEA predicted by this study were higher than those predicted by Wang and Mauzerall (2004), who projected for the period 1990–2020 a change between +5 and +15 ppb (April, July, and October) under the B2 emission scenario. Larger increases in ozone concentration predicted in the present study may be a result of several factors. Use of the meteorological conditions of the high-ozone month of January 2006 for the simulations may have been an important factor. Also, ozone chemistry is nonlinear, and ozone formation in an area is largely determined by the VOC/NO_x ratios. These ratios are not the same in different parts of the simulation domain, hence leading to the future nonuniform ozone changes over the domain.

The simulated monthly plume patterns imply that the regional transport of O_3 within CSEA is likely and may cause elevated concentrations in remote areas downwind of intensive emission regions. This in turn can cause damage to human health and ecosystems, including loss of agricultural crop yields in the areas surrounding large cities.

6.3.1.2 Maximum Hourly Ozone

The spatial distributions of ozone in the domain at the time when the simulated hourly zone was maximum in the months of January 2006 and 2020 (A2 and B2) are presented in Figure 6.7. In 2006, the maximum simulated hourly concentration of 167 ppb occurred on January 13 at 1300 local standard time (LST) in a grid inside BMR, which was in the same range as the hourly maximum observed O_3 of 152 ppb on the same day at a station in BMR. In January 2020, the simulated maximum hourly ozone under the A2 scenario would be around 187 ppb, higher than that under B2 of 168 ppb.

The spatial distribution of the maximum ozone concentrations in both scenarios was consistent with the results of monthly average patterns presented. The locations of the highest concentrations would still be in BMR. The areas with hourly ozone levels greater than 100 ppb (shown by a bold line in Figure 6.7) would be larger in January 2020 for both scenarios than in January 2006. In addition, under A2, hourly ozone levels of above 100 ppb would be observed at the southern region of Vietnam. As the areas of high ozone levels expand in the future, more potentially adverse effects on human health and crops are expected.

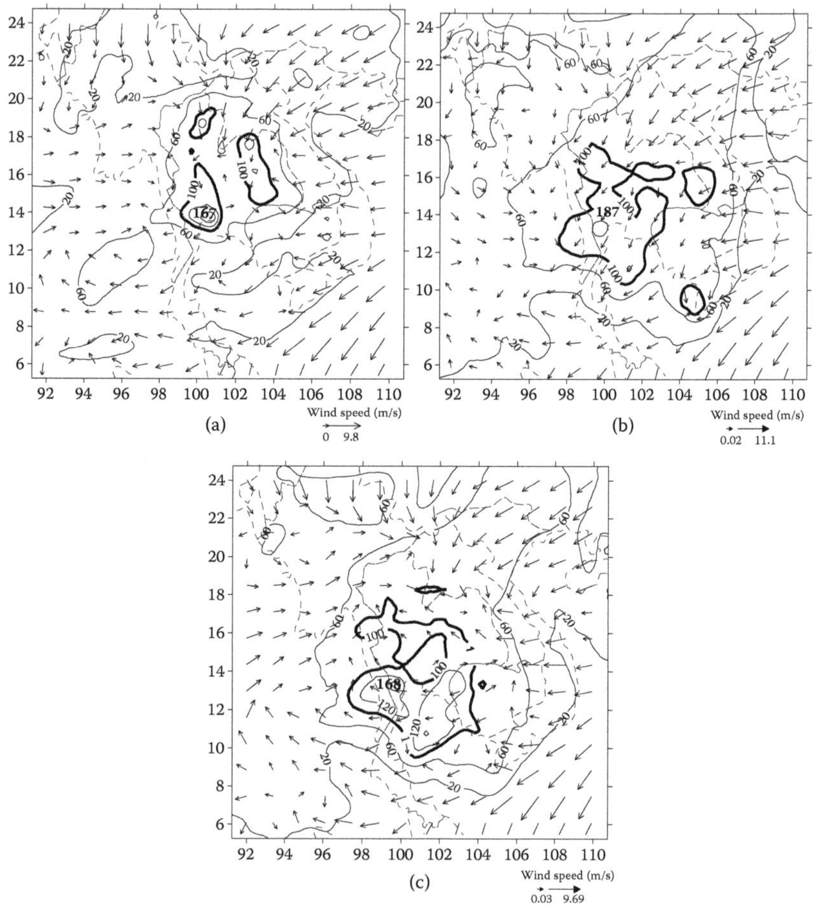

FIGURE 6.7
Maximum simulated hourly ozone (ppb) for January 2006 and 2020 in CSEA under different emission scenarios (a) January 2006; (b) January 2020 (A2); and (c) January 2020 (B2) (bold line borders areas with hourly O_3 above 100 ppb).

6.3.2 Projected AOT-40

The potential impacts of ground-level ozone on agricultural crops in the region caused by current and future ozone levels were assessed using the AOT-40 exposure metric. Due to the limitation in the observation data over CSEA, even for the year 2006 the AOT-40 determination was still made using the ozone simulation results. Note that the AOT-40, which was developed using a dose-response relationship established in Europe, was used in this study because such relationships have not yet been established for Asian conditions. Thus, it is assumed that plants and cultivars grown in Asia would respond to ozone similarly to those grown in Europe. This is an uncertainty

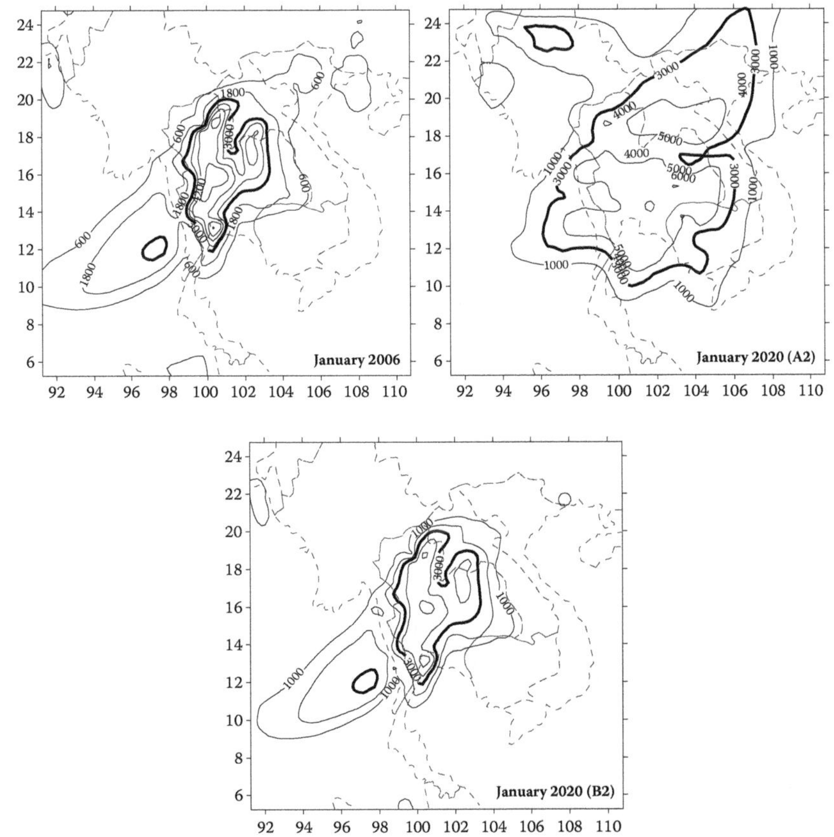

FIGURE 6.8
Simulated AOT-40 map for January 2006 and 2020 in CSEA (bold line borders the areas with monthly AOT-40 above 3,000 ppb.h).

in studies using the AOT-40 metric because other factors such as climate, local agricultural practices, and crop life cycle may alter plant responses to ozone (Emberson et al., 2010). Nevertheless, the AOT-40 metric is still useful information for the future changes of the relative spatial magnitude of risk across the region as well as the potential damage that may be likely to occur due to the ozone exposure.

In 2006, the highest AOT-40 values were obtained over the BMR and other provinces in the central region of Thailand, consistent with the simulated distribution of monthly ozone levels (Figure 6.8). The central and northern regions of Thailand would have the highest monthly AOT-40 values, ranging from 4,000 to 5,500 ppb.h. Note that these AOT-40 values, for only 1 month (January 2006) were already well above the WHO recommended AOT-40 of 3,000 ppb.h for protection of crops for the entire growing season that normally lasts 3 months. Assuming the AOT-40 for the growing season would

TABLE 6.3
Summary of Simulation Results for CSEA

Parameters	Simulation Scenarios		
	2006	2020 A2	2020 B2
Domain maximum monthly average ozone (ppb)	65	80	70
Difference in January average: 2020–2006 (ppb)		−15 to +51	−5 to +45
Domain maximum hourly ozone in January (ppb)	167	187	168
Domain maximum AOT-40 in January (ppb.h)	5,400	7,000	5,500
Areas with January AOT-40 ≥ 3,000 ppb.h, 1,000 km^2	242	1,125	321

be three times as much as for January, then the values obtained for the central part of Thailand were comparable to the 3-month accumulated values reported for China of 12,000–15,000 ppb.h (Aunan et al., 2000) and in most regions of Europe of 15,000–25,000 ppb.h (Zlatev et al., 2001). The AOT-40 spatial pattern in January 2006 was consistent with the ozone distribution patterns showing the influence of the NE monsoon. The simulated spatial pattern of AOT obtained in this study was similar to that of a previous study for 2004 by Nghiem (2008).

As expected, the simulated AOT-40 values for January 2020 would be significantly higher than that in January 2006. The maximum January AOT-40 values under A2 were almost 1.3 times than that in January 2006. The exposure risk with the monthly AOT-40 above 3,000 ppb.h extended to all countries in CSEA, including Laos, Cambodia, and Vietnam and the Yunnan Province of China. Under the B2 scenario, the maximum January AOT-40 values were not much higher than that in 2006 (i.e., 5,500 ppb.h vs. 5,400 ppb.h), and the areas with high AOT-40 would be mainly in Thailand. Thus, under the A2 scenario, by 2020, the model predicted that there would be larger areas in several CSEA countries that would be affected by ozone exposure. The estimated affected areas (with monthly AOT-40 ≥ 3,000 ppb.h) increased from 242,000 km^2 (January 2006) to 1,125,000 km^2 (under A2) and 321,000 km^2 (under B2). Crop yield loss as well as health effects are expected to be more serious in the future if necessary measures are not taken. A summary of the simulation results is given in Table 6.3.

6.3.3 Toward Ozone Pollution Mitigation

Although the data limitations prevent a more confident quantitative assessment of the magnitude of crop losses due to ozone injury to crop plants in CSEA, the potential economic loss could be substantial. In the European Union in 2000 an estimated 6.7 billion euros were lost due to the impacts of ozone on crops. Global economic losses from reduced yield were estimated to be $14–26 billion for rice, soybean, maize, and wheat combined (Royal-Society, 2008). In addition, the modeled high-ozone concentrations in areas of high agricultural production in Thailand, Laos, Cambodia, and Vietnam under the A2 scenario suggest that prudent policy measures are required to reduce ozone.

The emissions inventory (Table 6.1) shows that the main sources of ozone precursors were emissions from transportation, residential combustion, and open burning of biomass, and the major national sources were Thailand and Vietnam. Methane emissions from rice cultivation in CSEA are also likely to be a significant contributor to ozone formation.

SEA is one of the major biomass-burning regions of the world (Streets et al., 2004). It is estimated that biomass burning could contribute 8–18 ppb of ozone pollution, and its impact could be 10–30% on ozone concentrations (Fu et al., 2011). Some sources can be addressed by the use of low-cost measures to reduce emissions. For example, the burning of agroresidue, especially rice straw, is a traditional practice in much of SEA. However, these crop wastes can be productively used, for example, to make biofuels or mushroom-growing media, as discussed in Chapter 13 of this book. Alternatively, farming practices involving the use of agricultural wastes to improve soil organic matter, structure, and nutrient concentrations are also well established in many parts of the world. To make this change would require a cultural change, supported by infrastructure development, education programs, and demonstration projects and backed by regulation and enforcement measures. Other sources of biomass-burning emissions include forest fires associated with land clearing or controlled burn of forests and savanna woodland. Solid waste open burning is also a substantial source of air pollution including ozone precursors in SEA, expecially in periurban areas where the waste collection system is still limited.

Changes in transport emissions are usually more expensive policy options than changes to agricultural biomass burning in SEA (U.S. Agency for International Development [USAID], 2010) and include options to (a) introduce cleaner fuels and (b) tighten vehicle emissions standards and effectively implement inspection and maintenance programs. In addition, a range of measures highlighted in Chapter 1 of this book, including the use of cleaner fuel stove systems, would help to reduce the exposure to indoor air pollution greatly and reduce ozone emission precursors.

These control options achieve reductions in emissions of ozone precursors; the cost effectiveness and net benefits of the individual measures should be considered in the design of control strategies to maximize the reductions achievable within a specified budget. The practicality, cost effectiveness, and net benefits of a particular control option all vary depending on the particular local context.

An adaptation approach that can offer relatively low-cost gains is to integrate ozone tolerance into crop breeding and selection programs. There is some evidence that adaptation to increasing concentrations of ozone is already occurring but not formally recognized as farmers select crops and varieties that are productive and profitable under local conditions. Accordingly, ozone-sensitive varieties and species are being replaced by more ozone-tolerant crops in the Mekong region (ADB/MU, 2010).

In addition to its role as an air pollutant, ozone is a major greenhouse gas and the third-most-important contributor to human-induced climate change after carbon dioxide and methane (IPCC, 2007). By taking actions to reduce the emission of ozone precursors to protect food security, biodiversity, and human health and well-being, policy makers are also addressing climate change, that is, the cobenefit as discussed in Chapter 1.

6.4 Summary and Conclusions

Intensive anthropogenic emissions in SEA, especially from fossil fuel burning and biomass open burning, in combination with the tropical meteorological conditions are responsible for current high levels of surface ozone, especially in the dry months. Lack of ozone-monitoring data in most parts of the domain prevents any comprehensive assessment of its complex formation and impacts. Comprehensive photochemical modeling systems can be used in this situation to assist in the analysis of the present and future regional ozone air quality issue. Based on the simulation results by CMAQ/MM5, which had been previously set up and successfully evaluated for the domain, the ground-level ozone in the region was already high in the base year of 2006. The ozone levels would be increased substantially in 2020, more under the A2 than the B2 scenario, with the high values simulated not only for northern Thailand but also for other countries in the region. There is a high potential for regional transport of ozone, with the present and future ozone plumes extended from the major urban areas to the surrounding regions following the direction of the monsoon. Agricultural crops in the region would be under threat of high-ozone exposure based on the projected future monthly AOT-40 levels, which exceeded, over large parts of CSEA, the WHO recommended maximum value for the entire crop growing season.

A refined emission database and emission projections are required to improve the simulation results. Other important emission sources (e.g., shipping and updated methane emission from natural wetlands) should also be included in the emission input data. The simulations should also be done for the representative concentration pathways (RCPs) scenarios as well as realistic emission scenarios developed for the domain, taking into account the technology and policy interventions of each country. A longer simulation period, preferably 1 year, should be used to obtain monthly variations in ozone concentrations in the SEA domain and to improve impact assessment using AOT-40 estimated for three consecutive months. Potential economic loss due to ground-level ozone pollution in SEA in the future calls for ozone precursor emission reduction. Nonburning alternatives for crop residues and municipal solid waste and minimization of forest fires should be considered a priority. In addition, an adaptation approach that integrates ozone tolerance into crop breeding and selection programs can offer relatively low cost gains for SEA.

References

An D.D. 2005. Photochemical smog modeling for air quality management in the Hanoi metropolitan region, Vietnam, master's thesis, Asian Institute of Technology, Pathumthani, Thailand.

Asian Development Bank and Murdoch University (ADB/MU). 2010. *Risks and Adaptation to Climate Change in BCI Pilot Sites in PRC, Thailand and Vietnam* (Technical Report), edited by Murray F., Asian Development Bank and Murdoch University, Manila, Philippines.

Aunan K., Berntsen T.K., and Seip H.M. 2000. Surface ozone in China and its possible impact on agricultural crop yields, *Journal of Human Environment*, 29(6): 294–301.

Burnett R.T., Smith-Doiron M., Stieb D., Raizenne M.E., Brook J.R., Dales R.E., Leech J.A., Cakmak S., and Krewski D. 2000. Association between ozone and hospitalization for acute respiratory diseases in children less than two years of age, *American Journal of Epidemiology*, 153(5): 444–452.

Byun D.W., and Ching J.K.S. 1999. *Science Algorithms of the EPA Model-3 Community Multi-Scale Air Quality (CMAQ) Modeling System*, National Energy Research Laboratory, Research Triangle Park, NC.

EEA. 2003. European Environment Agency. Tropospheric ozone in the European Union—"The consolidated report." Topic report no. 8/1998, November 2003. http://reports.eea.eu.int/TOP08-98/en/page001.html

Emberson L., Ashmore M., and Murray F. 2003. *Air Pollution Impacts on Crops and Forests: A Global Assessment*, Imperial College Press, London.

Emberson L., Buker P., Engardt M., van Thienhoven A.M., Agrawal M., Zunckel M., and Hicks K. 2010. Assessing air pollution impacts to agriculture: a framework methodology for assessing the risks caused by ground-level ozone (O_3) in South Asia and southern Africa. http://www.sei-international.org/apcen/downloads/.

Fu J.S., Hsu N.C., Gao Y., Huang K., Li C., Lin N.H., and Tsay S.C. 2011. A regional chemical transport model to identify the influences of biomass burning during 2006 BASE-ASIA, *Atmospheric Chemistry and Physics Discussion*, 11: 3071–3115.

Fu J.S., Jang C., Cheng C., and He K. 2004. Attachment 8: project for the MCNC/U.S. Environmental Protection Agency: ICAP 2 Project-Task 8 pilot project of national/regional air quality modeling assessment in China (final report). http://www.cep.unc.edu/empd/projects/ICAP%20/index.shtml.

Fuhrer J., Skirby L., and Ashmore M.R. 1997. Critical levels for ozone effects on vegetation in Europe, *Environmental Pollution*, 97: 91–106.

Hogsett W.E., Weber J.E., Tingey D.T., Herstrom A.A., Lee E.H., and Laurence J.A. 1997. An approach for characterizing tropospheric ozone risk to forests, *Environmental Management*, 21: 105–120.

Horowitz L.W., Walters S., Mauzerall D., Emmons L., Rasch P., Granier C., Tie X., Lamarque J., Schultz M., and Brasseur G. 2003. A global simulation of tropospheric ozone and related tracers: description and evaluation of MOZART, *Journal of Geophysical Research*, 108, doi: 10.1029/2002JD002853.

Huy T.N.Q. 2008. Photochemical smog modeling for ozone air quality management in Ho Chi Minh City, Vietnam, master's thesis, Asian Institute of Technology, Pathumthani, Thailand.

Intergovernmental Panel on Climate Change (IPCC). 2007. *The Physical Science Basis. Contribution of Working Group I to the Fourth Assessment Report of the Intergovernmental Panel on Climate Change*, edited by Solomon S., Qin D., Manning M., Chen Z., Marquis M., Averyt K.B., Tignor M., and Miller H.L., Cambridge University Press, Cambridge, UK, 291–319.

Intergovernmental Panel on Climate Change (IPCC)SRES. 2000. *Special Report on Emissions Scenarios*, edited by Nakicenovic, N., and Swart, R., Cambridge University Press, Cambridge, UK.

Kato N., and Akimoto H. 2007. Anthropogenic emissions of SO_2 and NO_x in Asia: emission inventories, *Atmospheric Environment*, 26A: 2997–3017.

Kim Oanh N.T., and Permadi D.A. 2009. Photochemical smog modeling for ozone air quality management. In *Modelling of Pollutants in Complex Environmental Systems* (Vol. 1), edited by Hanrahan G., ILM, Hertfordshire, UK, 291–319.

Lee Y.C., Calori G., and Carmichael G.R. 2002. Ozone episodes in urban Hong Kong 1994–1999, *Atmospheric Environment*, 36: 1957–1968.

Lippmann M. 1991. Health effects of tropospheric ozone, *Environmental Science and Technology*, 25: 1954–1966.

Long D.H. 2006. Quality assurance and quality control for ambient air pollution data produced by automatic monitoring network in Hanoi, Vietnam, master's thesis, Asian Institute of Technology, Pathumthani, Thailand.

Nakicenovic N., Victor N., and Morita T. 1998. Emissions scenarios database and review of scenarios, *Mitigation and Adaptation Strategies for Global Change*, 3: 95–120.

Nghiem L.H. 2008. Photochemical modeling for simulation of ground-level ozone over the continental Southeast Asian region to assess potential impacts on rice crop, doctoral dissertation, Asian Institute of Technology, Pathumthani, Thailand.

Nghiem L.H., and Kim Oanh N.T. 2008. Evaluation of the mesoscale meteorological model (MM5)-community multi-scale air quality model (CMAQ) performance in hindcast and forecast of ground-level ozone, *Journal of Air and Waste Management Association*, 58: 1341–1350.

Novak K., Skelly J.M., Schaub M., Kräuchi N., Hug C., Landolt W., and Bleuler P. 2003. Ozone air pollution and foliar injury development on native plants of Switzerland, *Environmental Pollution*, 125: 41–52.

Orendovici T., Skelly J.M., Ferdinand J.A., Savage J.E., Sanz M.J., and Smith G.C. 2003. Response of native plants of northeastern United States and southern Spain to ozone exposures; determining exposure/response relationships, *Environmental Pollution*, 125: 31–40.

Pollution Control Department (PCD). 2010. Air quality monitoring stations in Thailand. http://aqnis.pcd.go.th/station/allstation.htm, accessed 2010.

Pochanart P., Kreasuwun J., Sukasem P., Geeratithadaniyom W., Tabucanon M.S., Hirokawa J., Kajii Y., and Akimoto H. 2001. Tropical tropospheric ozone observed in Thailand, *Atmospheric Environment*, 35: 2657–2668.

Royal-Society. 2008. *Ground-Level Ozone in the 21st Century: Future Trends, Impacts and Policy Implications*. Science Policy REPORT 15/08, The Royal Society, London, UK.

Russell A., and Dennis R. 2000. NARSTO critical review of photochemical models and modeling, *Atmospheric Environment*, 34: 2283–2324.

Sankovski A., Barbour W., and Pepper W. 2000. Quantification of the IS99 emission scenario storylines using the atmospheric stabilization framework (ASF), *Technological Forecasting and Social Change*, 63(2–3): 263–287.

Shiu C.J., Liu S.C., Chang C.C., Chen J.P., Chou C.C.K., Lin C.Y., and Young C.Y. 2007. Photochemical production of ozone and control strategy for Southern Taiwan, *Atmospheric Environment*, 41(40): 9324–9340.

Streets D.G., Bond T.C., Carmichael G.R., Fernandes S.D., Fu Q., He D., Klimont Z., Nelson S.M., Tsai N.Y., Wang M.Q., Woo J.H., and Yarber K.F. 2003. An inventory of gaseous and primary aerosol emissions in Asia in the year 2000, *Journal of Geophysical Research*, 108, doi: 10.1029/2002JD003093.

Streets D.G., Bond T.C., Lee T., and Jang C. 2004. On the future of carbonaceous aerosol emissions, *Journal of Geophysical Research*, 109, doi: 10.1029/2004JD004902.

Takigawa M., Akimoto H., Takahashi M., and Kobayashi K. 2009. Future projection of surface ozone and its impact on crop yield loss over East Asia in 2020, paper presented at the 10th annual WRF users workshop, Boulder, CO, June 2009.

Tu J., Xia Z.G., Wang H., and Li W. 2007. Temporal variations in surface ozone and its precursors and meteorological effects at an urban site in China, *Atmospheric Research*, 85(3–4): 310–337.

U.S. Agency for International Development (USAID). 2010. *Black Carbon Emission in Asia: Source, Impact, and Abatement Opportunities*, United States Agency for International Development Regional Development Mission for Asia, Bangkok.

Wang X., and Mauzerall. D.L. 2004. Characterizing distributions of surface ozone and its impact on grain production in China, Japan and South Korea: 1990 and 2020, *Atmospheric Environment*, 38: 4383–4402.

Weisel C.P., Cody R.P., and Lioy P.J. 1995. Relationship between summer time ambient ozone levels and emergency department visits for asthma in central New Jersey, *Environmental Health Perspective*, 103: 97–102.

World Health Organization (WHO). 2000. *Air quality guidelines for Europe*, World Health Organization Regional Office for Europe, Copenhagen, Denmark.

Zhang B.N., and Kim Oanh N.T. 2002. Photochemical smog pollution in the Bangkok metropolitan region of Thailand in relation to O_3 precursor concentrations and meteorological conditions, *Atmospheric Environment*, 36(26): 4211–4222.

Zhang Q., Streets D.G., Carmichael G.R., He K.B., Huo H., Kannari A., Klimont Z., Park S., Reddy S., Fu J.S., Chen D., Duan L., Lei Y., Wang L.T., and Yao Z.L. 2009. Asian emissions in 2006 for the NASA INTEX-B emission, *Atmospheric Chemistry and Physics*, 9: 5131–5153.

Zlatev Z.I., Dimov I., Ostromsky T., Geernaert G.I., Tzvetanov I., and Bastrup-Birk A. 2001. Calculating losses of crops in Denmark caused by high ozone levels, *Environmental Modeling and Assessment*, 6: 35–55.

7

Synoptic Climatological Approach to Forecasting Air Pollution Potentials

Nguyen Thi Kim Oanh

CONTENTS

7.1 Introduction

The complexity of air pollution problems is associated with emission, dispersion, transformation, and removal of pollutants from the atmosphere and is always a challenge in developing effective air quality management. Dispersion modeling is a common tool used to analyze past and current pollution situations and to project future pollution under various emission and meteorology scenarios (details in Chapters 5 and 6). Application of relevant dispersion models is particularly required to study secondary pollutants, which are formed in the atmosphere in a series of complex chemical reactions, such as ozone and secondary particles. These air quality dispersion models are widely available (Kim Oanh and Permadi, 2009; Russell and Dennis, 2000), but they require detailed input data of source emission and meteorology. Lack of suitable emission data with necessary temporal (hourly) and gridded spatial distributions is the common issue in many areas in Asia and therefore presents an obstacle for dispersion modeling.

An alternative method for pollution potential prediction, namely, the synoptic climatological approach, has been proven to be a useful tool in air quality management in many instances (Eder et al., 1994; Heidorn and Yap, 1986; Kathleen, 2002; Kim Oanh et al., 2005; Kim Oanh and Leelasakultum, 2011; Lam and Cheng, 1998; and references therein). Specifically, this tool is useful for analyzing air pollution in complex mountainous topography where atmospheric dispersion processes are complicated and meteorology-induced pollution episodes are likely to occur (Kim Oanh et al., 2005).

This chapter presents general concepts of the synoptic climatological approach to air pollution potential forecasting. A step-by-step application procedure of this approach is provided and illustrated with concrete case studies.

7.2 Air Pollution Climatology

7.2.1 Meteorology and Air Pollution Dispersion

Day-to-day fluctuations in ambient air pollution levels in a location are a function of variations in meteorological conditions and source emission strength. Meteorological conditions affect ambient air pollution in a number of ways. Meteorology plays important roles in dispersion, regional and long-range transport, transformation, and removal processes of air pollutants from the atmosphere. Emission rates of air pollutants are also related to meteorological conditions directly (e.g., evaporative emission and biogenic emission) or indirectly (e.g., increase in energy demand for space heating or cooling, increase in open biomass burning in drier weather). Knowledge of pertinent climatological data of a region is useful to analyze air pollution potentials to plan effectively for air quality management.

Atmospheric systems are commonly divided into macroscale (global), synoptic (large), and meso- and microscales. The general atmospheric circulation (long-term average motions) defines the macroscale systems with the global structure of the established mean zonal winds over the earth. The shift of the intertropical convergence zone (ITCZ) to the north in summer and to the south in winter results in the seasonal reversal of wind directions or monsoons. In Asia, the northeast (winter) and southwest (summer) monsoons are prevalent over most parts of the continent, as discussed in Chapter 1. Macroscale systems have a large horizontal dimension (above 10,000 km) and potentially can affect the long-range transport of air pollution.

The synoptic-scale systems include fronts (separating different air masses), cyclones (low-pressure areas), and anticyclones (high-pressure areas), which have a horizontal dimension of a few hundred km to 10,000 km. They are also called synoptic weather systems as they govern day-to-day weather conditions over a region. These systems are normally present for a few days, and during this period, they affect processes of pollutant dispersion and removal. Cyclones and fronts typically have strong winds and good vertical mixing to dilute air pollutants. The upward motions in these systems form clouds, which subsequently precipitate out and effectively remove pollutants from the atmosphere (wet removal). Synoptic systems of the most concern from an air pollution point of view are the stationary or quasistationary anticyclones, also called stagnating anticyclones. The downward motion in these high-pressure systems induces subsidence temperature inversions, which may persist for a few days. The typical weather pattern of stagnating anticyclone systems with light wind and clear sky also enhances formation of radiative inversions at nighttime, which are broken the next morning. Overall, stagnant atmospheric conditions associated with stationary high-pressure systems can induce air pollution episodes.

Mesoscale systems include those with a horizontal dimension generally ranging from a few kilometers to a few hundred kilometers. They are associated with established airflow circulations in some specific geological and topographical configurations, such as coastal areas (sea-land breezes), mountainous areas (mountain-valley winds), and urban areas surrounded by countrysides (urban-induced heat island circulation). Pollutants released from a source, if trapped in these systems, are dispersed following airflows in these local circulations. In certain circumstances, very high pollution levels may be built up, such as those in the coastal (shoreline) fumigation caused by sea-land breeze circulation (Zannetti, 1990).

The microscale systems include those processes with a horizontal dimension of less than a few kilometers and vertical confinement in the planetary boundary layer (PBL). These include, for example, building effects, surface roughness, temperature, and wind profiles in PBL. Since most emission sources release air pollution in this lower layer of the atmosphere, the microscale processes are the most important in determining local transport of pollution by wind and its diffusion by the turbulence within the PBL. The dispersion patterns of an elevated plume, commonly referred to as the plume shapes (fanning, coning, looping, trapping, lofting, and fumigation) reflect the influence of microscale processes in the PBL.

All atmospheric systems or processes of different scales are complexly interrelated to influence air pollution in a location. For example, pollutants released from low-lying urban sources, if trapped in a microscale circulation such as a building wake, would first disperse following typical flows in the wake. The pollutants will further disperse following other microscale circulations, such as street channels, and participate in mesoscale processes such as heat island-induced thermal circulation. These micro- and mesoscale circulations are more pronounced when processes of larger scales are relatively weak. If there is a tropical cyclone or a front present, then pollutants will be washed out from the atmosphere by precipitation. In contrast, if there is a stationary anticyclone present, then stagnant atmosphere may lead to meteorology-induced air pollution episodes.

Thus, through micro- and mesoscale processes pollutants released from a source are dispersed and mixed in the lower troposphere and simultaneously undergo transformation and removal processes. More conservative pollutant species follow motions in larger-scale systems to circulate over long distances or the whole globe, in some instances, causing long-range transboundary transport phenomena. Pollutants released from intensive sources such as massive forest fires in Southeast Asia (SEA) in 1997 (Asian Development Bank [ADB], 1999) or dust storms originating from the Gobi Desert and arid areas (Kim et al., 2003, and references therein) under suitable conditions can travel many thousands kilometers, as discussed in Chapter 1.

7.2.2 Air Pollution Climatology and Air Pollution Potential

Climatology is the study of climate, which in a narrow sense can be defined as weather conditions averaged over a period of time. Traditionally, to obtain climatological data a typical averaging period of over 30 years is used (Arya, 1999, and references therein). Accordingly, climatology can be considered a compilation of statistical information about the long-term average weather patterns.

Air pollution climatology deals with statistical relationships between an aggregate of weather elements and ambient air pollution levels. Such relationships are established based on long-term records of relevant meteorology variables and air quality measurements at a location. Air pollution climatology is normally expressed in terms of statistics of weather elements that have strong implications for atmospheric dispersion and can be assembled to forecast air pollution potentials. Examples of air pollution climatological parameters include the statistical information on wind (windroses), atmospheric stability, solar radiation, maximum mixing depth, air pollution roses, and so on (Schnelle and Dey, 2000; Stern, 1976). Scott and Diab (2000), for example, used the ventilation coefficient (product of mixing depth and vertically averaged wind speed) as a combined indicator of the atmospheric vertical and horizontal dispersion capacity to develop a forecast scheme of air pollution potentials.

7.3 Synoptic Climatological Approach

7.3.1 General Description

The word *synoptic* is derived from a Greek word meaning "seen together." Synoptic meteorology is the study of large-scale atmospheric processes and weather prediction utilizing synoptic weather information, such as synoptic charts and synoptic weather observations.

The synoptic climatological approach to air pollution prediction takes into account the relationship between the totality of meteorological variables with their complex interrelations comprising an air mass and the air pollution levels in a location. As air pollution levels normally respond to this totality rather than to arbitrarily selected meteorological variables (Lam and Cheng, 1998), the synoptic climatological approach shows a clear advantage over a simple regression analysis of time series data, for example.

In the synoptic climatological approach, the synoptic systems and their associated meteorological variables or weather types over a region are identified and analyzed in relation to air pollution levels using long-term data records. The approach has been widely applied to analyze and forecast air pollution potentials (Chen et al., 1992; Davis and Kalkstein, 1990a; Kalkstein and Corrigan, 1986; Kim Oanh et al., 2005; Kim Oanh and Leelasakultum, 2011; Lam and Cheng, 1998, and references therein). Basically, the approach consists of three steps: (a) classification of meteorological conditions into homogeneous groups representing the large-scale synoptic situations, (b) analysis of air pollution characteristics associated with each pattern, and (c) development of statistical models to forecast the pollutant levels for synoptic patterns with high pollution potentials. The procedure is shown with concrete steps in Figure 7.1.

Based on this approach, the forecasting procedure for air pollution potentials involves identification of polluted meteorology patterns that, when present, may induce high ambient air pollution. If a polluted pattern is identified on a forecast day, then a warning signal can be issued so that necessary measures can be taken to reduce emissions or exposure of the population. Further, statistical models can be used to estimate pollution levels.

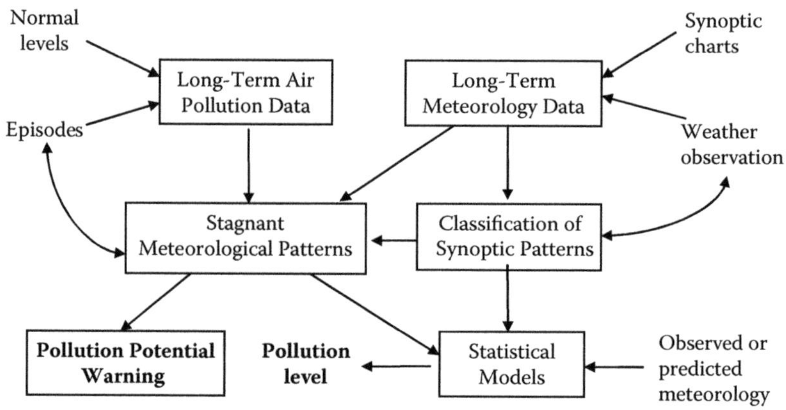

FIGURE 7.1
Synoptic climatological approach to air pollution prediction.

7.3.2 Synoptic Meteorology Classification

Methods used for classification and identification of large-scale synoptic patterns can be divided into two groups: (1) qualitative and semiquantitative and (2) quantitative or automated classification. The first group belongs to the traditional approach, which principally relies on flow patterns (or pressure fields) and often involves a visual inspection of synoptic weather maps and a visual comparison with a limited number of a priori established patterns for a study area (Eder et al., 1994; Heidorn and Yap, 1986; Kallos et al., 1993; Scott and Diab, 2000; Triantafyllou, 2001). This is the most straightforward way but requires intensive knowledge of synoptic meteorology and may be subjective. To overcome this difficulty, quantitative methods (the second group) are used; these involve development of a quantitative or automated synoptic classification scheme. Davis and Kalkstein (1990a, 1990b) provided a review of these quantitative methods that they referred to as the "air-mass based" to distinguish from the qualitative methods, which were referred to as the "weather type" or "flow pattern."

7.3.2.1 Temporal Synoptic Index and Spatial Synoptic Index Automated Classifications

When the weather data collected from a single station are used to develop an automated classification scheme, it is referred to as the temporal synoptic index (TSI) (Davis and Kalkstein, 1990a, 1990b). The TSI was used in a number of air pollution studies (Davis, 1991; Eder et al., 1994; Kalkstein and Corrigan, 1986; Lam and Cheng, 1998; Sanchez et al., 1990). When the procedure is applied for data collected from a spatial network of surface and upper air weather stations, it is called the spatial synoptic index (SSI). Davis and Kalkstein (1990a) used the approach to produce daily maps indicating areas of the United States under the influence of similar synoptic situations in a year. An automated classification, TSI or SSI, can use weather data from several observations during a day, which helps to account for daily variations in pattern identification.

Given the right selection of meteorological variables and stations, the SSI can combine both the air-mass-based and the flow pattern approaches. This approach may also account for the features associated with the development and movement directions of regional synoptic pressure systems, such as horizontal pressure gradients and temporal pressure changes (tendency). The SSI therefore is expected to produce better automated meteorology classification schemes, but it also greatly increases the number of data points (a few thousand) to be handled.

7.3.2.2 **Development of an Automated Classification Scheme**

The quantitative approach utilizes a long-term time series of a large number of meteorological variables for the synoptic meteorology classification; hence, data reduction techniques are required. Commonly, this classification is done using the principal component analysis (PCA) and K-means clustering technique available in the SPSS packages (Kim Oanh et al., 2005; Kim Oanh and Leelasakultum, 2011; Lam and Cheng, 1998; Winijkul, 2002). PCA is performed on the symmetric matrix of correlation coefficients of the original meteorological data. After rotation, the most important principal components are selected (based on a cutoff eigenvalue), which should collectively explain a high portion, normally above 70–80%, of the total variance of the original meteorology dataset. Subsequently, principal component scores, representing the projection of the original meteorological data onto the extracted principal components, are calculated. The final step in the synoptic meteorological classification scheme development involves clustering the obtained component scores to produce synoptic patterns. A two-stage clustering technique (an average linkage followed by the K-means technique) is normally used as it allows a reclassification of days after they have been grouped into a cluster to refine the cluster solutions.

7.3.2.3 **Selection of Meteorology Variables**

Selection of the meteorology stations and the pertinent meteorological variables is essential for the success of automated classification scheme development. Generally, air mass properties can be indicated by several meteorological variables, including air temperature, dew point, cloud cover, sea-level pressure, and pressure change. Air temperature and dew point are the most important indicators defining the air mass thermodynamic properties. These parameters describe the heat and moisture properties of the atmosphere. Pressure and wind are the indicators of air mass hydrodynamic properties, while cloud cover is an indication of moisture aloft (Eder et al.,

1994). Other important parameters are also considered for meteorology classification, for example, horizontal pressure gradients between meteorology stations, pressure tendency (to reflect development of synoptic systems), mixing heights, and solar radiation.

Commonly, available observation data for all meteorological variables are used in the initial analysis, and the final selection of variables is made based on the trial-and-error method. First, the raw meteorological dataset should be examined to remove significant outliers that may distort the results (Hair et al., 1998). The outliers are commonly identified by examining the distribution of the data after they are converted to the standard scores (with a mean of 0 and a standard deviation of 1). The most suitable classification scheme is normally selected by examining a number of trial meteorological datasets. The identified synoptic patterns should meet certain criteria: (a) Days belonging to the same identified pattern should display similar synoptic situations on synoptic charts; (b) typical synoptic situations shown in synoptic charts should be distinctly different between the patterns; and (c) means and ranges of meteorology variables (weather type) of each pattern should be consistent with the synoptic situations shown in synoptic charts. An overall procedure for the development of automated meteorology classification schemes is presented in Figure 7.2.

7.4 Case Studies

The case studies presented in this chapter demonstrate the applicability of the synoptic climatological approach to air pollution potential forecasting. The first case study deals with the SO_2 episode potential in the Maemoh Valley in winter months (November–January), while the second case study aims at prediction of PM_{10} (particles with an aerodynamic diameter less than 10 μm) during the haze period in Chiangmai (February–April). Both study areas are located in complex mountainous areas in northern Thailand. The third case study was conducted to analyze the ozone air quality in the Bangkok metropolitan region (BMR) in relation to representative meteorological conditions during high-ozone months, November–May. All three case studies employed the procedure presented in Figure 7.1.

The SSI approach was used to develop the automated meteorology classification scheme. Number and locations of regional weather stations used for the classification varied between the case studies, as shown in Figure 7.3. These stations, selected using the trial-and-error method, appear to be the best to reflect major developmental features of the synoptic systems that typically occur in the study areas during the periods under consideration. Characteristic levels and spatial distributions of selected pollutants in the study areas in each pattern were analyzed. A stepwise regression was used

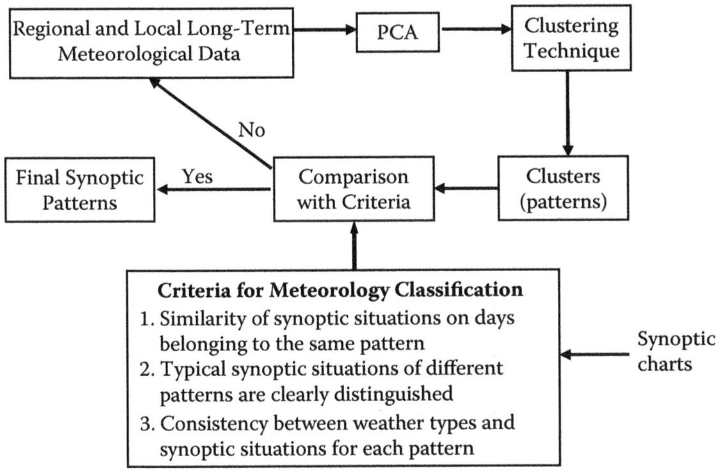

FIGURE 7.2
Steps in development of automated synoptic meteorology classification scheme.

to develop air pollution prediction models, separately for each synoptic pattern. The stepwise regression method has been applied widely in synoptic climatological air pollution studies (Eder et al., 1994; Lam and Cheng, 1998; Wolff et al., 1986) due to its ability to identify sequentially the optimum subset of independent variables to be incorporated into the regression equation. Finally, an operational procedure for air pollution potential forecasting is proposed that is elaborated in Section 7.4.4.

7.4.1 Forecast of SO_2 in Maemoh Valley

7.4.1.1 Background

The Maemoh Valley of Lampang Province is located in northern Thailand. The valley is about 15–20 km wide and 50 km long and aligned from the northeast to southwest. Mountainous complex terrain borders the valley on three sides and leaves the southwest side open. The Maemoh power plant, located in the valley, is the largest coal-fired power plant in Thailand. In the past, the plant burned lignite with a high sulfur content. Consequently, during winter (November–January) ambient SO_2 levels were reported often to exceed the Thailand hourly National Ambient Air Quality Standards (NAAQS) of 780 μg/m^3 and the interim hourly standard for the Maemoh Valley of 1,300 μg/m^3. SO_2 pollution episodes in this complex mountain-valley terrain are likely meteorologically induced and associated with plume fumigation due to inversion breakup (Sargent and Lundy, 1994).

During the winter months, northern Thailand is generally influenced by a high-pressure ridge extending from the Siberian high-pressure system. The NE monsoon prevalent during this period brings in a cool and dry weather

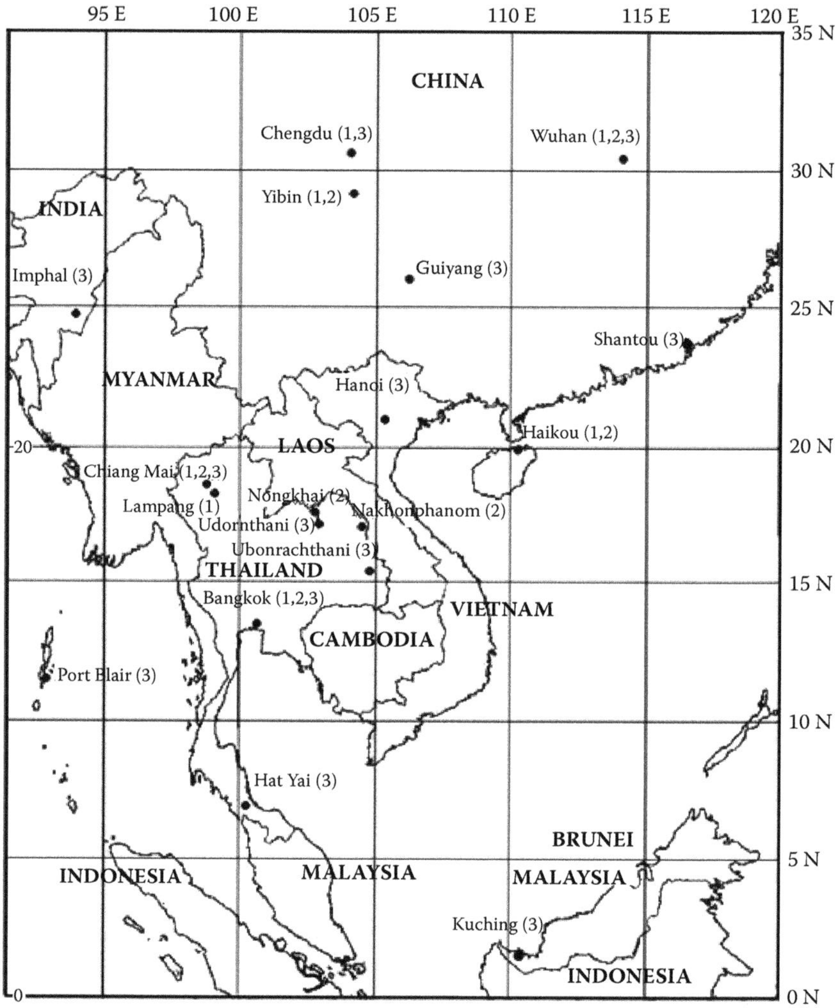

FIGURE 7.3
Locations of weather stations used in the meteorology classification. Numbers in parentheses indicate the case study: (1) SO_2 in Maemoh; (2) haze in Chiangmai; and (3) ozone in BMR.

type to the area. Stagnant conditions occur over the valley, and multiple strong inversion layers are often observed at nighttime. This is the result of a combination of the elevated subsidence inversion and the surface radiative inversion typical for stationary anticyclone systems. At night, the plumes released from stacks, 80–150 m high, are trapped in this multiple-layer inversion. In the morning, when the sun rises, the lower part of the inversion breaks up, and fumigation occurs with accumulated SO_2 rapidly mixed down, which causes a sharp increase in the ground-level concentrations.

The pollutant dispersion in the Maemoh Valley is further complicated by the mountain-valley topography. Large-scale meteorological conditions appear to control the potential for fumigation events, whereas mesoscale conditions control the onset time of the events, SO_2 levels, and plume placement (Kowalewsky, 1997). The fumigation was observed at various distances from the emission source and lasted for a few hours. The SO_2 levels during the episode were observed at different times of a day (i.e., not only in the morning as in classic fumigation), but a few were also observed at late afternoon, and some lower peaks were observed virtually all daylight hours. The resulting fumigation patterns are extremely complex and present a challenge to dispersion modeling for real-time prediction of SO_2 levels in the valley (Kim Oanh et al., 2005).

7.4.1.2 Governing Synoptic Patterns during Winter Months

Automated meteorology classification was done based on the data collected at 0100 LST (Thailand local standard time) daily in November–January of 1995–2000, from seven surface stations (denoted with "1" in Figure 7.3) plus the Chiangmai upper air station. The final meteorology dataset used for the classification comprised 13 variables, including most weather elements at the local station of Lampang plus wind at 850 mb in Chiangmai and pressure tendencies and horizontal pressure gradients between the selected surface stations. The synoptic meteorology during the November–January period was classified into six patterns that best met the classification criteria. These patterns (Figure 7.4) basically present different developmental stages of a high-pressure ridge influencing the study area during winter. The averages and ranges of each meteorological variable are also remarkably different between the patterns, and the weather type in each pattern appears to be consistent with the pressure field (Kim Oanh et al., 2005). The occurrence frequencies of the patterns are also distinguished, with the first three patterns occurring more often during November and the last three occurring more during January (Table 7.1). Patterns I and III had the highest occurrence frequencies during the period, 24% and 21%, respectively.

7.4.1.3 SO_2 Levels Associated with Synoptic Patterns and Prediction Models

Characteristics of the ambient SO_2 pollution, monitored at 12 ambient stations in the valley during the winter months of 1995–1997, including levels and temporal and spatial distributions, were examined for each pattern and showed clear differences between the patterns (Kim Oanh et al., 2005). Pattern V was the most polluted pattern (Figure 7.5), with extremely high SO_2 pollution levels that can be linked to poor dispersion conditions typically observed in a strong stationary ridge present over the region (Figure 7.4). Pattern I, ranked the second highest in SO_2 levels, was characterized by a moderate extension of a high-pressure ridge over northern Thailand with relatively clear sky and light wind. Pattern VI, ranked third in SO_2 levels, was a

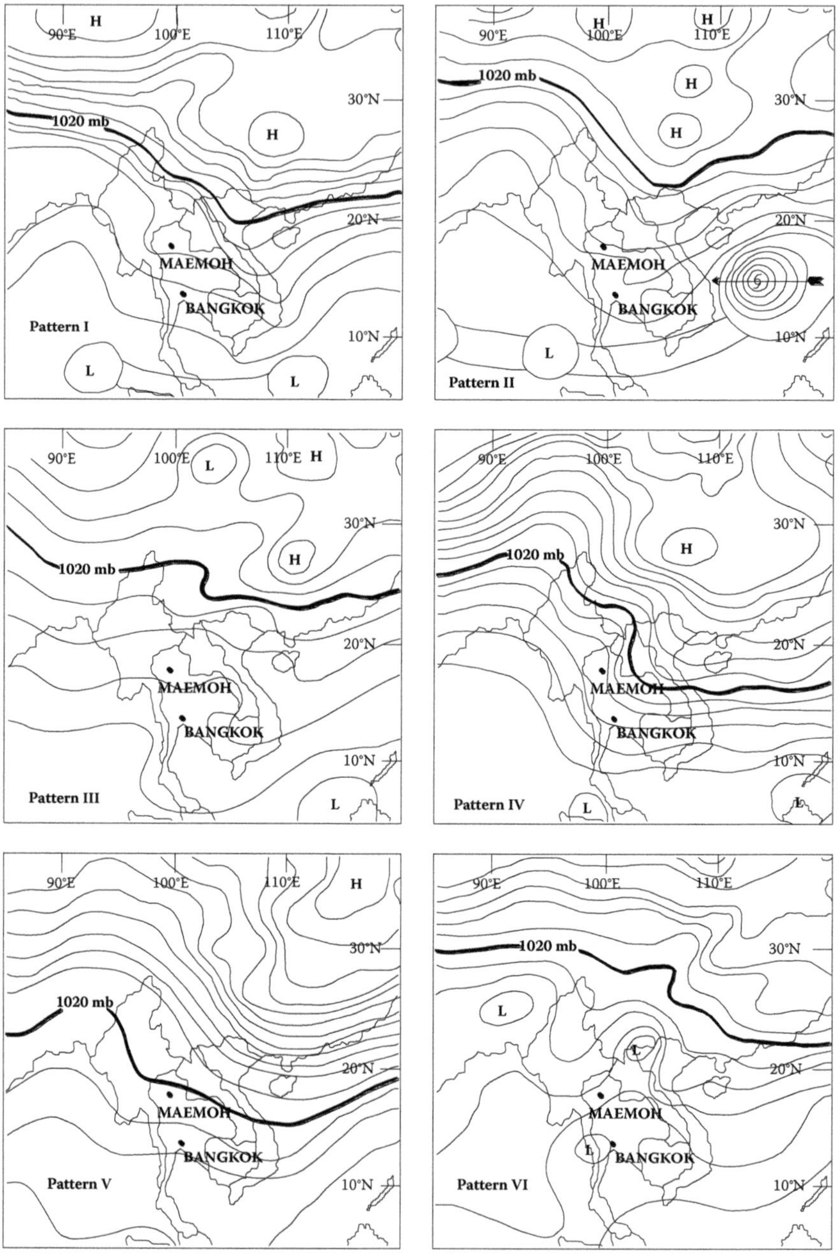

FIGURE 7.4
Examples of regional surface synoptic maps for the six synoptic patterns governing over northern Thailand during November–January. (*Source:* Kim Oanh N.T., Chutimon P., Ekbordin W., and Supat W. 2005. Meteorological pattern classification and application for forecasting air pollution episode potential in a mountain-valley area, *Atmospheric Environment,* 39(7): 1211–1225. Courtesy of AE.)

TABLE 7.1

Occurrence Frequency (%) of Synoptic Patterns by Month (1995–2000)

	Synoptic Patterns						
Month	**I**	**II**	**III**	**IV**	**V**	**VI**	**Total, Days (100%)**
November	45	26	26	2	1	1	109
December	20	3	17	21	36	2	89
January	9	1	21	25	15	29	137
Total	24	10	21	16	16	13	335

Source: Extracted from Piromyaporn C. 2000. Development of a synoptic climatological approach to predict ambient SO_2 concentrations at the Mae Moh Valley, master's thesis, Asian Institute of Technology, Pathumthani, Thailand.

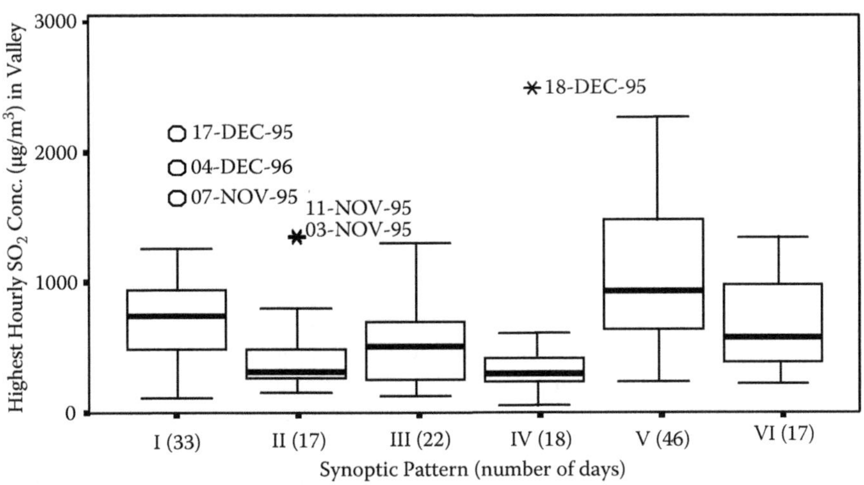

FIGURE 7.5

Box plots showing statistics of highest hourly SO_2 levels measured at 12 stations in Maemoh Valley during November–January 1995–1997 associated with six synoptic patterns. *Note:* Number of days for a pattern during the period is in parentheses. The bold horizontal line across the box is the 50th percentile, while the lower and upper hinges are the 25th and 75th percentiles, respectively. Whiskers are in the 10th and 90th percentile. An open circle (○) or an asterisk (*) denotes possible outliers. (Adapted from Kim Oanh N.T., Chutimon P., Ekbordin W., and Supat W. 2005. Meteorological pattern classification and application for forecasting air pollution episode potential in a mountain-valley area, *Atmospheric Environment,* 39(7): 1211–1225. Courtesy of AE.)

transitional pattern with several thermal lows over the study area. Clear sky (cloud cover was the lowest of all patterns) and dry weather (second-lowest dew point) observed in this pattern are favorable for formation of radiative inversions, consequently a high potential for the plume fumigation. Lower SO_2 levels were observed in the valley in the other three patterns, which had conditions conducive to good dispersion or wet removal of air pollution. Synoptic charts of pattern III typically showed a weak high-pressure system

TABLE 7.2

Stepwise Regression Models to Predict the Highest Hourly SO_2 Levels in Maemoh Valley

Pattern	Equation	R^2	Relative Error, %
I	SO_2 = 1781+ (18 × P_SO_2) – (264 × WS100) – (73 × Ta10)	0.66	34
V	SO_2 = 55957 – (231 × WS100) – (37 × PG1) – (54 × P_Lampang)	0.61	33

Source: Extracted from Kim Oanh, N.T., Chutimon, P., Ekbordin, W., and Supat, W. 2005. Meteorological pattern classification and application for forecasting air pollution episode potential in a mountain-valley area. *Atmospheric Environment*, 39(7): 1211–1225. (Courtesy of AE.)

Note: SO_2: The highest 1-h level in the valley on a given day when the pattern is identified (μg/m^3).
P_SO_2: 24-h average SO_2 of the previous day over the valley (μg/m^3).
Ta10 (air temperature at 10 m, °C) and WS100 (wind speed at 100 m, m/s): observation data at 0100 LST at the Maemoh meteorological tower.
PG1: horizontal pressure gradient between Lampang and Yibin (China) (mb).
P_Lampang: Sea-level pressure at Lampang (mb).

over China (no ridge extending to northern Thailand). Pattern II exhibited a weather type with high moisture content and cloudiness, with a tropical cyclone appearing sometimes off the coast of Vietnam. Pattern IV was associated with a high-pressure ridge that approached the region with strong horizontal pressure gradients; hence, the resulting strong wind could effectively dilute the air pollution (Kim Oanh et al., 2005).

Prediction models (Table 7.2) were developed to forecast the highest hourly SO_2 occurring in the Maemoh Valley (a single maximum per day) for the two most polluted synoptic patterns (patterns I and V). Relative error, the ratio between the root mean-squared error (RMSE) and the average pollution level in a pattern, was 33–34%. The models produced a reasonable agreement between the observed and predicted values but underestimated peak values of the data series (Figure 7.6), which is commonly reported for statistical models (Eder et al., 1994; Lam and Cheng, 1998), partly due to the inherent averaging effects.

Note that these prediction models were developed based on the data collected during the periods of November–January in 1995–1997. The models do not explicitly include the SO_2 emission rates; hence, they are applicable only for the emission rates of this data period. The models need to be reexamined for applications in other years if the emission rates change (e.g., after implementation of emission reduction measures). The revisited models should be tested with a long-term independent dataset before use for operational forecasts.

7.4.2 Forecast of PM_{10} during Haze Period in Chiangmai

7.4.2.1 Background

Chiangmai City, a large city in northern Thailand, experiences haze episodes with high PM_{10} levels around February–April each year. The most

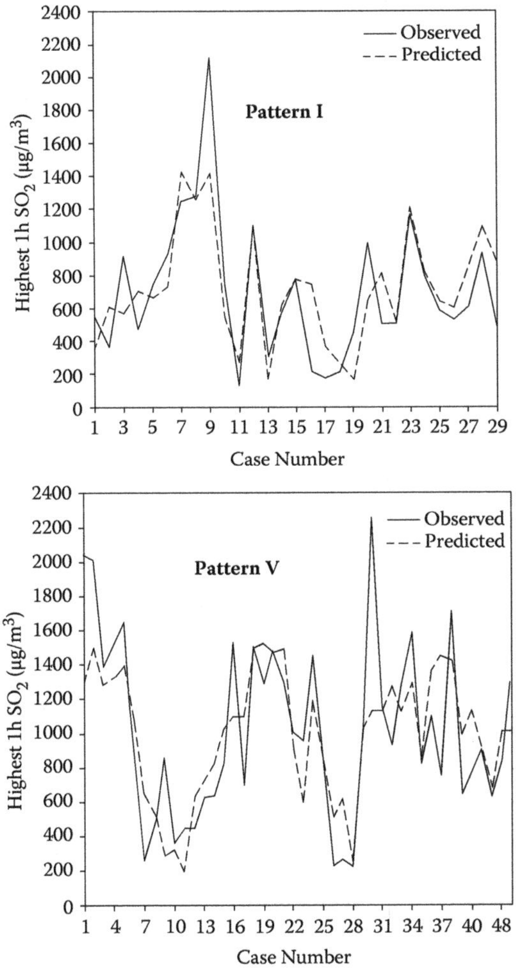

FIGURE 7.6
Observed versus predicted highest 1-h SO_2 in Maemoh Valley for patterns I and V based on data compiled for November–January 1995–1997. Case number: days in the pattern numbered following the appearance order during the period. (*Source:* Kim Oanh N.T., Chutimon P., Ekbordin W., and Supat W. 2005. Meteorological pattern classification and application for forecasting air pollution episode potential in a mountain-valley area, *Atmospheric Environment,* 39(7): 1211–1225. Courtesy of AE.)

severe PM_{10} pollution was observed in March 2007, when the whole city was blanketed with haze for a few weeks, and the 24-h PM_{10} level peaked at 396 μg/m³, which was significantly higher than the levels, generally within 150–300 μg/m³, observed in other years during 2001–2008 (Kim Oanh and Leelasakultum, 2011). The number of respiratory patients, based on data collected from 23 Chiangmai public hospitals, in March 2007 (21,336) was much higher than that in March of other years (Chiangmai Provincial Public Health Office [CMPHO], 2008). The onset of severe haze episodes such as that in March 2007 is caused by a combination of many factors. A thorough analysis of the underlying causes and development of a predictive model are necessary for episode management. Due to unavailability of accurate emission input data for dispersion modeling, this case study used the synoptic climatological approach for the haze prediction.

7.4.2.2 Governing Synoptic Patterns during Haze Months

As compared to the previous case study of the Maemoh SO_2 pollution, this case study covered a different period of the year (i.e., February–April) but the same study area. The regional stations used in this case (denoted with "2" in Figure 7.3) are quite similar to those in the previous case. Observation data recorded at 0700 LST for 8 years (2001–2008) from the selected surface stations and the upper air station of Chiangmai were included in the data screening (Kim Oanh and Leelasakultum, 2011). The final meteorological dataset used for the automated meteorology classification consisted of 17 surface meteorological variables (all data recorded at Chiangmai, sea-level pressure at stations in Thailand and in China, horizontal gradients of sea-level pressure between Chiangmai and every other station in the network) plus the morning mixing height in Chiangmai. Four clusters corresponding to four distinguished synoptic patterns were produced (Figure 7.7). As expected, the typical patterns in the February–April period appeared to be different from those obtained in the previous case study for November–January. In fact, this case study considered the hottest months in the area; hence, the thermal lows appeared frequently and more pronouncedly to the west of Chiangmai.

Pattern II had the highest occurrence frequencies and was mostly seen in March and April (Table 7.3), while pattern IV had the highest occurrence frequency in February. The other two patterns had a low occurrence frequency (about 10%) during the period. Pattern I occurred more in April, while pattern III was evenly distributed over the 3 months.

7.4.2.3 PM_{10} Levels Associated with Synoptic Patterns and Monthly Emission

7.4.2.3.1 PM_{10} Levels in Different Synoptic Patterns

PM_{10} levels measured at two automatic air quality monitoring stations in Chiangmai during February–April of 2001–2008 were analyzed in relation to the synoptic patterns. Average levels and ranges of PM_{10} were clearly distinguished between the four patterns (Figure 7.8). The lowest PM_{10} levels were observed in pattern I. Synoptic charts of pattern I typically showed a presence of ITCZ across SEA. The weather conditions in Chiangmai were characterized by high cloudiness, high mixing height and dew point, and strong wind, which all enhanced pollution dispersion, hence lowering PM_{10} levels. Pattern III, ranked second lowest in PM_{10} levels, was observed with

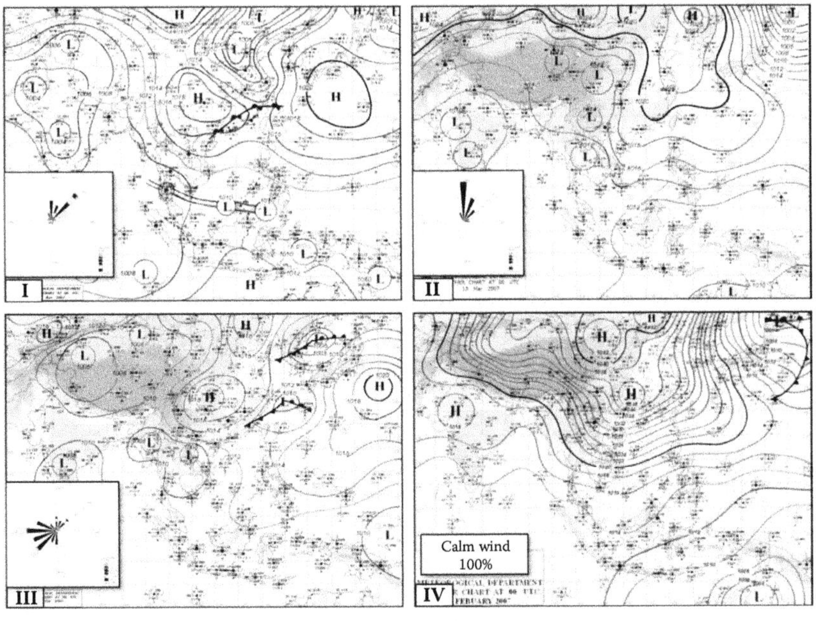

FIGURE 7.7
Examples of regional surface synoptic charts at 07:00 LST with inserted windroses at Chiangmai. (*Source:* Kim Oanh N.T., and Leelasakultum K. 2011. Analysis of meteorology and emission in haze episode prevalence over mountain-bounded region for early warning, *Science of the Total Environment*, 409: 2261–2271.)

TABLE 7.3
Occurrence Frequency (%) of Four Synoptic Patterns in February–April (2001–2008)

	Synoptic Patterns					
Month	**I**	**II**	**III**	**IV**	**Unclassified**	**Total, Days (100%)**
February	1	36	13	49	0	226
March	9	53	14	23	1	248
April	21	58	14	7	1	240
Total	11	49	13	26	1	714

Source: Extracted from Leelasakultum K. 2009. The Chiangmai haze episode in March 2007: cause investigation and exposure assessment master's thesis, Asian Institute of Technology, Pathumthani, Thailand.

the highest wind speed in Chiangmai of all patterns, which also indicates good pollution dispersion.

The highest 24-h PM_{10} levels were observed in pattern II, followed by pattern IV. The high levels of PM_{10} in these two patterns can be explained by the stagnant air conditions associated with the synoptic situations. On synoptic charts, pattern II was typically characterized by the presence of a vast thermal low with several cells observed over the areas west to Thailand, generally over

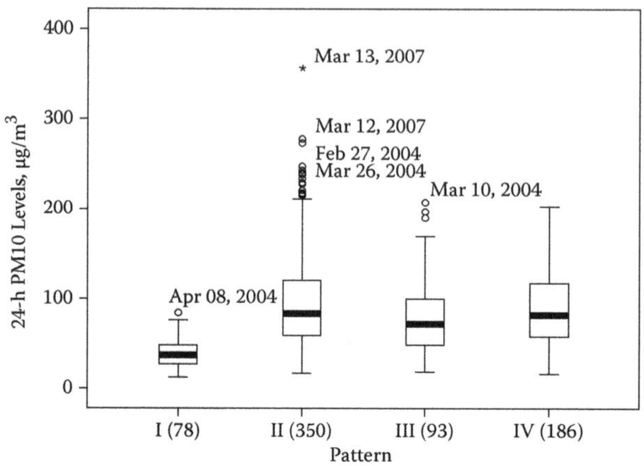

FIGURE 7.8
Box plots showing statistics of average daily PM_{10} in Chiangmai associated with different synoptic patterns, February–April 2001–2008 (see note in Figure 7.5). (Kim Oanh N.T., and Leelasakultum K. 2011. Analysis of meteorology and emission in haze episode prevalence over mountain-bounded region for early warning, *Science of the Total Environment*, 409: 2261–2271.)

SEA, and central China, and in most cases, a thermal low was seen over northern Thailand. In addition, a ridge from a high pressure centered in northeastern China was seen extending toward SEA and the Pacific Ocean. This pattern was also characterized by the lowest sea-level pressure at four stations in Thailand and quite low pressure gradients between Chiangmai and other stations in the network. In particular, clear sky (cloudiness of 1/10), light wind (averaged at 0.2 m/s), low mixing height, and low dew point (dry weather) associated with this pattern could enhance the radiative inversion formation at night (Kim Oanh and Leelasakultum, 2011). All these meteorological conditions were favorable for a buildup of high air pollution levels in the study area. Similar to pattern II, pattern IV was also characterized by a strong stationary pressure ridge deeply extending to SEA. However, the sea-level pressure at all stations in Thailand and China was the highest of all patterns, remarkably different from pattern II. Weather in Chiangmai in pattern IV was also typically characterized by clear sky and dry air (second-lowest dew point) conditions and low wind (100% calm wind observed in Chiangmai at 0700 LST). Thus, pattern IV in principle also enhanced radiative and subsidence inversion formation, and together with low wind, the pollution dispersion would be limited both vertically and horizontally (Kim Oanh and Leelasakultum, 2011).

In terms of stagnant air conditions, pattern II and IV appeared to be similar. However, the surface and upper air temperatures in pattern II (observed more in hotter months of March and April) were higher than pattern IV (occurred mainly in February). Most strong haze episode days in 2004 and 2007 were identified as outliers in pattern II (Figure 7.8). All the days when PM_{10} exceeded the 24-h NAAQS (120 μg/m^3) in March 2007 belonged to pattern II. Distinctly high PM_{10} levels in pattern II coinciding with its prevalence in March and April suggest that increases in source emission strength in these months should also be considered.

7.4.2.3.2 Monthly Emission Variation

Monthly fractions of major sources in northern provinces surrounding Chiangmai were determined for 2007 and showed that the most remarkable increases in the emission intensity were in March. The sources with higher emission strength in this month included biomass open burning (hotspot counts); seasonal industries (tobacco curing, brick kilns); and agricultural residue field burning (based on the monthly rice production). The most remarkable increase observed during March and early April 2007 was found for the number of hotspots (indicating biomass open burning) counted on the area surrounding Chiangmai and over the forest area located to the west of the city. Traffic emission should be an important contributor to the air pollution in the region, but the emission intensity remained relatively constant during the year (Kim Oanh and Leelasakultum, 2011).

The coincidence of the stagnant meteorology (pattern II) and the increase in the emission intensity, especially biomass open burning, should be the main cause for the highest PM_{10} levels in March 2007 (Figure 7.9). Backward trajectories constructed on the days of pattern II typically show the air masses originating from the west region (Bangladesh, India, Myanmar, and some from Andaman seas), passing Myanmar and the northern territory of Thailand before arriving at Chiangmai. High PM_{10} levels in Chiangmai were observed when air masses moved slowly and took a longer time to pass over adjacent areas with dense hotspots before arriving at the city (Kim Oanh and Leelasakultum, 2011).

7.4.2.4 PM_{10} Prediction Models

Haze episodes with high PM_{10} levels in Chiangmai most likely occur in synoptic pattern II; hence, identification of this synoptic pattern is useful for haze warning. A stepwise regression model was developed to predict the 24-h PM_{10} in Chiangmai on the days of pattern II (Equation 7.1). The model was developed using the data collected during February–April of 3 years (2007–2009). The daily hotspot counts, which represent the important biomass-burning emission source, were also included in the initial dataset for the stepwise regression analysis but were automatically dropped. This was most probably due to a strong correlation between the hotspot counts and other variables that appeared in the model (i.e., visibility and previous-day 24-h PM_{10}) (Kim Oanh and Leelasakultum, 2011).

$$PM_{10} = -4044 + 0.57\ P_{PM_{10}} - 0.008\ \text{Vis} - 1.95\ \text{Hum} + 4.26\ \text{SLP} \qquad (7.1)$$

where PM_{10} is the 24-h PM_{10} ($\mu g/m^3$) on the forecast day in Chiangmai; P_{PM10} is the 24-h PM_{10} ($\mu g/m^3$) on the previous day, averaged for two stations in Chiangmai; Vis is the visibility (m); Hum is the relative humidity (%); and

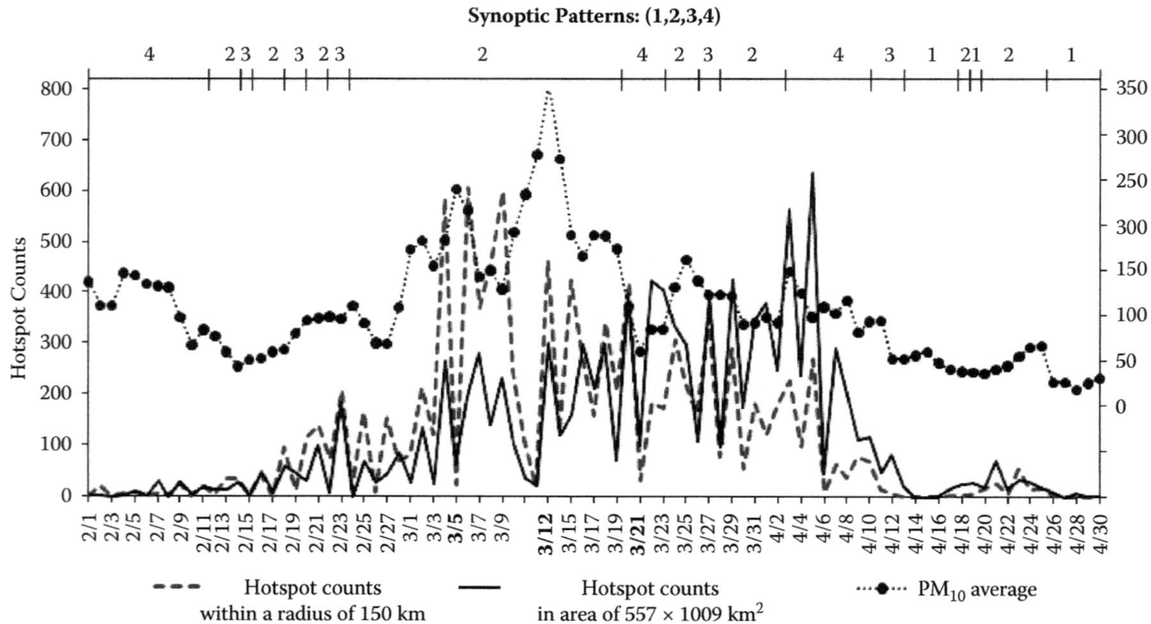

FIGURE 7.9
Major features of the haze period in February–April 2007. (*Source:* Kim Oanh N.T., and Leelasakultum K. 2011. Analysis of meteorology and emission in haze episode prevalence over mountain-bounded region for early warning, *Science of the Total Environment*, 409: 2261–2271.)

SLP is the sea-level pressure (mb) observed in Chiangmai at 0700 LST on the forecast day.

The model performed well with the dataset used for model development (2007–2009), with $R^2 = 0.87$ and a RMSE of 22 μg/m^3, as well as for a 4-year independent dataset, with R^2 of 81% and RMSE of 23 μg/m^3. The model was also evaluated to assess its forecast skill in terms of correct and incorrect predictions of the PM_{10} levels above or below the NAAQS (120 μg/m^3) following the category forecast verification statistics of the U.S. Environmental Protection Agency (USEPA, 2003). The results showed that the model was able to achieve an accuracy $A = 90.4\%$, which meets the minimum requirement of 90% for the initial implementation of the national air quality forecasting capacity by USEPA (Davidson et al., 2004).

7.4.3 Surface Ozone in Bangkok Metropolitan Region

7.4.3.1 Background

Photochemical smog pollution is an issue of concern in BMR, Thailand, where high levels of ozone are often recorded during the dry season. Since around the year 2000, the air quality in BMR has generally improved, but ozone pollution remains a problem. Hourly ozone levels in BMR were reported to exceed the Thailand NAAQS (100 ppb) at 0.3% of total hourly measurements in 2009 (Pollution Control Department [PCD], 2010). The period of high O_3 levels in BMR is from November to May (Nghiem and Kim Oanh, 2008; Zhang and Kim Oanh, 2002).

Formation, destruction, and accumulation of ozone in polluted urban areas are complex and involve a number of physical and chemical processes and their interactions. Ozone formation strongly depends on meteorological conditions. In the Midwest and Northwest of the United States, for example, the variability in surface ozone levels was found to be largely induced by meteorology, which explains as much as 40–60% of the variance in raw O_3 data (Wolff et al., 2001).

Several photochemical smog models have been applied to simulate the ozone pollution in BMR, as highlighted in Chapter 5. Lack of an accurate and updated emission database is the weakest link in three-dimensional (3D) model applications not only for BMR but also for other areas in Asian developing countries. The synoptic climatological approach may be an alternative to analyze and forecast the potential of ozone pollution in an urban area. This case study was designed to test this hypothesis.

7.4.3.2 Synoptic Patterns Governing BMR during High-Ozone Months

Meteorological data collected from 13 regional surface weather stations (denoted with "3" in Figure 7.3) during November–May of 9 years (1992–2000) were included in the initial dataset. This large number of stations was used

TABLE 7.4

Mean and Standard Deviation of the Meteorological Variables in Six Synoptic Patterns for BMR (1992–2000)

Meteorology		Synoptic Pattern					
Variable	Unit	I	II	III	IV	V	VI
CLC455	tenths	5.8 ± 2.0	6 ± 1.8	5.8 ± 2.3	5.9 ± 2.0	8.9 ± 1.27	4.7 ± 1.85
Ta455	°C	26.3 ± 2.0	27 ± 1.7	23.7 ± 2.3	26.3 ± 1.7	26.5 ± 1.84	21.7 ± 2.31
RAIN455	mm	1 ± 4.6	0.4 ± 2.2	0.1 ± 0.9	0.2 ± 0.9	13.8 ± 20.37	0
WSP455	knot	1.1 ± 1.7	0.8 ± 1.4	3.5 ± 2.3	0.9 ± 1.5	0.9 ± 1.7	0.4 ± 1.1
SUN455	hour	8.5 ± 1.9	8.4 ± 1.7	7.1 ± 2.7	8.1 ± 2.0	3.7 ± 2.5	8.4 ± 1.2
SLP455	mb	1010 ± 2.0	1010 ± 2.1	1013 ± 2.4	1010 ± 2.1	1009 ± 2.3	1014 ± 2.2
MOI455	g/kg	23.6 ± 2.9	24.5 ± 2.0	18.4 ± 2.9	23.7 ± 2.5	24.2 ± 2.51	17.9 ± 3.2
Td455	°C	23.4 ± 2.6	24.2 ± 1.8	18.2 ± 3.1	23.4 ± 2.2	23.8 ± 2.18	17.4 ± 3.4
VIS455	km	7.9 ± 2.7	7.9 ± 2.5	10.0 ± 2.4	8 ± 2.6	8.7 ± 2.6	6.0 ± 2.6
PGD333	mb	0.4 ± 1.71	0.5 ± 1.2	–2.1 ± 1.7	0.3 ± 1.4	–0.7 ± 1.65	–1.6 ± 1.62
PGD820	mb	2 ± 3.8	3 ± 3.3	9.2 ± 3.3	6.3 ± 3.8	3.2 ± 4.0	6.2 ± 2.6
PGD294	mb	11.1 ± 5.7	5.3 ± 5.1	15.7 ± 4.8	15.3 ± 4.7	6.9 ± 6.4	11.7 ± 4.4
PGD816	mb	8.3 ± 5.7	4.1 ± 5.4	15.1 ± 4.9	13.9 ± 4.8	6.2 ± 6.1	9.9 ± 4.8
PGD316	mb	5.5 ± 3.0	7 ± 2.8	10.8 ± 2.6	9.1 ± 3.1	6 ± 3.4	8.6 ± 2.4
PGD413	mb	0.7 ± 1.8	1.1 ± 1.6	–2.4 ± 2.1	0.6 ± 1.7	1.7 ± 2.2	–2.3 ± 1.8
24PC294	mb	6.7 ± 5.5	–2.8 ± 3.5	–1.2 ± 3.7	2.6 ± 4.2	–0.6 ± 4.4	–1 ± 3.7
24PC816	mb	6.1 ± 5.2	–3.2 ± 3.8	–0.9 ± 3.7	3.4 ± 4.0	–0.8 ± 3.8	–1.5 ± 4.1
24PC333	mb	1.7 ± 1.7	0.1 ± 1.3	0.2 ± 1.3	–0.3 ± 1.1	–0.4 ± 1.3	–0.4 ± 1.6
24PC455	mb	1.1 ± 1.4	–0.2 ± 1.1	0.9 ± 1.2	–0.3 ± 1.0	0 ± 1.2	–0.5 ± 1.1

Source: Extracted from Winijikul (2002)

Note: The first nine variables are from the Bangkok surface station (48455) listed in the order of appearance: cloud cover, air temperature, rainfall, wind speed, daily duration of sunshine, sea-level pressure, moisture content, dew point, and visibility. Other variables include the horizontal pressure gradients between Bangkok and stations 43333 (Port Blair), 48820 (Hanoi), 56294 (Chengdu), 57816 (Guiyang), 59316 (Shantou), 96413 (Kuching), and the last 24-h pressure change at stations 56294, 57816, 43333, and 48455.

to capture major developmental features of large-scale synoptic systems during these months, such as high-pressure systems in China and thermal lows to the west of Thailand. The final dataset used for the classification was composed of 19 meteorology variables recorded at 0700 LST, which represented the thermodynamic and hydrodynamic characteristics of an airmass as well as features of synoptic pressure system development. This dataset include most weather elements recorded at the Bangkok station plus the horizontal pressure gradients between Bangkok and each selected station and pressure tendencies at several stations, as shown in Table 7.4. Six synoptic meteorology patterns governing weather over BMR during the high-ozone months were identified. Examples of synoptic charts for two patterns (V and VI) are shown in Figure 7.10.

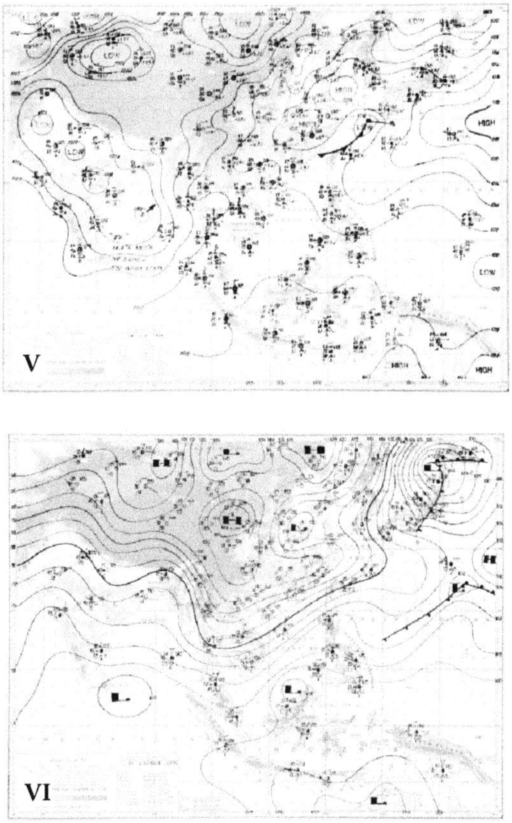

FIGURE 7.10
Typical surface synoptic charts of patterns V and VI governing BMR in November–May period. (Synoptic charts were provided by the Thailand Meteorology Department.)

These patterns were clearly different in terms of the averages and ranges of the selected meteorology variables (Table 7.4) and the frequencies of occurrence in different months (Table 7.5). Overall, patterns II (27%) and IV (23%) had the highest frequencies of occurrence during the study period. Pattern II occurred more frequently during March–April, while pattern IV appeared more during November–March. Other patterns have lower occurrence frequencies (9–16%), and their monthly distributions were also clearly distinguished. Pattern VI, for example, did not occur in April–May, whereas pattern V was most prevalent during these 2 months.

7.4.3.3 Ozone Levels and Spatial Distribution in Relation to Synoptic Patterns

The hourly ozone monitoring data collected at 15 air quality monitoring stations in BMR over 3 years (1998–2000) were analyzed. The highest 1-h ozone

TABLE 7.5

Occurrence Frequency (%) of Synoptic Pattern Governing over BMR during November–May Period of 1992–2000

	Synoptic Patterns						
	I	II	III	IV	V	VI	Total, Days (100%)
November	10	17	20	19	16	18	232
December	5	7	30	23	4	32	240
January	7	17	9	29	3	36	211
February	6	28	12	33	2	18	190
March	9	44	8	31	7	2	210
April	14	49	2	17	19	0	191
May	13	31	2	9	46	0	191
Total	9	27	12	23	13	16	1465

Source: Extracted from Winijkul (2002).

TABLE 7.6

Summary Statistics of Daily Highest 1-h Ozone over BMR (Mean ± 1 SD) Associated with Six Synoptic Patterns, 1998–2000

Pattern	No. of Days	Ozone, ppb	Days with O_3 > 100 ppb	High-Ozone Areas in BMR
I	46	82 ± 28	9 (20%)	East
II	162	76 ± 31	33 (20%)	East
III	77	82 ± 22	13 (17%)	East and northwest
IV	150	77 ± 30	31 (21%)	East
V	90	74 ± 25	9 (10%)	Northwest
VI	61	94 ± 24	20 (33%)	South and southwest

Source: Extracted from Winijkul E. 2002. Development of synoptic climatological model for forecasting photochemical smog potential in Bangkok metropolitan region. Master's thesis, Asian Institute of Technology, Pathumthani, Thailand.

Note: In parentheses are percentages of days with hourly ozone exceeding NAAQS of 100 ppb.

occurred in BMR in each day (one value per day) during the study period, and its spatial distribution in each pattern (Table 7.6) and the average of daily highest 1-h ozone levels measured at 15 stations (Figure 7.11) were examined. Except for pattern VI, ozone levels in the other five patterns were not quite well distinguished, although spatial distributions over BMR were somewhat different between the patterns (Table 7.6). Patterns II and V had not only lower ozone levels but also several outliers, which indicated the abnormally high values (Figure 7.11).

Overall, the highest ozone levels were associated with pattern VI, which also had the highest percentage of days with 1-h ozone exceeding NAAQS, 20 days of 61 or 33%, during the 3-year period. The meteorological conditions typical for this pattern are conducive to ozone formation and accumulation.

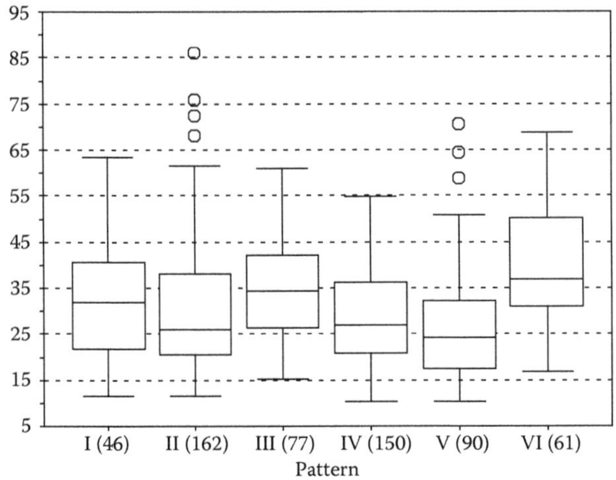

FIGURE 7.11
Box plots showing the statistics of the average of daily maximum 1-h ozone levels measured at 15 monitoring stations (1998–2000) in BMR associated with six synoptic patterns (see note in Figure 7.5). (Adapted from Winijkul, 2002.)

A high-pressure ridge was normally observed in this pattern (Figure 7.10), reaching deep to SEA. The ridge was in the stagnating or weakening stage, indicated by negative 24-h pressure tendencies (Table 7.4). Light wind (lowest wind speed of all patterns), clear sky (lowest cloudiness), long daily sunshine hours, dry weather (no rain, lowest dew point and moisture content), and the highest sea-level pressure in Bangkok were some distinguished features of this pattern. The spatial distribution showed that in pattern VI higher ozone levels mostly occurred on the south to southwest parts of BMR, which are downwind of the city under the influence of the NE monsoon. Detailed analysis of the ozone levels in the other five patterns showed that, despite the distinct differences between the synoptic patterns in terms of the regional pressure fields and the ranges of meteorological variables, the difference in the ozone levels between the patterns was not sufficiently large to develop a forecast scheme. Specifically, although pattern V (lowest ozone) and pattern VI (highest ozone) had clearly distinguished synoptic pressure patterns (i.e., low-pressure systems vs. high-pressure systems over the region; Figure 7.10), and weather conditions in the region (i.e., rainy vs. dry conditions in BMR; Table 7.4) the difference in ozone levels was relatively small, and the ranges overlapped (Table 7.6). Winijkul (2002) also tried with the TSI approach using the data collected at the Bangkok station for the meteorology classification and showed even fewer differences in ozone levels and spatial distribution between the patterns as compared to the case with SSI presented here.

Surface ozone, as a secondary pollutant, may be more complexly related to the regional and local meteorology than the primary pollutants such as SO_2 and haze considered in the previous two case studies. More important,

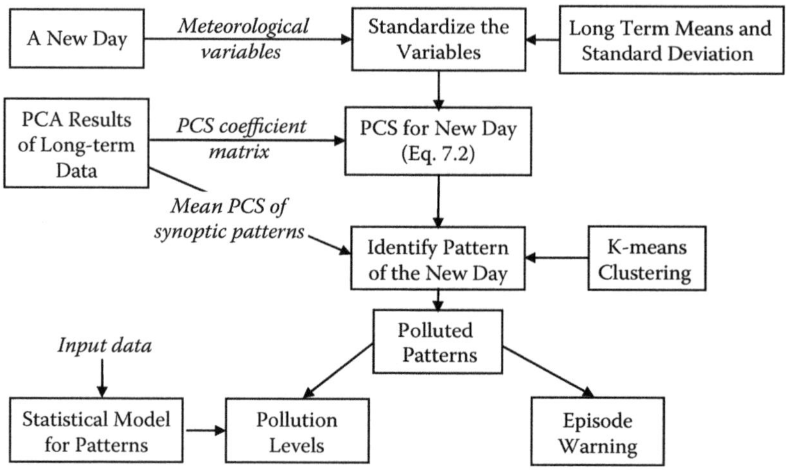

FIGURE 7.12
Proposed operation forecasting procedure for meteorology-induced air pollution episodes.

ozone levels in a city are expected to be strongly dependent on traffic emission, which varies significantly during a week. Thus, effects of weekdays and weekends should be considered when analyzing ozone levels within each synoptic pattern. Further research is still required to develop a warning signal for ozone episodes in Bangkok if the synoptic climatological approach is chosen. Other statistical tools, such as the neural network techniques, can also be applied for prediction of surface ozone air quality in urban areas and may be a good alternative when detailed emission input data are not available (Nghiem and Kim Oanh, 2009). Overall, application of 3D photochemical models is highly recommended for ozone air quality management as discussed in Chapters 5 and 6.

7.4.4 Procedure for Air Pollution Episode Warning and Prediction

Operational procedures for routine air pollution episode warning can be developed after the automated synoptic meteorology classification is successfully made, that is, the identified meteorology patterns exhibit large differences in air pollution levels, such as for SO_2 and PM_{10} in the first two case studies. A proposed operational procedure is shown in Figure 7.12, which consists of (a) identifying a synoptic pattern for a new day using the automated classification scheme; (b) issuing a warning signal if a highly polluted pattern is identified; and (c) predicting pollutant levels in the study area using evaluated regression models.

The meteorological variables used for classification of a new day into a synoptic pattern are those appearing in the final dataset used for the meteorology classification. Either observations or forecasted data available for this day can be used to identify the synoptic pattern. The long-term means and standard deviation of the variables are determined using the long-term meteorology data series and are used to standardize the variables. The principal component scores (PCSs) of the new day are calculated using Equation 7.2:

$$\mathrm{PCS} = \sum_{i=1}^{p} W_i X_i \tag{7.2}$$

where W_i are the PCS coefficients taken from the PCS coefficient matrix, X_i are standardized variables of the new day, and p is the number of meteorological variables.

The mean component scores for each synoptic pattern are those obtained by PCA during automated scheme development. Based on the K-means clustering principle, the new day under consideration is classified into a pattern that has the PCS closest to the PCS obtained for the day.

For example, in the case of haze in Chiangmai, if a new day is classified into synoptic pattern II during March, a warning signal for unfavorable meteorological conditions can be issued. Further, 24-h PM_{10} level in the city can be estimated using Equation 7.1. The haze-warning forecast scheme can provide a "haze alert" in advance so that emission reduction measures can be timely implemented. In particular, the open burning of solid waste and agriculture fields and prescribed forest fires should be minimized during the period. The warning signal can help the population avoid unnecessary exposure to episode air pollution levels (i.e., to restrict outdoor activities, especially for the aged and infirm and children).

7.5 Summary

Day-to-day variations in ambient air quality are determined by fluctuations in meteorology and source emission intensity. The most important role of meteorology is in the dispersion, transformation, and removal of air pollutants from the atmosphere, which under certain circumstances may lead to formation of meteorology-induced air pollution episodes. Air pollution climatology relates meteorological conditions to ambient air pollution levels in a location using long-term data records. The synoptic climatological approach classifies the meteorological conditions into homogeneous groups or patterns and is widely applied to forecast air pollution potentials. An automated synoptic meteorology classification scheme can be developed using the PCA and clustering techniques. Selection of regional weather stations and meteorological variables, which is done by a trial-and-error method, is the most important step for successful development of an automated classification scheme. The case studies showed good results in applying the synoptic climatological approach in forecasting SO_2 and PM_{10} episodes in the mountainous areas in northern Thailand where meteorology-induced pollution episodes have been reported. The synoptic climatological approach may be an appropriate alternative in these two cases, especially when no detailed emission data are available for 3D dispersion modeling. The regression models developed in these case studies should be revisited regularly to reflect changes in the source emission intensity following, for example, successful implementation of some emission reduction measures. The synoptic patterns developed for the BMR during the high-ozone months are also distinctly different in terms of pressure fields and the typical ranges of meteorological variables. However, the ozone levels during the study period were not well distinguished between the patterns. This suggests that, for a secondary pollutant such as ozone in urban areas, further investigations on emission variations (weekday-weekend) should also be included in the analysis. When the required emission input data are lacking, other statistical methods such as the artificial neural network may be used for ozone prediction. However, the complexity of the ozone air quality issue should be analyzed by the 3D photochemical smog modeling tool, which in turn requires detailed emission input data.

References

Arya S.P. 1999. *Air Pollution Meteorology and Dispersion*, Oxford University Press, New York.

Asian Development Bank (ADB). 1999. *Planning for Fire Prevention and Drought Management Project*, Asian Development Bank, Jakarta, Indonesia.

Chen T.C., Van Loon M.H., Wu K.D., and Yen M.C. 1992. Changes in the atmospheric circulation over the Pacific-North American area since 1950, *Journal of Meteorology*, 70: 1137–1146.

Chiangmai Provincial Public Health Office (CMPHO). 2008. Chiangmai Provincial Public Health Office, Strategy Division. http://www.chiangmaihealth.com/cmpho_web.

Davidson P.M., Seaman N., Schere K., Wayland R.A., Hayes J.L., and Carey K.F. 2004. National air quality forecasting capability: first steps toward implementation, paper presented at the Sixth Conference on Atmospheric Chemistry: Air Quality in Megacities, Seattle, January, 2004.

Davis R.E. 1991. A synoptic climatological analysis of winter visibility trends in the mid-eastern United States, *Atmospheric Environment*, 25B: 165–175.

Davis R.E., and Kalkstein L.S. 1990a. Development of an automated spatial synoptic climatological classification, *International Journal of Climatology*, 10(8): 769–794.

Davis R.E., and Kalkstein L.S. 1990b. Using a spatial climatological classification to assess changes in atmospheric pollution concentrations, *Journal of Physical Geography*, 11: 320–342.

Eder B.K., Davis J.M., and Bloomfield P. 1994. An automated classification scheme designed to better elucidate the dependence of ozone on meteorology, *Journal of Applied Meteorology*, 33: 1182–1199.

Hair J.F., Anderson R.E., Tatham R.L., and Black W.C. 1998. *Multivariate Data Analysis* (5th ed.), Prentice Hall, Upper Saddle River, NJ, pp. 164–165.

Heidorn K.C., and Yap D. 1986. A synoptic climatology for surface ozone concentrations in southern Ontario, *Atmospheric Environment*, 20: 695–703.

Kalkstein L.S., and Corrigan P. 1986. A synoptic climatological approach for geographical analysis—assessment of sulfur dioxide concentrations, *Annals of the Association of American Geographers*, 76(3): 381–395.

Kallos G., Kassomenkos P., and Pielke R.A. 1993. Synoptic and mesoscale weather conditions during air pollution episodes in Athens, Greece, *Boundary-Layer Meteorology*, 62: 163–184.

Kathleen V.S. 2002. A synoptic climatological evaluation of surface ozone concentrations in Lancaster Country, Pennsylvania. Center for Climatic Research–University of Delaware. http://www.udel.edu/SynClim/ozone.html.

Kim K.H., Choi G.H., Kang C.H., Lee J.H., Kim J.Y., Yound Y.H., and Lee S.R. 2003. The chemical composition of fine and coarse particles in relation with the Asian dust events, *Atmospheric Environment*, 27(6): 753–765.

Kim Oanh N.T., Chutimon P., Ekbordin W., and Supat W. 2005. Meteorological pattern classification and application for forecasting air pollution episode potential in a mountain-valley area, *Atmospheric Environment*, 39(7): 1211–1225.

Kim Oanh N.T., and Leelasakultum K. 2011. Analysis of meteorology and emission in haze episode prevalence over mountain-bounded region for early warning, *Science of the Total Environment*, 409: 2261–2271.

Kim Oanh N.T., and Permadi D.A. 2009. Photochemical smog modeling for ozone air quality management. *Modelling of Pollutants in Complex Environmental Systems* (Vol. 1), edited by Hanrahan G., ILM, Hertfordshire, UK, 291–319.

Kowalewsky K.J. 1997. *Development and Implementation of a Mathematical Model for Simulating the Meteorology and Air Pollution Situation in the Mea Moh Valley District*, Report submitted by Radian International, Ltd. to Electricity Generating Authority of Thailand, Bangkok, Thailand.

Lam K.C., and Cheng S. 1998. A synoptic climatological approach to forecast concentrations of sulfur dioxide and nitrogen dioxide in Hong Kong, *Environmental Pollution*, 101: 183–191.

Leelasakultum K. 2009. The Chiangmai haze episode in March 2007: cause investigation and exposure assessment master's thesis, Asian Institute of Technology, Pathumthani, Thailand.

Nghiem L.H., and Kim Oanh N.T. 2008. Evaluation of the mesoscale meteorological model (MM5)-community multi-scale air quality model (CMAQ) performance in hindcast and forecast of ground-level ozone, *Journal of Air and Waste Management Association*, 58: 1341–1350.

Nghiem L.H., and Kim Oanh N.T. 2009. Comparative analysis of maximum daily ozone levels in urban areas predicted by different statistical models, *ScienceAsia*, 35: 276–283.

Piromyaporn C. 2000. Development of a synoptic climatological approach to predict ambient SO_2 concentrations at the Mae Moh Valley, master's thesis, Asian Institute of Technology, Pathumthani, Thailand.

Pollution Control Department (PCD). 2010. Thailand national ambient air quality standard. http://www.pcd.go.th, accessed 2011.

Russell A., and Dennis R. 2000. NARSTO critical review of photochemical models and modeling, *Atmospheric Environment*, 34: 2283–2324.

Sanchez M.L., Pascual D., Pamos C., and Perez I. 1990. Forecasting particulate pollutant concentrations in a city from meteorological variables and regional weather patterns, *Atmospheric Environment*, 24A: 1509–1519.

Sargent and Lundy. 1994. *Development of System Wide Emission Control Strategies Applicable to Mae Moh Power Plant*. Report submitted by Sargent & Lundy Co., Ltd. to Electricity Generating Authority of Thailand, Bangkok, Thailand.

Schnelle K.B., and Dey P.R. 2000. *Atmospheric Dispersion Modeling Compliance Guide*, McGraw-Hill, New York.

Scott G.M., and Diab R.D. 2000. Forecasting air pollution potential: a synoptic climatological approach, *Journal of Air and Waste Management Association*, 50: 1831–1842.

Stern A.C. 1976. *Air Pollution (3rd ed.). Vol. Air Pollutants, Their Transformation and Transport*, Academic Press, New York.

Triantafyllou A.G. 2001. PM_{10} pollution episodes as a function of synoptic climatology in a mountainous industrial area, *Environmental Pollution*, 112: 491–500.

U.S. Environmental Protection Agency (USEPA). 2003. Guidelines for developing an air quality (ozone and $PM_{2.5}$) forecasting program, U.S. Environmental Protection Agency, Research Triangle Park, NC. http://www.epa.gov/airnow/aq_forecasting_guidance-1016.pdf.

Winijkul E. 2002. Development of synoptic climatological model for forecasting photochemical smog potential in Bangkok metropolitan region. Master's thesis, Asian Institute of Technology, Pathumthani, Thailand.

Wolff G.T., Dunker A.M., Rao S.T., Porter P.S., and Zurbenko I.G. 2001. Ozone air quality over North America: Part I—review of reported trends, *Journal of Air and Waste Management Association*, 51(2): 273–282.

Wolff G.T., Siak J.S., Chan T.L., and Korsog P.E. 1986. Multivariate statistical analysis of air quality data and bacterial mutagenicity data from ambient aerosols, *Atmospheric Environment*, 20: 2231–2241.

Zannetti P. 1990. *Air Pollution Modeling: Theories, Computational Methods and Available Softwares*, Computational Mechanics Publications, Southampton Boston.

Zhang B.N., and Kim Oanh N.T. 2002. Photochemical smog pollution in the Bangkok metropolitan region of Thailand in relation to O_3 precursor concentrations and meteorological conditions, *Atmospheric Environment*, 36(26): 4211–4222.

Section III

Air Pollution Control Techniques

8

Development of Air-Cleaning Device Using Chemically Impregnated Activated Carbon for Interior Vehicle Application

Nguyen Thi Kim Oanh and Sukanya Kiatphuangchai

CONTENTS

8.1 Introduction

The problem of exposure to toxic air pollutants is of increasing concern in many urban areas, especially in Asian developing countries. High levels of traffic-related fine particles and gaseous air pollutants, such as volatile organic compounds (VOCs) and nitrogen oxides (NO_x), along busy streets may substantially increase the exposure of commuters in open transport means, pedestrians, and urban residents. Pollutants can also enter and become trapped in the vehicle interior and cause adverse health effects for commuters in closed air-conditioned vehicles. The common practice of setting low or even closing off the fresh air intake while driving in urban areas leads to a high risk of exposure to accumulated pollutants inside vehicles. Measurements of in-car air environment revealed significant levels of VOCs, SO_2, and NO_2 while traveling polluted streets (International Center for Technology Assessment [ICTA], 2000; Kiatphuangchai, 2008; Kongpran, 2011). Suitable control devices can be developed to clean the intake fresh air or the interior air of vehicles to reduce exposure while running in polluted urban streets.

8.1.1 Exposure to Street Air Pollution

Traffic is a major source of air pollution in urban areas worldwide (Colvile et al., 2001). In urban areas of Asian developing countries, high and uncontrolled growths of vehicle fleets in conjunction with slow increases in road surface areas and improper traffic management often lead to traffic congestion, hence further deteriorating air quality. In Bangkok, the capital city of Thailand, over the period from 1994 to 2006, the vehicle fleet grew at a rate of 5% per annum (Developing Integrated Emissions Strategies for Existing Land-Transport [DIESEL], 2008). In Hanoi, the motorcycle fleet alone increased by 16% per year during 2000–2008, reaching above 2 million units by 2008 (Hanoi's People Committee [HPC], 2008). Pollutants emitted from vehicles accumulate at high concentrations along streets before dispersing to other locations inside and outside cities. Narrow streets bordered with high buildings in most urban centers limit atmospheric dispersion and enhance pollution buildup at high levels in streets and at roadsides (Kim Oanh et al., 2008; Truc and Kim Oanh, 2007). Commuters as well as those working and dwelling along busy roads are thus at risk of exposure to hazardous air pollution levels (Saksena et al., 2008).

Toxic fine particles, gases (such as CO, NO_x, SO_2, and VOCs) and semi-VOCs are emitted from road transport vehicles. Except for SO_2, which is emitted proportionally to the sulfur content of fuels, emission of other pollutants is a function of combustion conditions. CO, VOCs, and semi-VOCs are the products of incomplete combustion. Gasoline, for example, contains thousands of VOCs, including paraffins, olefins, and aromatics (Noll, 1999). Gasoline-powered vehicle exhaust contains organic compounds of unburned and partially burned fuel, which all are toxic to human health. NO_x is the product of complete combustion formed at high temperatures with sufficient oxygen available and commonly refers to the sum of NO and NO_2. Fresh vehicle exhaust contains mainly NO (about 90% of NO_x). In the ambient air, NO is converted to NO_2 and various nitrogen-containing compounds. The ambient concentration of NO is normally far below 0.5 ppm;

hence, it is not considered to be of significant toxicity to human health (Wark et al., 1998). NO_2, the most prevalent form of nitrogen oxides in the atmosphere, is regulated in most countries. Short-term exposure to elevated NO_2 levels can result in adverse respiratory effects for both healthy people and people with asthma (U.S. Environmental Protection Agency [USEPA], 2009). Further, VOCs and NO_x participate in photochemical smog reactions to form surface ozone and toxic organic particles, as detailed in Chapters 1 and 6.

8.1.2 Levels of Pollutants Inside and Outside Vehicles

It is commonly perceived that urban commuters such as motorcyclists and bicyclists in open vehicles and pedestrians are among the most exposed to street air pollution, while those commuters in enclosed air-conditioned vehicles are less exposed. However, in many cases the quality of air inside a vehicle is reported to be similar to that of outside (ICTA, 2000).

Kiatphuangchai (2008) measured pollutants while traveling in an air-conditioned car in the Bangkok metropolitan region (BMR) and showed that levels of pollutants simultaneously monitored inside and outside the vehicle were almost the same (Table 8.1). In general, higher levels of pollutants, both inside and outside a car, were measured while driving a busy street passing the Bangkok city center as compared to that obtained when driving on a highway in a suburban area. The ranges of pollutants were 10–30 ppb for benzene and 16–42 ppb for toluene, while other VOCs presented in Table 8.1 were only 1–3 ppb in both areas. The range for NO_2 was from a dozen parts per billion in the suburb area to 400 ppb in the city center, while that of SO_2 was from around 100 ppb in the city center to a maximum of 780 ppb in the suburb area. Higher SO_2 in the suburb area was most likely due to a large number of diesel trucks traveling on the highway.

TABLE 8.1

Levels of Pollutants (2-h Average) Inside and Outside a Car Traveling in BMR (Average ± 1SD), ppb

	Suburb Area		City Center	
Pollutants	**Inside**	**Outside**	**Inside**	**Outside**
Benzene	14.9 ± 5.8	16.6 ± 5.8	11.8 ± 4.1	22.7 ± 6.1
Toluene	26.3 ± 9.6	23.8 ± 11.5	29.4 ± 5.9	33.0 ± 5.2
Ethylbenzene	0.8 ± 0.9	0.7 ± 0.2	1.2 ± 1.0	2.1 ± 0.5
m,p-Xylene	0.6 ± 0.5	1.5 ± 0.6	1.5 ± 1.2	4.1 ± 1.8
o-Xylene	0.5 ± 0.7	0.8 ± 0.9	1.1 ± 0.7	2.8 ± 0.9
NO_2	14.4 ± 9.3	20.4 ± 11.3	102 ± 10.5	169 ± 151
SO_2	286 ± 229	471 ± 259	116 ± 30	96 ± 64

Source: Data from Kiatphuangchai S. 2008. Removal of gaseous air pollutants using chemically impregnated activated carbon, master's thesis, Asian Institute of Technology, Pathumthani, Thailand.

Note that the monitoring was done while the fresh air intake was closed, following a common practice of local drivers, primarily to save energy. The vehicle doors were kept closed during each monitoring period (about 2 h). Therefore, except for the air exchange due to infiltration and leaks, virtually the same air was allowed to circulate inside the running vehicle; hence, the pollutants generated or trapped inside the vehicle could build up to high levels.

If fresh air intake for vehicles is allowed while driving but without purification, people traveling in closed vehicles for a long time may be in contact with high street air pollution levels. On the other hand, if the fresh air intake valve is closed, the circulating air inside the vehicle can also have toxic pollutants that accumulate to high levels. Exposure to high levels of toxic air pollutants while commuting in closed vehicles may be a problem of concern, especially during congestion.

This chapter presents the experiment setup to examine the VOC and NO_2 adsorptivity of KOH-impregnated activated carbon (K-IAC) versus virgin activated carbon (AC). A principal design of an AC filter system to remove VOC and acidic gases simultaneously is proposed for application for interior vehicle air cleaning.

8.2 Chemically Modified AC for Air Pollution Control

Application of appropriate cleaning devices to remove air pollution from fresh air intake or from circulating air inside a vehicle can efficiently reduce the exposure risk for commuters. Various control technologies are available, but most are pollutant specific. For example, adsorption is a common control technology for VOCs. Theoretically, a VOC with molecular weight above 45 can be removed efficiently by adsorption (Noll, 1999). AC is considered a suitable adsorbent as it has high removal capacity for VOCs, even at a low VOC concentration. The AC adsorption efficiency for VOCs is normally above 95% (USEPA, 1995). Regeneration of AC beds can reduce the operating cost of the control device (USEPA, 2009). AC can be prepared from a wide range of carbonaceous raw materials, including various agricultural by-products. The term *activated* refers to a treatment process to increase the porosity of carbon extremely, resulting in large internal surface area. Original (unmodified/virgin) AC retains gaseous molecules on its surface by physical intermolecular (van der Waals) attraction. This essentially creates an internal "charge," which enables gaseous contaminants (adsorbate) to precipitate out of a contaminated airstream onto the AC surface (adsorbent). This physical attraction force in AC is particularly suitable for removing a number of VOCs, including benzene, toluene, and xylenes. However, the AC is not effective in retaining many inorganic acidic gases, including NO_x and SO_2 (Muller and Stanley, 2006).

To remove NO_x and SO_2, AC is normally modified using various chemicals (impregnates) to retain target contaminants effectively (Lee et al., 2002, 2003).

TABLE 8.2

Impregnates and Target Contaminants for Activated Carbon Adsorption

Impregnate	Target Contaminants
Phosphoric acid (H_3PO_4) or citric acid ($C_6H_8O_7$)	Ammonia, amines, *n*-methyl-2-pyrrolidone (NMP)
Iodine (I_2)	Mercury vapor
Potassium hydroxide (KOH)	H_2S, SO_2, NO_x, Cl, HCl, HF, BF_3

Source: Data from Circul-Aire-Inc. 2011. Contaminant and applications handbook, Multi-Mix Chemical Media. http://circulaire.cn/html_en/product_pdf/multimix/Contaminant_Applications_Handbook.pdf, accessed July 2011.

Retention of gaseous contaminants on the chemically modified AC involves the formation of a chemical bond between the gases and the surface of the AC adsorbent, which is termed *chemisorption*. Chemical modification of AC involves treatment using acidic or basic compounds, that is impregnation with foreign materials, which induce a number of advantages and disadvantages for subsequent pollutant adsorption (Yin et al., 2007). Selection of the right chemical for impregnation, therefore, can improve the adsorptivity of a selected pollutant. Examples of impregnates with their target contaminants are presented in Table 8.2.

Chemical impregnation is commonly done via a dry-feed process in which the chemical (impregnate) solution is added to the powdered original adsorbent materials, ensuring uniform distribution, followed by carbonization at high temperatures (Muller and Stanley, 2006). The process of AC impregnation enhances its inherent properties by distributing selected chemicals on its surface, which eventually changes the primary removal mechanism from physical adsorption to chemisorption. The impregnation can normally improve the attachment capacity of selected individual pollutants on the adsorbent surface. The impregnated AC is therefore pollutant specific, which may actually reduce removal efficiency for other nontargeted pollutants. Selection of optimum chemicals and operating conditions for AC adsorption beds is important to improve the removal efficiency of a mixture of air pollutants commonly found in source emissions or in the environment.

8.3 Experimental Design

8.3.1 AC Impregnation

Granular activated carbon (GAC), made from coconut shell, underwent the preparation and impregnation steps shown in Figure 8.1, which largely followed the method of GAC impregnation by KI presented by Joshi (2003). GAC

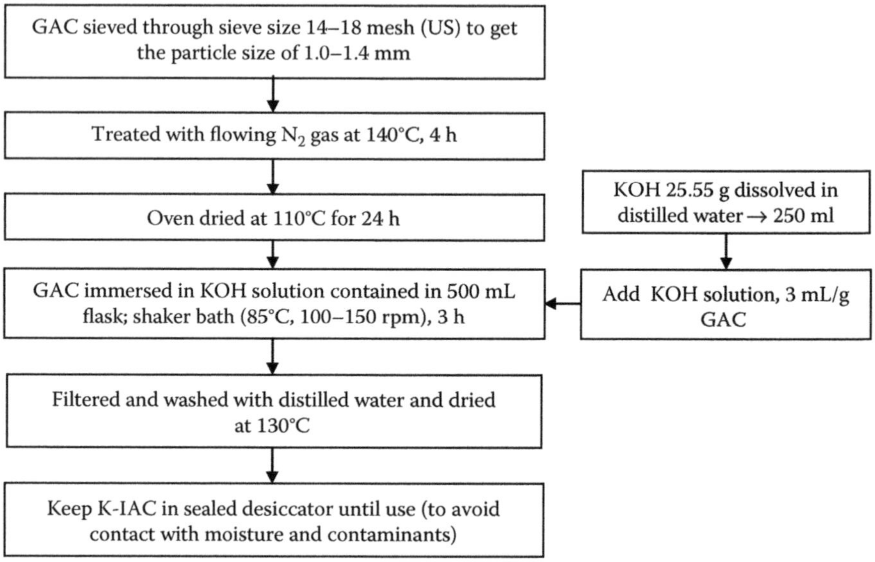

FIGURE 8.1
Impregnation of GAC with KOH.

was sieved through a 14- to 18-mesh (1.0- to 1.4-mm) sieve and treated with flowing N_2 for 4 h at 140°C. Next, the GAC was oven dried at 110°C for 24 h. The GAC was then placed in a 500-mL Erlenmeyer flask, and a KOH solution in distilled water (10%) was added at about 3 mL per 1 g GAC, just enough to submerge the GAC. The flask was then placed in a constant-temperature (85°C) shaker bath and shaken at 100–150 rpm for 3 h. After that, the GAC slurry was filtered, briefly washed by distilled water, and dried at 130°C. Prepared K-IAC was kept in a sealed container and placed in a desiccator to avoid contact with moisture and airborne contaminants that might reduce adsorbent function.

A scanning electron microscope (SEM) with magnification of 1,000 and 10,000 was used (in three repetitions) to examine the surface of virgin GAC and the K-IAC. The surface morphology textures of the two adsorbents shown in Figure 8.2 were clearly different. The virgin GAC surface was homogeneous in composition but contained a number of impurities (Figure 8.2[a]). The K-IAC (Figure 8.2[b]) had crystal particles uniformly distributed over the external surface. Note that these crystals deposited on the surface of K-IAC may partially block the AC pores. Therefore, the impregnation with KOH may lower the surface-to-volume ratio of the adsorbent (K-IAC), as well as the attraction force between gaseous pollutants and AC, hence causing a reduction in physical adsorption capacity (Lee et al., 2004; USEPA, 1995).

Energy dispersive x-ray (EDX) spectroscopy analysis was done to determine the chemical composition of GAC and K-IAC adsorbents. The results showed remarkable increases in potassium and oxygen content (in %) of the

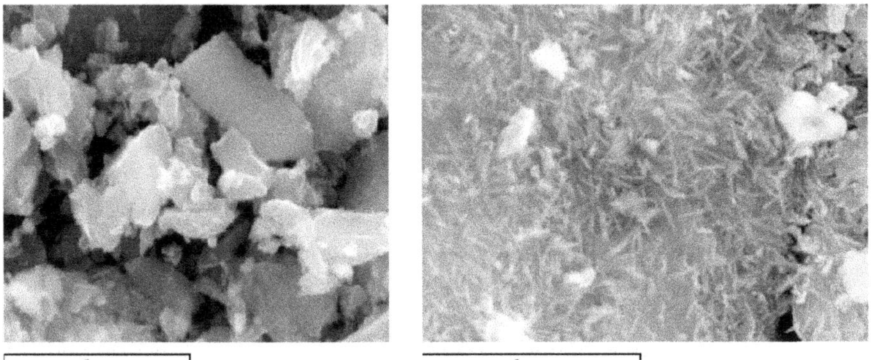

FIGURE 8.2
SEM images of virgin GAC and K-IAC at ×10,000.

TABLE 8.3
Summary of EDX Analysis of Virgin and K-IAC

		Weight%	
Element	**Symbol**	**Virgin GAC**	**K-IAC**
Carbon	C	84.5	56.8
Oxygen	O	13.3	28.7
Aluminum	Al	0.6	5.0
Silicon	Si	0.7	1.0
Sulfur	S	0.5	0.4
Potassium	K	0.03	7.3
Calcium	Ca	0.03	0.0
Iron	Fe	0.3	0.8
Zinc	Zn	0.1	0.0
Total		100	100

K-IAC compared to the virgin GAC (Table 8.3). This, together with the SEM images, confirmed the presence of KOH in the adsorbent. The increase in K and O contents in K-IAC was accompanied by a relative reduction of the percentage C content. However, an increase in Al content was also observed, which was not expected. This may be due to impurities in the KOH solution used for impregnation. Note that the K content in the K-IAC obtained in this study was 7.3% by weight, which is slightly lower than that (9.96%) reported for K-IAC used by Lee et al. (2002).

TABLE 8.4

Synthetic Airstreams Used in Different Adsorption Experiments

	Range of Inlet Concentrations	
Experiments	**Virgin GAC**	**K-IAC**
1. Benzene containing air (5 runs)		
Benzene, ppb	4–300	20–52
2. VOC and NO_2 mixture (3 runs)		
NO_2, ppm	55–77	55–77
Benzene/toluene, ppb	10–14	10–14
Xylenes/ethylbenzene, ppb	Not detected	Not detected

8.3.2 Synthetic Gas Streams

Two types of synthetic gas streams were prepared: (a) airstream containing only benzene and (b) airstream containing a mixture of VOCs and NO_2. The second type was used in the experiments to represent the actual street air pollution closely, in which many pollutants coexist and hence may induce competitive adsorption (Lillo-Rodenas et al., 2006). The levels of contaminants in the synthetic inlet gases are presented in Table 8.4.

To prepare the benzene-containing airstream, a known volume of benzene solution (99.7% purity) was injected into a glass bottle. The bottle was heated slightly to enhance benzene evaporation. Zero air was passed through to purge the benzene vapor to a 30-L Tedlar bag at a flow rate of about 3 L/min. The benzene concentrations of the generated synthetic gas were from 20–50 ppb, which was in the common concentration range found in the streets of Bangkok (Table 8.1), to 120–300 ppb. To prepare the airstream containing a mixture of pollutants (benzene, toluene, and NO_2), first the VOC vapors were generated and purged (using zero air) into a Tedlar bag. During the experimental runs, the VOC mixture was introduced from the Tedlar bag into a mixing chamber where NO_2 was added (Figure 8.3). The NO_2 inlet concentrations were in the range of 50–80 ppm, in fact higher than the street NO_2 levels reported for Bangkok (Table 8.1). This was done in anticipation of the presence of other acidic gases (NO, SO_2) in the real conditions, which can also be removed by the K-IAC, as discussed further in the chapter.

8.3.3 Experiment Setup

8.3.3.1 Adsorption Column

Two identical columns were used in each experiment, one with virgin GAC (column A) and another with K-IAC (column B), as shown in Figure 8.3. The columns were made from a stainless steel tube with an inner diameter of 1.9 cm and a length of 20 cm. The details of the column configuration are given in Table 8.5. Each column was packed with three sections of media (GAC or

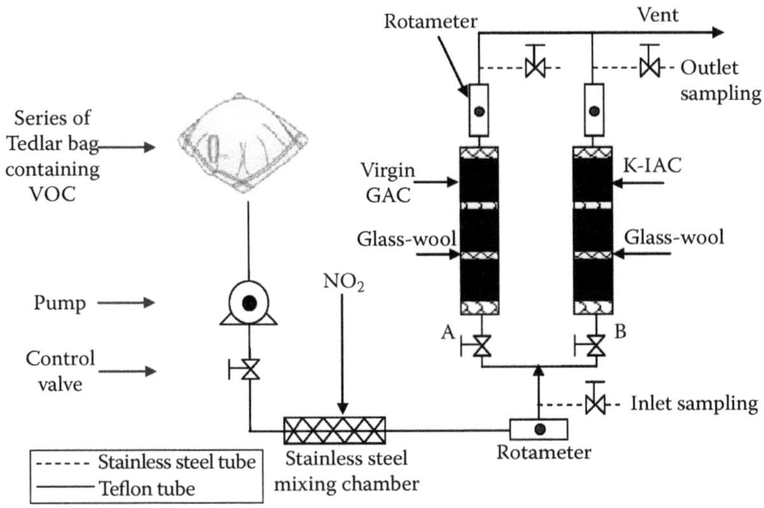

FIGURE 8.3
Experimental setup showing VOC and NO_2 inlet and two identical adsorption columns containing GAC and K-IAC.

TABLE 8.5
Adsorption Column Details

	Columns	
Parameters	**A (GAC)**	**B (K-IAC)**
Size, ID (cm) × H (cm)	1.9 × 20	1.9 × 20
Height of media H, cm	6.0	6.0
Volume of media, cm^3	17.0	17.0
Mass of media, g	8.2574	8.2565

K-IAC) separated by glass wool (Figure 8.3). Fresh (not regenerated) adsorbents were used in all experiments. Before use, the packed columns were purged with He for about 10 min and were tested to ensure there were no leaks.

8.3.3.2 Experimental Conditions

8.3.3.2.1 Benzene Adsorption

The benzene-contaminated air stored in a Tedlar bag was drawn through the adsorption columns using a pump with a flow rate of 3 L/min. This resulted in a 17.6-cm/s linear velocity of the airstream passing the column. To provide a continuous influent to the adsorption columns, a series of Tedlar bags (30 L each) containing the synthetic air was used. The resulting gas contact time in the bed was estimated at 0.34 s. The difference in the outlet flow rate

TABLE 8.6

Sampling and Analytical Methods for NO_2 and VOCs

Sampling Tube	Gas Flow to Tube, L/min	Extraction Method	Analytical Method
VOCs			
SKC-coconut–shell: 6-mm diameter, 70-mm length	0.2	Method 1501 (NIOSH, 2003)	Gas chromatography with flame ionization detection (GC-FID)
NO_2 Gas			
TEA-IMS tube	0.2	Method ID 182 (OSHA, 1991)	Ion chromatography

(2.6 L/min) and the inlet flow rate (3 L/min) corresponded to a drop in the linear velocity of 2.3 cm/s. Five experimental runs were performed. Each run was for about 1–2 h and was stopped before breakthrough occurred. All the experiments were performed at ambient temperatures, 29–32°C.

8.3.3.2.2 NO_2 and VOC Mixture Adsorption

Three experimental runs were conducted. The airstream containing the mixture of VOCs (benzene and toluene) and NO_2 was fed into the adsorption columns (Figure 8.3). Other setup details were similar to those for the benzene adsorption experiments presented.

8.3.4 Sampling and Analysis of Air Pollutants

Each sample of the inlet or outlet gas from the adsorption columns was collected in a Tedlar bag over a time period for subsequent analysis for VOCs (National Institute for Occupational Safety and Health [NIOSH], 2003) and NO_2 (Occupational Safety and Health Administration [OSHA], 1991), as detailed in Table 8.6. A known volume of sampled gas from a Tedlar bag was transferred to a designated charcoal tube using a pump (at 0.2 L/min), which was subsequently extracted for pollutant analyses.

8.4 Results and Discussion

8.4.1 Benzene Adsorption

Table 8.7 presents the results of benzene adsorption by virgin GAC and K-IAC adsorbent columns in various experimental runs. The results indicated the adsorptivity (adsorption capacity) of the media before the breakthrough occurred. The virgin GAC had better adsorptivity for benzene than

TABLE 8.7

Adsorption Capacity of Virgin GAC and K-IAC for Benzene in Air after the First Hour of Column Operation

Virgin GAC			K-IAC		
Run No.	**Inlet, ppb**	C_{out}/C_{in}	**Run No.**	**Inlet, ppb**	C_{out}/C_{in}
1	301	0.07	1	22.8	0.40
2	120	0.08	2	20.0	0.48
3	11	0.05	3	41.4	0.34
4	7.8	0.05	4	33.1	0.52
5	4.2	0.12	5	52.0	0.34
Average and SD		0.07 ± 0.03	Average and SD		0.41 ± 0.08

the K-IAC. The outlet-to-inlet concentration ratios C_{out}/C_{in} for GAC were 0.07 ± 0.03, equivalent to a reduction of 93% ± 3% of the inlet benzene concentration, as compared to that of 0.41 ± 0.08 (a benzene reduction of 59% ± 8%) for the K-IAC column. The results for virgin GAC are reported for a wide range of inlet concentrations, from 4 to 300 ppb. Note that measurements for the experiments with low inlet concentrations (run 5) may have been affected by the uncertainty of the monitoring method. Overall, the adsorptivity obtained for virgin GAC was comparable to the reported range for VOC adsorption (USEPA, 1995).

The lower adsorptivity by K-IAC may be attributed to a reduction in the physical adsorption capacity (which mainly works for benzene) by the impregnation. The presence of KOH on its surface led to a reduction in surface-to-volume ratio (partially blocked the original pores) and a reduction in the attraction force between benzene and AC (USEPA, 1995).

8.4.2 Adsorption of VOCs and NO_2 in Mixture

8.4.2.1 Volatile Organic Compounds

Adsorptivity of VOCs in this case was similar to the case for the synthetic air containing benzene alone. The average C_{out}/C_{in} was 0.07 for benzene and 0.08 for toluene (Table 8.8), which corresponded to a reduction of 93% and 92%, respectively, of the inlet VOC concentrations by virgin GAC. The corresponding average values for K-IAC were 0.33 and 0.30 (67% and 70% of VOC removal efficiency, respectively), that is, a lower VOC adsorption capacity by K-IAC than by virgin GAC.

8.4.2.2 Nitrogen Dioxide

Adsorptivity of NO_2 was significantly improved on K-IAC (C_{out}/C_{in} = 0.24 ± 0.03, or a removal efficiency of 76%). This was significantly above that of 3–4% on the virgin GAC (C_{out}/C_{in} = 0.97 ± 0.01), as seen in Table 8.8. Lee et al.

TABLE 8.8

Adsorption Capacity of Virgin GAC and K-IAC for VOCs and NO_2 over the First Hour of Operation of Experimental Beds

	Benzene			Toluene			NO_2		
Run No.	Inlet, ppb	GAC C_{out}/C_{in}	K-IAC C_{out}/C_{in}	Inlet, ppb	GAC C_{out}/C_{in}	K-IAC C_{out}/C_{in}	Inlet, ppm	GAC C_{out}/C_{in}	K-IAC C_{out}/C_{in}
1	13.2	0.07	0.36	10.2	0.07	0.34	76.45	0.97	0.20
2	13.4	0.06	0.36	13.8	0.10	0.21	54.48	0.97	0.27
3	10.9	0.08	0.26	9.5	0.07	0.35	60.16	0.96	0.26
Average and SD		0.07 ± 0.01	0.33 ± 0.06	—	0.08 ± 0.02	0.30 ± 0.08	—	0.97 ± 0.01	0.24 ± 0.03

(2002) also reported the C_{out}/C_{in} ratios of about 0.4–0.5 for the K-IAC adsorption of NO_2-air (air containing NO_2 at 150 ppm). For the NO-NO_2-rich air inlet (122 ppm NO and 787 ppm NO_2), the C_{out}/C_{in} for NO_x was about 0.35 at the completion of the reaction (Lee et al., 2003).

According to Lee et al. (2002), the enhanced removal efficiency for NO_2 by K-IAC may be explained by the chemical reaction between KOH and NO_2:

$$2KOH + 3NO_2 \rightarrow 2KNO_3 + NO + H_2O \tag{8.1}$$

The KNO_3 crystal produced in this reaction would attach on the K-IAC surface with an ionic bond, which is formed as K^+ cations pull NO_2. Thus, when K-IAC is used, the chemisorption process occurs; hence, the removal efficiency of NO_2 is enhanced. Equation 8.1 also shows that for every three molecules of NO_2 removed, one molecule of NO is produced. The NO produced has the possibility of being readsorbed to other available selective adsorption sites after being converted into NO_2 in the presence of oxygen (Lee et al., 2004). The primary reason for the increase in adsorptivity is the availability of many selective adsorption sites on K-IAC with the K^+ and OH^- ions, which can react with the adsorbate. Lee et al. (2002) suggested that K-IAC is suitable for use in various industrial applications for NO_x emissions control.

K-IAC is expected to remove SO_2 as well, which mostly follows the same mechanism as for NO_2. Lee et al. (2003) showed that the adsorbed SO_2 was predominantly found on the external surface, producing mainly K_2SO_4 and H_2SO_4 and K_2SO_3. Thus, using K-IAC can simultaneously remove NO, NO_2, and SO_2. However, the difference in adsorption affinity between NO_x and SO_2 leads to different results for their individual and simultaneous adsorption (Lee et al., 2003).

8.4.3 Toward Design of an AC Filter for Interior Vehicle Air Cleaning

8.4.3.1 Automobile Air-Conditioning Ventilation System

A questionnaire survey conducted by Kiatphuangchai (2008) showed that there are two types of automobile air-conditioning systems in Thailand:

TABLE 8.9

Information on Automobile Air-Conditioning Systems Obtained from Survey

Car				Total Flow Rate (m^3/min)		Ventilation System
Year	**Brand**	**Model**	**cc**	**Min**	**Max**	
1978	Mercedes Benz	/8	2,000	1.8	4.5	Manual
1994	Toyota	Corolla	1,600	3.8	10.9	Manual
1995	Mercedes Benz	C-Class	2,000	1.1	8.7	Automatic
2002	Honda	Accord	2,300	7.1	18.5	Automatic
2006	Honda	Civic	1,800	2.8	7.8	Manual
2007	Honda	Jazz	1,500	5.4	7.2	Manual
2007	Toyota	New Vios	1,500	3.1	10.4	Manual
2007	Toyota	Vigo	2,500	6.4	8.4	Manual
			Average	3.94	9.55	
			Average of all	6.74 m^3/min		

Source: Data from Kiatphuangchai S. 2008. Removal of gaseous air pollutants using chemically impregnated activated carbon, master's thesis, Asian Institute of Technology, Pathumthani, Thailand.

manual and automatic, as presented in Table 8.9. Generally, a manual air-conditioning system is installed in common personal cars, pickups, vans, and trucks, whereas an automatic system is available in relatively new and luxury cars. In the manual system, users can manually adjust to obtain the desired amount of fresh air intake into the cabin. However, due to high on-road air pollution levels and to save fuel, most people choose to close the fresh air intake valve while driving in urban areas. Thus, mostly the same air is circulated in the cabin; hence, pollutants generated/trapped indoors can build up to a high level until the doors are opened again. This practice would also increase the levels of CO_2 and other human bioeffluents as well, especially on a long trip.

An automatic air-conditioning system operates based on the cabin temperature measured by a sensor. Some systems allow an additional amount of fresh air intake into the cabin when the temperature is lower than a desired level. Such systems are commonly installed in new or more luxury cars. However, some car models still allow users to close off the fresh air intake system totally; hence, the situation becomes similar to the manual system mentioned.

8.4.3.2 Proposed AC Filter Design

The experimental results in this study suggested that an air-cleaning device should contain both virgin GAC and K-IAC to remove VOCs and NO_2 simultaneously. As discussed, K-IAC is expected also to remove NO and SO_2 (Lee et al., 2004). A pleated activated carbon filter (ACF) is proposed (Figure 8.4).

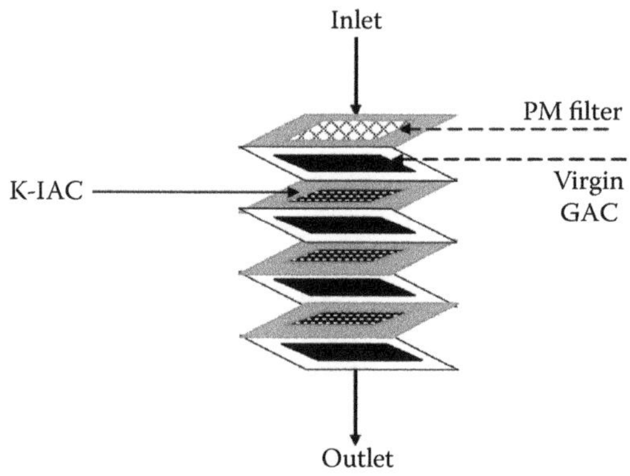

FIGURE 8.4
Proposed ACF with alternated virgin GAC and K-IAC beds. PM, particulate matter.

Such an ACF can be installed in the fresh air intake of the ventilation system of air-conditioned vehicles or in the vehicle cabin to treat the circulated air inside closed vehicles (personal cars, vans, light-duty pickups and buses, etc.).

ACFs containing multiple beds of alternating virgin GAC and K-IAC would also help to reduce the pressure drop across the system. The inlet flow of the ACF may be approximately 10% of the average flow of automobile air-conditioning systems, (i.e., 0.6–1 m^3/min), assuming 10% fresh air intake is set. There are hundreds of VOCs in street air, and the total level may reach a few parts per million, which will be competitively adsorbed on virgin GAC.

Similarly, street air contains a mixture of NO, NO_2, and SO_2; hence, there will be competitive adsorption on K-IAC.

Further study is required to determine the amount of adsorbents required for a desired operation period and for the typical air pollutant mixture in city streets. The adsorbent working capacity should first be determined. A pilot-scale device can be operated with actual street air intake and its performance evaluated. This will be the basis for sizing the actual device. In addition, due to general low levels of pollutants in airstreams being treated, nongenerative ACF systems may be considered, which further requires environmentally sound disposal of the adsorbents used. For example, after use the adsorbent beds may be incinerated in a high-temperature facility equipped with air pollution control, including NO_x removal devices.

8.5 Summary and Conclusions

This chapter presented information on the impregnation of GAC with KOH and further investigated the adsorption capacity of the virgin GAC and

K-IAC for benzene, toluene, and NO_2. The impregnation method was able to distribute KOH evenly on the surface of GAC. Lab-scale control devices conducted at ambient temperature (29–32°C) showed that VOC adsorptivity by virgin GAC was significantly higher than that of K-IAC, which may be attributed to the reduction in the physical adsorption capacity due to the impregnation. The opposite was observed for NO_2 adsorptivity that was negligible on virgin GAC but was significantly improved on K-IAC, which confirmed the chemisorption nature of the process. High levels of air pollutants measured both inside and outside vehicles traveling in urban areas suggested a high exposure risk to commuters despite sitting in closed air-conditioned cars. The proposed ACF with alternating beds of virgin GAC and K-IAC can retain VOCs, NO_x, and SO_2 simultaneously. Such a device can be used to clean the fresh air intake while traveling in an air-conditioned vehicle. It can also be used to clean the air circulating inside closed vehicles. Further research is required to prepare a detailed design of the control device for a typical mixture of multiple air pollutants in streets.

References

Circul-Aire-Inc. 2011. Contaminant and applications handbook, Multi-Mix Chemical Media. http://circulaire.cn/html_en/product_pdf/multimix/Contaminant_Applications_Handbook.pdf, accessed July 2011.

Colvile R.N., Hutchinson E.J., Mindell J.S., and Warren R.F. 2001. The transport sector as a source of air pollution [Review], *Atmospheric Environment*, 35(9): 1537–1565.

Developing Integrated Emissions Strategies for Existing Land-Transport (DIESEL). 2008. *Developing Integrated Emissions Strategies for Existing Land-Transport (DIESEL)*, UNDP/World Bank Energy Management Assistance Programme, Bangkok, Thailand.

Hanoi's People Committee (HPC). 2008. Hanoi Police Web site. http://www.anninhthudo.vn/Xa-hoi/De-cao-viec-xu-ly-nghiem-va-triet-de/338099.antd, accessed in 2011.

International Center for Technology Assessment (ICTA). 2000. *In-Car Air Pollution: Report No. 4 An Assessment of the Air Quality Inside Automobile Passenger Compartments; The Hidden Threat to Automobile Drivers*, International Center for Technology Assessment, Washington, DC.

Joshi M.U. 2003. Mercury control technology by chemically treated activated carbon using simulated flue gas, master's thesis, Asian Institute of Technology, Pathumthani, Thailand.

Kiatphuangchai S. 2008. Removal of gaseous air pollutants using chemically impregnated activated carbon, master's thesis, Asian Institute of Technology, Pathumthani, Thailand.

Kim Oanh N.T., Martel M., Pongkiatkul P., and Berkowicz R. 2008. Determination of fleet hourly emission and on-road vehicle emission factor using integrated monitoring and modeling approach, *Atmospheric Environment*, 89(3): 223–232.

Kongpran J. 2011. Monitoring of on-road and roadside air quality in Bangkok metropolitan region for assessment of vehicle emission and associated health risk, master's thesis, Asian Institute of Technology, Pathumthani, Thailand.

Lee Y.W., Choi D.K., and Park J.W. 2002. Performance of fixed-bed KOH impregnated activated carbon adsorber for NO and NO_2 removal in the presence of oxygen, *Carbon*, 40: 1409–1417.

Lee Y.W., Kim H.J., Park J.W., Choi B.U., Choi D.K., and Park J.W. 2003. Adsorption and reaction behavior for the simultaneous adsorption of NO-NO_2 and SO_2 on activated carbon impregnated with KOH, *Carbon*, 41: 1881–1888.

Lee Y.W., Park J.W., Jun S.J., Choi D.K., and J.E. Y. 2004. NO_x adsorption—temperature programmed desorption and surface molecular ions distribution by activated carbon with chemical modification, *Carbon*, 42: 59–69.

Lillo-Rodenas M.A., Fletcher A.J., Thomas K.M., Cazorla-Amoros D., and Linares-Solano A. 2006. Competitive adsorption of a benzene-toluene mixture on activated carbons at low concentration, *Carbon*, 44: 1455–1466.

Muller C., and Stanley B. 2006. Chemical filtration strategies for the control of airborne molecular contamination, part 2, *Controlled Environments Magazine*. http://www.cemag.us/articles.asp?pid=629.

National Institute for Occupational Safety and Health (NIOSH). 2003. *NIOSH Manual of Analytical Method. Hydrocarbons, Aromatic: Method 1501* (4th ed.), National Institute for Occupational Safety and Health, Cincinnati, OH.

Noll K.E. 1999. *Fundamentals of Air Quality Systems: Design of Air Pollution Control Devices*, American Academy of Environmental Engineers, Annapolis, MD, USA.

Occupational Safety and Health Administration (OSHA). 1991. *Nitrogen Dioxide in Workplace Atmospheres (Ion Chromatography) Method ID 182*, Occupational Safety and Health Administration, Utah, Branch of Inorganic Methods Development, OSHA Salt Lake Technical Center. Salt Lake, Utah.

Saksena S., Quang T.N., Nguyen T., Dang P.N., and Flachsbart P. 2008. Commuters' exposure to particulate matter and carbon monoxide in Hanoi, Vietnam, *Transportation Research Part D*, 13: 206–211.

Truc V.T.Q., and Kim Oanh N.T. 2007. Roadside BTEX and other gaseous air pollutants in relation to emission sources, *Atmospheric Environment*, 41: 7685–7697.

U.S. Environmental Protection Agency (USEPA). 1995. *Survey of Control Technologies for Low Concentration Organic Vapor Gas Streams*, Publication: EPA-456/R-95-003, USEPA, Research Triangle Park, NC.

U.S. Environmental Protection Agency (USEPA). 2009. Nitrogen dioxide. http://www.epa.gov/air/nitrogenoxides/health.html.

Wark K., Warner C.F., and Davis T.W. 1998. *Air Pollution: Its Origin and Control*, Addison Wesley Longman, Boston.

Yin Y.C., Aroua M.K., and Wan Daud W.M.A. 2007. Review of modification of activated carbon for enhancing contaminants uptakes from aqueous solutions, *Separation and Purification Technology*, 52: 403–415.

9

Application of Nanometallic Particles Coated on Porous Materials for VOC Destruction

Amornpon Changsuphan and Nguyen Thi Kim Oanh

CONTENTS

9.1 Introduction: VOC Emission and Control Technologies

Volatile organic compounds (VOCs) are defined as organic compounds that evaporate readily at room temperature, such as benzene, toluene, ethylbenzene, and xylenes (BTEX) and chlorinated solvents such as trichloroethylene (U.S. Environmental Protection Agency [USEPA], 2011). VOCs are easily absorbed and broadly metabolized in the human body (Agency for Toxic Substances and Diseases Registry [ATSDR], 2004). The U.S. Environmental Protection Agency (USEPA) has specified a number of VOCs as air toxics. In many urban areas of Asia, several anthropogenic sources, such as traffic, residential, and industry, release large amounts of VOCs into the atmosphere, which subsequently build up to high ambient levels. In the atmosphere with the presence of sunlight, VOCs participate in photochemical reactions with NO_x (nitrogen oxides) to form photochemical smog. High levels of fine particles and surface ozone present in the smog are health hazards. Ozone is a strong greenhouse gas and a phytotoxic pollutant that damages plants and affects agriculture and forestry, as highlighted in Chapter 6.

Awareness of the harmful effects of VOCs has brought about more stringent environmental regulations to minimize their levels in the air. Accordingly, control of VOC emissions at sources has become a major concern in many countries in Asia. For industrial sources, both cleaner production and end-of-the-pipe approaches are used (Chapters 10, 11, and 14). Among the end-of-the-pipe treatment technologies for VOCs, nanostructure materials, an emerging technology, have received a great deal of attention for their prospective applications. This chapter presents a brief overview of VOC emission sources, ambient levels, as well as the existing control technologies, with an emphasis on the application of nanostructure materials. A case study of using nanoscale zinc oxide (nZnO) particles coated on porous materials (adsorbents) for VOC removal is provided. Roles of adsorption and oxidation in removal of benzene, toluene, and ethylbenzene in the nZnO-coated adsorption systems were investigated in the presence of ultraviolet (UV) light and ozone.

9.1.1 Major Industrial Emission Sources of VOCs

The majority of anthropogenic VOC sources are associated with industrial processes that use solvents or other chemical agents as raw materials, additive agents, or catalysts (e.g., petrochemical processing, surface coating [paints], and printing [inks]). Petroleum industries have large VOC emissions from both the manufacturing process and fugitive sources. The USEPA (1995) estimated that in the United States annually, the petroleum industry released 369,058 tons (short ton) of VOCs, while the petrochemical industries producing organic chemicals and inorganic chemicals respectively released 201,888 and 52,091 tons of VOCs. Kihlman et al. (2009) indicated that the VOC emission from refinery industries in Sweden was about 509.6 kg/h or 4,500 tonnes (metric tons) per year.

High levels of VOCs are normally present in the airstreams from coating activities that apply decorative or protective materials in liquid or powder form to substrates. These coatings normally include general solvent-based paints, varnishes, lacquers, and water-based paints. VOCs emitted from the surface coating industry originate from solvents used in the paint formulations, solvents used to paint at the coating facility, or solvents used for cleanup (USEPA, 1981). Chang et al. (2002) investigated the VOC emissions from an automobile coating industry in Taiwan and found high levels in various airstreams, ranging from 480 ± 320 mg/m^3 in the primer coat booth to 1,960 ± 1,040 mg/m^3 in the topcoat oven.

The printing industry uses various production processes, which appear in the same basic sequence—imaging, prepress, printing, and postpress—and all release VOCs into the atmosphere. Emissions may result from the volatile adhesives used in binding and solvents found in some types of inks, coatings, and laminates used in the finishing process (USEPA, 2002a). High levels of VOCs may be observed around the industry. For example, the 4-month average levels of benzene in a large printing facility located in Bangkok, Thailand, reached 102 ± 28 ppm, while other BTEX levels were in the range 2–3 ppm (Thanacharoenchanaphas et al., 2007).

9.1.2 Ambient VOC Levels and Source Contributions

Studies showed that VOC concentrations in ambient air in industrial areas are normally higher than other areas. VOC levels measured at ambient sampling sites around a petrochemical complex and an oil refinery in Turkey, for example, were 4–20 times higher than those measured at a suburban site (Cetin et al., 2003). Likewise, the concentrations and daily variations of total VOCs measured at an industrial site in Ulsan, Korea, were higher than those at a downtown site (Na et al., 2001). In southeastern Michigan in the United States, higher VOC levels, both indoor and outdoor, were found in an industrial area as compared to urban and suburban areas, suggesting a health risk (Jia et al., 2008). In addition, higher VOC levels and more compounds were detected indoors as compared to outdoors, with the highest levels observed for benzene, toluene, p,m-xylenes, and carbon tetrachloride (medians from 1 to 2 μg/m^3). In Map Ta Put, the biggest industrial estate of refinery and petrochemical industries in Thailand (Pollution Control Department [PCD], 2011b), the levels of several VOCs (e.g., benzene, 1,3-butadiene and 1,2-dichloroethane) were reported to be significantly higher than the Thai VOC annual standards at several locations during 2006–2010 (PCD, 2011a).

Source apportionment for VOCs measured in ambient air is commonly performed using receptor modeling (see details of receptor modeling in Chapters 2 and 3), which also points to industry as an important contributing source in many places. Positive matrix factorization (PMF) modeling results showed that the anthropogenic sources were the major contributors to VOCs at two sites in the Los Angeles air basin. These included evaporative emission (31–34%), vehicle exhaust (22–24%), liquid/unburned gasoline (13–27%), and industry (15–17%), as compared to 1–3% from biogenic emission (Brown et al., 2007). A source apportionment study using chemical mass balance (CMB) model for 58 VOCs in the Pearl River Delta region, China (Liu et al., 2008), showed that vehicle exhaust was the largest source, contributing more than 50% to VOC pollution at three urban sites. Solvent usage was the biggest contributor to VOCs at an industrial site (33%), while the evaporation of solvents and coatings was the largest contributing source (31%) at a nonurban site located downwind of the industrial area.

9.1.3 Overview of VOC Control Techniques

Conventional end-of-the-pipe control technologies for VOC treatment include incineration, absorption, adsorption, condensation, and biological oxidation. Major features of these technologies are discussed next.

9.1.3.1 Incineration

Incineration is widely used for the destruction of a wide variety of VOCs. It is the most suitable for applications to treat a gas stream with a consistent flow rate and VOC concentration. There are three fundamental incineration technologies: thermal oxidation, thermal catalytic oxidation, and flare.

Thermal oxidation can be used to treat gas streams having VOCs either at a low concentration range (less than 10 ppm by volume) or at very high concentrations approaching 10,000 ppm. Supplementary fuels are required to maintain the combustion temperature of 500–1,000°C (USEPA, 2008). Thermal catalytic oxidation is widely used for destruction of VOC-laden gas streams (i.e., containing high VOC levels) and provides VOC destruction at lower temperatures (150–400°C). Both thermal and thermal catalytic oxidation technologies usually achieve the VOC destruction efficiency above 95% and often exceed 99% (Rafson, 1998). In principle, the inlet VOC concentrations should be kept below 25% of the lower explosive limit (Noll, 1999; USEPA, 2008). Flaring is an open combustion process used to burn combustible components (i.e., mostly hydrocarbons of waste gases from industrial operations) at high temperatures above 1,000°C (USEPA, 2002c). A flare system is suitable for treating gas streams containing high VOC levels and can achieve a destruction efficiency of more than 98% (USEPA, 2002c). All incineration technologies used for VOC treatment can produce air pollution (particulate matter and gases) if not properly designed and operated. A heat recovery system can be combined and usually installed in the thermal oxidation and thermal catalytic oxidation systems.

9.1.3.2 Absorption

Absorption is not widely used for VOC control. Instead, this technology is the most widely used to control inorganic gaseous emissions, especially for SO_2. A packed bed wet scrubber appears on the list as proven control technology for VOCs but is still not widely used because of the low control efficiency. Accordingly, only a few applications of scrubbers for removing VOCs were reported (Rafson, 1998).

9.1.3.3 Physical Adsorption

Physical adsorption is one of the most widely applied control technologies for VOCs. In the physical adsorption process, organics are selectively collected on the surface of a porous solid. Activated carbon (AC) is the most commonly used adsorbent for treatment of gas streams. Adsorption beds are generally used with two objectives in quite different situations: VOC recovery and VOC destruction. They are used for VOC recovery when a VOC-laden gas stream only contains one to three organic solvent compounds, and it is economical to recover and reuse these compounds. It is used for VOC destruction when the gas stream contains a large number of organic compounds at low concentrations, usually lower than 100 ppm, and it is necessary to preconcentrate these organics prior to thermal or catalytic oxidation. In an adsorption system, the temperature is kept below 55°C, and the relative humidity should remain below 50%. The adsorption removal efficiency is often high (i.e., in the range of 98% to 99%) (Noll, 1999).

Physical adsorption systems can be divided into nonregenerable (throw-away) and regenerable types. Nonregenerable systems are simple but find many applications in cases of low flow rate or low VOC concentrations (i.e., below 1 or 2 ppm), such as for odor control and indoor air cleaning. The adsorbent with adsorbed materials is discharged after use, which should be done in an environmentally friendly way. In a regenerable system, when the adsorbent is approaching saturation with organic vapor, a bed is isolated from the gas stream and desorbed. Low-pressure steam or hot gas is often used to remove the adsorbed VOCs. The concentrated stream from the desorption cycle is treated to recover the organic compounds. After desorption, the adsorption bed is returned to service (Zerbonia et al., 1995). Two or more beds are used in a system to ensure its continuous operation.

9.1.3.4 Condensation

Condensation is the technique that separates a volatile component from the remaining vapors through saturation and phase change mechanisms. Condensation systems can be divided into the nonrefrigerated and refrigerated types. In a nonrefrigerated condenser, the pressure is increased to condense the volatile vapors at a given temperature, so no refrigerants are required. It is widely used for raw material or product recovery in chemical processing industries (USEPA, 2002b). Refrigerated condensers use refrigerants (chilled water, brine solutions, ammonia, or chlorofluorocarbons) to cool and hence condense VOCs in a gas stream. Refrigerated condensers are widely used as pretreatment air pollution control devices for removing VOC contaminants from emission streams prior to other devices such as incinerators or adsorbers to reduce the overall cost of control systems (Noll, 1999). They can treat VOC-laden streams with a VOC concentration above 5,000 ppm and achieve a removal efficiency of above 90% (USEPA, 2002b).

Refrigerated condensers are divided into two types: contact and surface condensers. In a contact condenser, VOCs are in direct contact with the refrigerant, and the condensate is mixed with the refrigerant, so it cannot be readily recovered. In a surface condenser, VOCs are allowed to condense on the outside surface of tubes, while the refrigerant is circulated inside the tubes, and the separate condensate is recovered.

9.1.3.5 Biofiltration

The biofiltration technique uses living material to capture and biologically degrade pollutants. The system has many advantages, including low capital cost, low operating cost, low energy and maintenance requirements, high reliability, consistent destruction rates, and no hazardous combustion related by-products. The microbe consumes VOCs to form new cells; hence, no secondary disposal is required (Zerbonia et al., 1995). However, a biofiltration system also has some disadvantages, including a requirement for consistent loadings and the need for more space than other control techniques. There are three types of biofiltration, those that use a biofilter, biotrickling filter, and bioscrubber (Devinny et al., 1999). Further details on the application of biofiltration for VOC control is presented in Chapter 11.

9.1.4 VOC Control Using Metallic Particles and Nanoscale Metallic Particles

9.1.4.1 Metallic Particles

Metallic particles of concern in air pollution control are the particles of oxides of transition metals, such as Ti, Co, Fe, Cu, Mo, Rh, Zn, and so on. These particles normally are 100 nm or larger. Several studies investigated VOC removal using transition metal oxides in combination with other materials or chemicals, which showed that removal efficiency can be improved, mainly through catalytic oxidation.

Titanium dioxide (TiO_2) particles are commonly applied for VOC control in the combination of O_3 and UV or visible light. In the presence of ozone and UV light, the degradation rate of hexane could reach up to 50% and toluene up to 92.5% (Pengyi et al., 2003). With visible light, TiO_2 catalyst was also shown to have high activity for VOC degradation (Wang et al., 2007). In addition, doping of lanthanum improved the photocatalytic activities of TiO_2 under a low doping ratio (2.5 wt%) with both the decomposition rate of benzene and the formation rates of CO_2 and CO over doped TiO_2 film larger than on dopant-free TiO_2 film (Zhang et al., 2006).

In their study, Law et al. (2003) showed that the catalytic oxidation of VOCs in contaminated air on Co(II)-containing zeolites (NaX) worked best in the presence of reactive oxygen species (released from an ionizer). Another study (Brazlauskas and Kitrys, 2008) used CuO/zeolite (NaA, CaA, NaX, and CaX) as adsorbent-catalysts for oxidation of alcohol in air and reported a high oxidation rate at a temperature of 100–350°C.

9.1.4.2 Nanoscale Metallic Particles

Recent research has focused on applications of nanoscale metallic particles for VOC removal. These nanoparticles, having a diameter below 100 nm, can enhance or exhibit several exceptional properties (Rohit, 2008). Several types of nanoscale particles have been investigated for air pollution control, including $nTiO_2$, nZnO, $nCeO_2$, nFe_2O_3, and nFe_3O_4 (the prefix n here indicates the nanoscale), with the pollutants being treated including a wide range of common VOCs and dioxins in the air. Improvement in the treatment performance is influenced by their size, surface area structure, adsorption capacity, removal efficiency, and interparticle interactions (Li et al., 2006).

In photocatalytic oxidation of toluene under sunlight, $nTiO_2$ is effective, while with ozone added the destruction rate of n-octane was shown to increase in a laboratory-scale photoreactor (Wang et al., 2007). Menesi et al. (2008) studied photooxidation of toluene and ethanol vapors over $nTiO_2$-clay composites in recycling flow reactors and found a synergic effect at the solid-gas interface between the vapors and $nTiO_2$-clay composites.

Other types of nanoparticles are also investigated for VOC control purpose. Hernandez-Alonso et al. (2004) used cerium (IV) oxide (CeO_2) nanoparticles for toluene removal and reported the complete oxidation of the compound under UV light with a wavelength less than 400 nm. Kastner et al. (2008) showed that synthetic α-Fe_2O_3 and Fe_3O_4 nanoparticles had reaction rates 80 and 200 times higher, respectively, when compared with wood fly ash or AC in the catalytic ozonation of propanal at room temperature (23–25°C). Fe_xO_y nanocomposite was used in catalytic oxidation with ozone for treatment of polychlorinated dibenzo-p-dioxins (PCDDs) and polychlorinated dibenzofurans (PCDFs) in contaminated air at temperature of 120–180°C, which showed destruction and removal efficiencies of 94% and 91%, respectively (Wang et al., 2008).

Several researchers applied nZnO for removal of air contaminants. Chang et al. (2008) studied the oxidation of methanol and production of hydrogen over Au/Al_2O_3 and nZnO-promoted Au/Al_2O_3 ($Au/nZnO/Al_2O_3$) catalysts and showed that the addition of nZnO to Au/Al_2O_3 can increase the catalytic activity for a partial oxidation to produce hydrogen. Chen and Tang (2007) applied a granular $nZnO/Al_2O_3$ catalyst for trichloroethylene removal from a contaminated airstream and found that the removal efficiency increased to above 90% when oxygen in the feed gas was increased to 5% (at the optimum temperature and space velocity). nZnO can be used in combination with other nanoparticles; for example, Ismail (2008) successfully developed the ZnO/TiO_2-SiO_2 photocatalysts to oxidize trichloroethylene to HCl and CO_2 with the presence of UV light at a wavelength of 365 nm.

9.2 Experiment Setup for VOC Destruction on nZnO-Coated Adsorbents

The porous adsorbents including AC and zeolite were coated with nZnO and investigated for VOC removal in lab-scale experiments. The VOC removal efficiency of coated adsorbents was compared to that of virgin adsorbents under different experimental conditions (i.e., with UV or O_3, at different bed temperatures and gas hour space velocity [GHSV]). Roles of adsorption and catalytic oxidation mechanisms in VOC destruction were investigated.

9.2.1 Fundamental of VOC Destruction Mechanisms on Coated Adsorbents

Studies show that $nTiO_2$ and nZnO are able to adsorb and reflect UV light, hence promoting photodegradation (Rohit, 2008). Integration of UV light or ozone into TiO_2-coated adsorption systems can enhance the oxidation/decomposition of VOCs. The TiO_2 coated on AC in combination with UV light can act as the electron carrier, so it can prevent the recombination of electron-hole species, which leads to higher VOC oxidation (Li et al., 2006). A combination of UV plus O_3 on TiO_2-coated zeolite can improve the VOC removal efficiency due to the synergic effects among adsorption, ozonation, and catalytic reaction (Huang et al., 2009; Pengyi et al., 2003). However, when combined with UV light, the ozone molecules, if used in excessive amount, were shown to scavenge the hydroxyl radical generated from $nTiO_2$, hence reducing trichloroethylene oxidation (Shen and Ku, 2002).

Coating with nZnO particles is expected to improve the VOC adsorption capacity thanks to the Bronsted base behavior of nZnO. Chang et al. (2008) investigated this behavior of nZnO in cleavage of the O–H bond of methanol and oxidization of methanol in the presence of oxygen. However, nZnO can also easily adsorb and react with CO_2 and intermediate products and change their form to zinc carbonate and soluble carboxyl zinc, thus deactivating the coated absorbent (Yansheng et al., 2009).

The nZnO particles are considered a highly appealing alternative to $nTiO_2$ due to their similar chemicophysical characteristics and photodegradation mechanisms (Barreca et al., 2009). It is expected that application of nZnO-coated absorbents when integrating with UV or ozone also improves VOC removal efficiency in ways similar to that for $nTiO_2$.

9.2.2 Coating Adsorbents with nZnO

Two types of commercial-grade adsorbents (i.e., 13X zeolite and 12 × 40 mesh granular AC) were used. nZnO with an average size of about 63 nm and specific surface area of 15–25 m^2/g was obtained from Sigma Aldrich, USA. Coating nZnO on adsorbents was done following the method presented by Kalpana et al. (2006). nZnO was mixed with the respective adsorbent at a ratio of Zn to adsorbent weight of 0.5:1 in deionized (DI) water and stirred continuously for 3 h. The mixture was filtered and dried in the oven at 250°C for 1 h and slowly cooled to room temperature. The steps used in the coating processes are shown in Figure 9.1. For the coated zeolite adsorbent, the material was further sieved through a 14 × 80 mesh to obtain the desirable particle sizes.

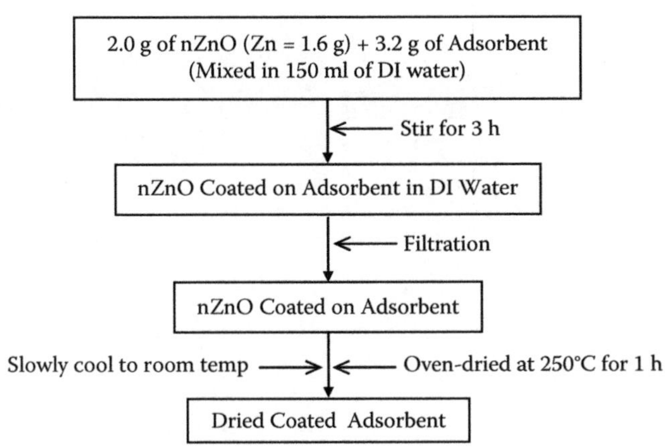

FIGURE 9.1
The procedure for coating of nZnO particles on adsorbents.

Morphologies of the virgin and coated adsorbents were characterized by scanning electron microscopy (SEM). The obtained SEM images in this case study were similar to those reported in previous studies for zeolite (Anandan et al., 2006) and AC (Sobana and Swaminathan, 2007). Aggregates of nZnO particles were attached to the surface of coated zeolite (Figure 9.2), whereas on the surface of virgin zeolite, individual particles of spherical shape were clearly seen. Similarly, SEM images of the coated AC also showed nZnO particles agglomerating on the adsorbent external surface, while on the surface of virgin AC a clear surface was shown.

Compositions of coated adsorbents were analyzed by a wavelength dispersive x-ray (WDX) technique, and the results showed that the coated zeolite contained Zn at 38.9% by weight, that is, higher than expected (33% assuming all nZnO was taken up), while the coated AC had lower Zn (28.3%). Higher Zn content in the coated zeolite may be a result of the exchange of the Zn ion with other elements present in the virgin zeolite. These elements are subsequently lost into the aqueous suspension; for example, a loss of 20% of sodium ions from zeolite due to the replacement by Zn ions was observed by Wark et al. (1997) while preparing nZnO- (and nCdO-) coated zeolite.

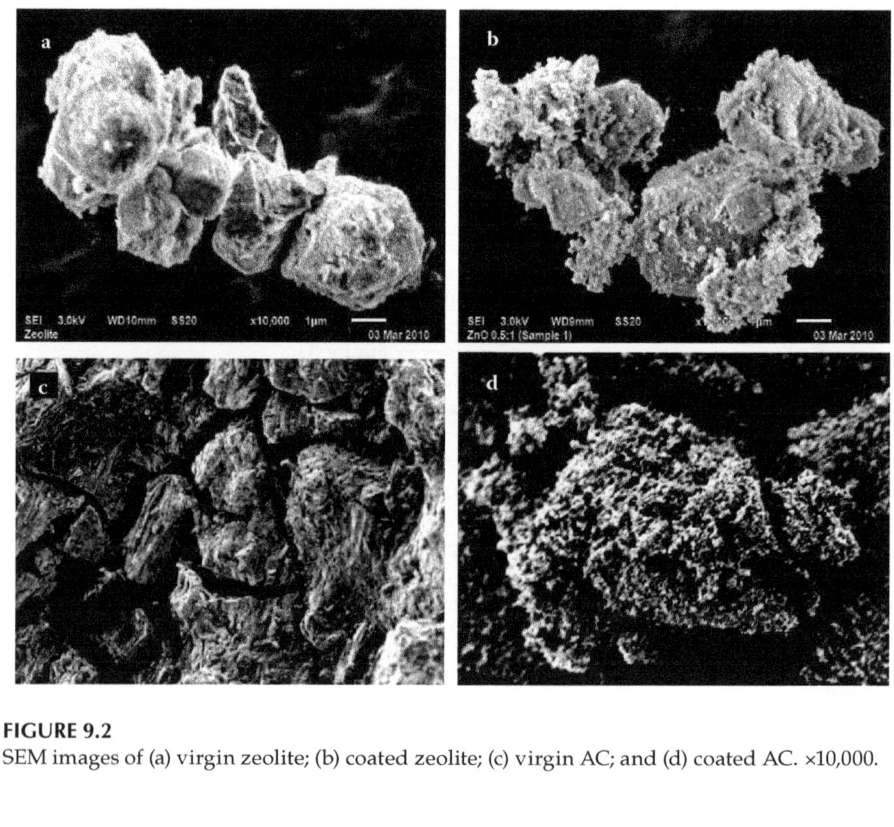

FIGURE 9.2
SEM images of (a) virgin zeolite; (b) coated zeolite; (c) virgin AC; and (d) coated AC. ×10,000.

9.2.3 Preparation of Synthetic Airstreams Containing VOCs

Two types of synthetic airstreams were prepared: (1) airstreams containing benzene alone (5 ppm) and (2) airstreams containing a mixture of benzene, toluene, and ethylbenzene (BTE), each at 5 ppm. The first type was used in benzene adsorption and catalytic oxidation at ambient temperature, while the second type was used in the study of VOC catalytic oxidation with UV light or ozone at different operational conditions. To prepare the benzene-containing airstream, a pump was used to purge clean air (with BTE levels below 3 ppb) passing through a heated impinger (150°C) containing a known amount of benzene into a Tedlar bag (25 L). Similarly, for the VOC mixture, the BTE liquids were placed in a heated impinger, and the vapors were diluted with clean air to get 25 L in each Tedlar bag.

9.2.4 Lab-Scale Adsorption System

The system had three sections and is schematically shown in Figure 9.3. The inlet section consisted of a gas supply system, a mixing tank, and an ozone generator. When ozone was injected, it was continuously generated and diluted with zero air to the desired concentrations before introducing into the system. A sampling port was provided at the inlet of the reactor. The reactor section was composed of the adsorbent beds, a temperature controller, a thermocouple, UV lamps, and bed-length selection valves. The adsorbent was packed in a Pyrex tube and separated into three sections connected in a series, each section with a maximum depth of 250 mm. The heater could heat the column up to 400°C. Three UV lamps (15 W, UV-C with a central

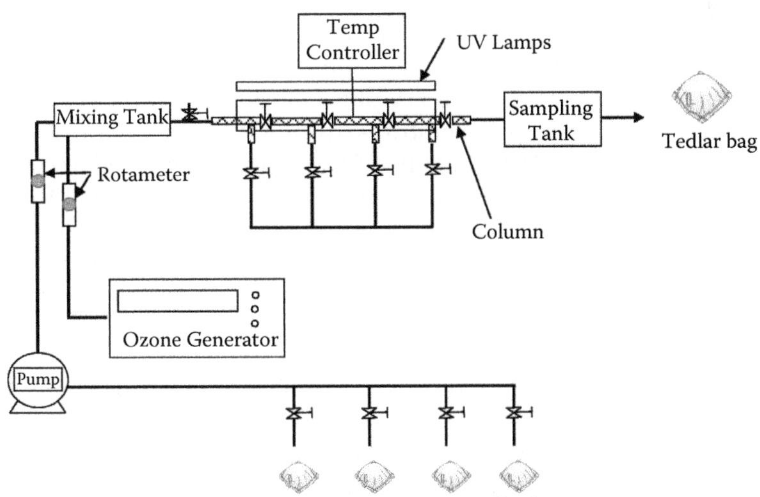

FIGURE 9.3
Scheme of the experimental setup.

wavelength of 254 nm and intensity of 58 W/m^2) were placed above the column (at a distance of around 5 cm) in one row to illuminate and enhance photooxidation. The outlet section had a sampling tank that was connected to a Tedlar bag (3 or 25 L each) to collect the effluent integrated over a period of time for subsequent VOC analysis. Except for the Pyrex tube, all other connection parts of the system were made of stainless steel.

9.2.5 Experimental Conditions

9.2.5.1 Benzene Adsorption and Catalytic Oxidation at Ambient Temperature

Experiments for benzene adsorption and catalytic oxidation at ambient temperature (~30°C) were conducted with both virgin and coated adsorbents. A total of five experiments were conducted with AC and five with zeolite (one virgin, one coated, and one each with UV, O_3, and UV plus O_3, respectively). Each experiment was repeated in at least three runs (replicates). The introduced ozone levels were in the range of 7–7.3 ppm. The benzene-containing airstream (stored in Tedlar bags) was passed through the column by a pump at a constant flow rate of 0.4 L/min for zeolite experiments and at 1.0 L/min for AC experiments.

9.2.5.2 VOC Mixture Catalytic Oxidation at Different Temperatures

The experiments were conducted with zeolite beds (virgin and coated) with UV light or ozone introduced at different temperatures and GHSVs. Catalytic VOC oxidation was studied after the bed media were saturated with the BTE mixture (purging the BTE-containing airstream through the system for 28 min). Five experiments were conducted at ambient temperature (virgin, coated, and each with UV, O_3, and UV plus O_3). Other experiments were conducted at five elevated bed temperatures (from 100°C to 400°C) and with four different GHSVs (from 818 to 6,136 h^{-1}). Ozone was introduced at concentrations between 14 and 16 ppm. Each experiment was replicated at least three times. Tedlar bags containing BTE were connected in a series, and a pump was used to pass the BTE-containing stream continuously into the column.

9.2.6 Sampling and Analysis

The inlet samples were collected directly from each Tedlar bag before introduction into the adsorption system. Each outlet sample was collected in a Tedlar bag over a time period (averaging time) of 3 to 10 min, and the bags were continuously replaced (with an interval of 3 to 10 min) to provide almost continuous VOC measurements. BTE were analyzed by a gas chromatograph (GC; HP 5890 series 11 Plus) fitted with a flame ionization detector (FID). A gas sample was collected (from Tedlar bags) and directly injected into the

GC/FID using a gastight syringe for the VOC analyses. For VOC quantification, two sets of standard curves were prepared covering two ranges: ppb levels (outlet samples) and ppm levels (inlet and outlet samples). Each curve was constructed using at least four VOC levels ($R^2 > 0.99$). The minimum detection limit of the analytical method was 3 ppb for each BTE compound.

Ozone in the inlet of the system was analyzed online by an O_3 analyzer (49C, Thermo Environmental Instruments Inc., USA). For quality assurance/quality control (QA/QC), the results were also checked with those produced by the classical method of KI absorption-colorimetry, which showed a good agreement.

9.3 Results and Discussion

9.3.1 Benzene Adsorption and Catalytic Oxidation at Ambient Temperature

Benzene adsorption was examined before the breakthrough occurred. The investigation of catalytic oxidation of benzene was done with UV light or ozone introduced into the adsorption system both before and after the breakthrough occurred.

9.3.1.1 Benzene Adsorption over Virgin and Coated Adsorbents

The benzene average removal efficiency (RE) for AC experiments were almost the same for all conditions presented in Table 9.1, that is, 98–99%. For zeolite experiments, the coated bed had higher RE (~98%) compared to the virgin bed (~94%). Other experimental conditions on the coated zeolite beds showed almost the same RE (98–99%), which were also similar to the results of the AC experiments.

TABLE 9.1

Benzene Removal Efficiency (%) on Virgin and Coated Zeolite and AC at Ambient Temperature

Zeolite Experiment Time (before Breakthrough): 21 min, GHSV = 818 h^{-1}		AC Experiment Time (before Breakthrough): 138 min, GHSV = 18,483 h^{-1}	
Experimental Conditions	**Average RE**	**Experimental Conditions**	**Average RE**
Virgin zeolite	94.2 ± 2.4	Virgin AC	99.5 ± 0.2
Coated zeolite	97.9 ± 0.9	Coated AC	98.3 ± 0.1
Coated zeolite + UV	98.4 ± 0.4	Coated AC + UV	99.0 ± 0.1
Coated zeolite + O_3	98.2 ± 1.0	Coated AC + O_3	99.0 ± 0.2
Coated zeolite + UV + O_3	98.9 ± 0.2	Coated AC + UV+ O_3	98.2 ± 0.3

Coating with nZnO particles is expected to improve the capacity of both zeolite and AC owing to the Bronsted base behavior of ZnO, as mentioned in Section 9.2.1. However, the improvement was shown only for the zeolite case, which suggests that there may be other causes involved. The coating may increase the zeolite surface area due to the presence of nZnO particles on the adsorbent surface but block some pores of AC. This is because the nZnO particles (~63 nm) were larger than the pore size of zeolite used in this study; hence, the particles were likely to be attached on the surface. For the AC used in this study, the pore sizes were from 1 to above 500 nm, so some of them may have been partially blocked by nZnO particles, hence reducing its physical adsorption. In the case of coated AC, the blockage of active pores may not be fully compensated by the Bronsted base property of nZnO, so the RE was slightly reduced.

It is also expected that introduction of UV light or O_3 can enhance benzene oxidation at the ambient temperature, but the effects were not clearly shown (Table 9.1). Note that the results are reported for the beds before breakthrough occurred, and it is likely that the physical adsorption predominantly controlled the benzene removal in these experiments.

9.3.1.2 Benzene Removal after Breakthrough

Breakthrough curves of zeolite and AC experiments (Figure 9.4) showed that the zeolite beds (28 g adsorbent in 750-mm bed depth, inlet flow rate of 0.4 L/min) had a short adsorption operation time (<40 min), while that for AC beds (8 g in 83-mm bed depth, inlet flow rate of 1.0 L/min) lasted much longer (~140 min). The benzene adsorption capacity of the zeolite was estimated at around 10^{-2} g/kg adsorbent, which is about 150–200 times lower than that of AC (~1.5–2.0 g/kg adsorbent).

The coated zeolite beds (with and without UV light or O_3) clearly showed a reduction in the adsorption capacity when the breakthrough occurred earlier (at 21 min) as compared to the virgin zeolite bed (39 min). However, before the breakthrough occurred, the coated zeolite beds had consistently lower outlet concentrations than the virgin zeolite. Competitive adsorption of CO_2 and intermediate products on nZnO particles (Yansheng et al., 2009) may be a reason for the earlier breakthrough of the coated beds. After the breakthrough, the C_{out}/C_{in} ratio of coated zeolite beds with O_3 and UV plus O_3 stabilized at 0.3–0.4 (RE of 60–70%). This suggests that after the breakthrough point, when the beds were saturated and adsorption virtually stopped, the benzene oxidation continued enhancing RE in the presence of O_3 and UV plus O_3. The coated bed with and without UV appeared quite similar (Figure 9.4[a]) and showed that at ambient temperature the UV effect on benzene oxidation was not pronounced.

The AC experiments showed low outlet benzene levels before breakthrough points in all cases. The virgin AC bed had the longest operation time (180 min), while that for the coated bed varied, with a minimum of about 140

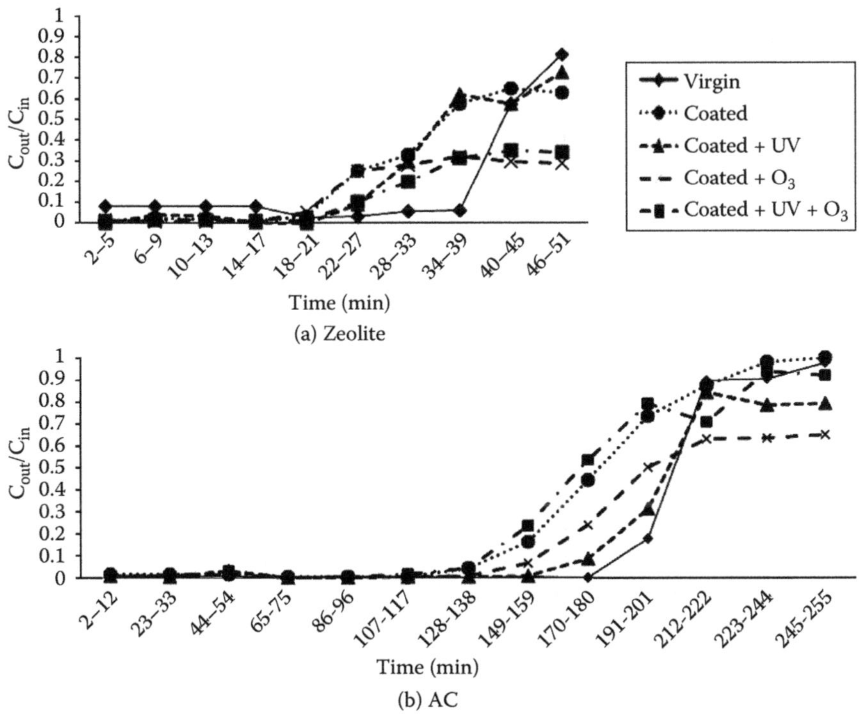

FIGURE 9.4
Breakthrough curves of (a) zeolite experiment and (b) AC experiment (C_{out}/C_{in} = outlet-to-inlet concentration ratio).

min. The blockage of AC pores by nZnO particles and competitive adsorption of intermediate and final products on coated beds, as mentioned, may be main reasons the operation time of the coated AC beds was shortened. At the end of the experiments (~255 min), the highest outlet benzene levels were observed for both the coated AC and virgin AC beds. The coated AC bed with UV light behaved quite similarly to the virgin AC bed was a long operation time (~160 min), and the outlet concentration increased sharply after the breakthrough. The experiments with the coated AC bed, coated plus O_3, and coated plus UV plus O_3 showed a reduction in the operation time, but the outlet concentrations increased more slowly than for the other two cases. Thus, photodegradation and oxidation under the presence of UV light or O_3 continued to destroy benzene after the adsorbents were saturated.

Both zeolite and AC experiments showed that the coated beds with ozone had stabilized C_{out}/C_{in} ratios lower than for other cases. In fact, the experiments with UV plus O_3 did not produce the highest RE as was expected. This may be explained by the presence of excessive ozone that scavenged hydroxyl radicals (generated by ZnO under UV) (Shen and Ku, 2002). Note that ozone was detected at the outlet of the adsorption system (at part-per-billion levels), indicating the presence of excessive ozone in the system. However, the differences in behavior of the systems when UV, O_3, or UV plus O_3 were introduced still need further research for a better explanation.

9.3.2 Catalytic Oxidation of a VOC Mixture on Zeolite

The catalytic oxidation study was conducted after the zeolite beds were saturated with the BTE mixture by purging the BTE-containing airstream through the beds for 28 min. This was about the time for coated zeolite beds to have breakthrough when only benzene alone at 5 ppm was used (Figure 9.4); hence, it should be well beyond the breakthrough point for the BTE mixtures of 5 ppm each. UV light or ozone was introduced to study the VOC oxidation as physical adsorption was largely eliminated after this point.

9.3.2.1 VOC Oxidation at Ambient Temperature Compared to 100°C Bed Temperature

The results of BTE conversion (oxidation) efficiency under various experimental conditions are presented in Table 9.2. At ambient temperature, the conversion efficiency of BTE was the highest for the coated bed with ozone, which was similar to the case presented for benzene. The second highest was for coated beds with UV plus O_3. Other cases (virgin zeolite, coated zeolite, and coated zeolite with UV) had low average conversion efficiencies, generally within 1–2%.

Incorporation of O_3 or UV plus O_3 can improve the BTE conversion efficiency on coated zeolite beds by 10–20%. The improvement in efficiency was generally lower than the case of benzene alone (Figure 9.4), which may be attributed to a shorter residence time in the BTE experiments (GHSV = 6,137 h^{-1}) than in the benzene experiments (GHSV = 818 h^{-1}).

At 100°C bed temperature, the BTE conversion efficiency on the coated beds was improved significantly. The effects of UV alone (photodegradation)

TABLE 9.2

Averaged Conversion Efficiency (±1 SD) of BTE on Zeolite Beds (%) after Breakthrough

Experimental Conditions	Ambient Temperature Time: 120 min, GHSV = 6,136 h^{-1}			100 °C Bed Temperature Time: 120 min, GHSV = 6,136 h^{-1}		
	B	T	E	B	T	E
Virgin zeolite	NA	0.4 ± 0.6	1.1 ± 1.9	3.9 ± 1.0	NA	NA
Coated zeolite	2.1 ± 1.8	2.4 ± 2.6	2.2 ± 2.0	9.2 ± 5.4	5.1 ± 3.5	4.5 ± 2.1
Coated zeolite + UV	0.8 ± 1.2	0.0 ± 0.0	0.8 ± 1.7	16.2 ± 2.1	9.6 ± 2.3	10.9 ± 3.7
Coated zeolite + O_3	23.2 ± 5.7	12.1 ± 2.7	10.4 ± 4.7	21.0 ± 3.2	13.9 ± 2.3	17.0 ± 1.7
Coated zeolite + UV + O_3	18.4 ± 8.0	8.3 ± 5.1	6.4 ± 2.6	24.9 ± 6.8	17.5 ± 8.6	31.0 ± 9.2

Note: NA, not analyzed

and UV plus O_3 on coated beds were clearly seen at this high temperature; that is, there was an improvement in average conversion efficiency by about 8–18%. The highest improvement was shown for the case of UV plus O_3 for all three compounds. However, the effects of the temperature increase in the experiment with ozone alone were not so clearly shown; that is, there was almost the same range for benzene and toluene, and an increase by 7% for ethylbenzene. The effects of UV plus O_3 were less than additive for benzene and toluene and more than additive for ethylbenzene. Note that some excessive ozone was detected in the outlet gas stream of the experiments at 100°C temperature (~60 ppb) and at ambient temperature (~120 ppb), so the scavenging effects on hydroxyl radicals may be expected in both cases. The ozone scavenging effect appeared to be less pronounced in the experiments at 100°C; hence, the UV+O_3 combination could produce the highest RE for all three pollutants (Table 9.2).

Thus, overall the nZnO, UV light, and O_3 enhanced the catalytic oxidation of BTE at higher temperature, which is similar to results of other studies, such as those of Brazlauskas and Kitrys (2008), Wang et al. (2008), Chang et al. (2008), and Chen and Tang (2007). Further increase in bed temperature was done to investigate this effect.

9.3.2.2 Effects of Bed Temperature

The performance of coated zeolite beds with UV plus O_3 at a GHSV of 818 h^{-1} was examined at five different bed temperatures, from 100°C to 400°C. Higher oxidation (conversion) efficiency was observed at higher temperature for all three compounds, with the maximum average conversion efficiency of 70.5% for benzene, 91.4% for toluene, and 99.8% for ethylbenzene (Table 9.3). The most significant improvement was for the temperature change from 100°C to 200°C, by about 20% for benzene and toluene and about 40% for ethylbenzene. A further increase in temperature (i.e., from 200°C to 400°C) could improve the efficiency by 10% for benzene, 25% for toluene, and only 5% for ethylbenzene. Higher temperature enhanced the complete oxidation of VOCs, but the energy consumption was also considerable. The minimum temperature to achieve 90% conversion efficiency for ethylbenzene was 200°C and for toluene was 300°C. For benzene, the

TABLE 9.3

BTE Conversion Efficiency at Different Bed Temperatures (%)

Bed Temperature (°C)	B	T	E
100	39.5 ± 8.8	44.8 ± 9.5	55.5 ± 1.1
200	60.7 ± 3.9	66.3 ± 5.4	93.6 ± 3.2
250	63.1 ± 2.7	74.6 ± 3.8	97.4 ± 0.7
300	65.6 ± 4.2	90.1 ± 5.5	99.5 ± 0.5
400	70.5 ± 2.4	91.4 ± 5.8	99.8 ± 0.1

TABLE 9.4

Conversion (±1 SD) on Coated Zeolite Beds at Different GHSV at 100°C (%)

GHSV (h^{-1})	B	T	E
6,136	24.9 ± 6.8	17.5 ± 8.6	31.0 ± 9.2
2,453	36.2 ± 3.6	37.7 ± 4.0	47.1 ± 3.6
1,227	36.0 ± 7.8	47.4 ± 0.4	51.8 ± 8.3
818	39.5 ± 8.8	44.8 ± 9.5	55.5 ± 1.1

maximum conversion was only 70.5% at the highest temperature of the experimental range (400°C).

High conversion efficiency of BTE in the coated beds in the presence of UV light or ozone indicated an advantage of the control technology over the conventional adsorption. The oxidation process effectively eliminated the need for desorption (bed regeneration) to ensure continuous operation of the system. However, other factors, including cost, should be considered when scaling up the device.

9.3.2.3 Effects of Gas Hour Space Velocity at 100°C Bed Temperature

The experiments were conducted on the coated zeolite beds with UV plus O_3 at 100°C at four different GHSVs (by changing the flow rate and bed depth), and the results are shown in Table 9.4. The conversion efficiency of all three compounds increased by 15–27% when GHSV reduced from 6136 h^{-1} to 818 h^{-1}, which should be attributed to the corresponding increase in the residence time to better complete the VOC oxidation reactions.

9.4 Summary

High levels of VOCs are normally observed in industrial areas; these are mainly linked to various industrial activities. Due to VOC toxicity and other harmful effects, the emission is being increasingly regulated. Various end-of-the-pipe VOC control techniques are available. Recent research has focused on applications of nanoscale metallic particles, which have several exceptional properties to enhance VOC removal.

This case study investigated the effects of nZnO particles coated on AC and zeolite adsorbents under various experimental conditions for removal of benzene and a BTE mixture. The surface of coated adsorbents was examined by SEM, and the compositions were characterized by WDX, which showed that the nZnO particles were properly incorporated into the adsorbents. The experiments were conducted to study the adsorption and oxidation of the

VOCs in airstreams at different operational conditions (i.e., with UV and ozone introduced and at varying bed temperatures and GHSVs).

For the benzene experiments, at ambient temperature, before the breakthrough point when the physical adsorption was the main removal mechanism, the effects of UV light or ozone on the RE were marginal on both virgin and coated beds of zeolite and AC. Only the coated zeolite bed was shown to enhance the physical adsorption of benzene, which was most likely attributed to the increase in the surface area by the coated nZnO particles. After breakthrough occurred, considerable improvements in VOC removal in the presence of O_3 alone and UV plus O_3 were observed, which confirmed the effects of oxidation. The coated beds with introduced ozone appeared to yield the highest benzene removal efficiency in the experiments with both (AC and zeolite) coated adsorbents at ambient temperature.

The experiments with BTE on the zeolite beds were performed after the beds were saturated with VOCs to investigate the oxidation efficiency. At ambient temperature, the highest efficiency was observed for the coated bed with O_3 alone, while at 100°C bed temperature the coated bed with UV plus O_3 yielded the highest conversion efficiency. The most significant improvements in the conversion efficiency of all three compounds were observed for the temperature increase from 100°C to 200°C. The minimum temperature to achieve 90% conversion efficiency under the experimental conditions was 200°C for ethylbenzene and 300°C for toluene, while for benzene the maximum efficiency was only 70.5% at 400°C. The decrease in the GHSV was shown to increase the conversion efficiency, which was attributed to the corresponding increase in residence time to better complete oxidation reactions.

The effective VOC oxidation on nZnO-coated beds in various experiment conditions showed a potential for the application of the technology in a continuous operation mode as the need for desorption/bed regeneration could be largely eliminated. However, other factors, including the cost of the installation and operation of such a system, should also be considered.

Further study should be focused on improvement of the efficiency of catalytic oxidation by varying the Zn-to-zeolite ratios. Removal of other VOCs should also be investigated. The intermediate products should be analyzed to examine their competitive adsorption effects. The complexity of VOC removal mechanisms when integrating UV, O_3, and UV plus O_3 should be investigated in depth to find the optimum design and operational conditions to achieve maximum VOC removal.

References

Agency for Toxic Substances and Diseases Registry (ATSDR). 2004. *Interaction Profile for: Benzene, Toluene, Ethylbenzene, and Xylene (BTEX)*, U.S. Department of Health and Human Service, Public Health Service, Agency for Toxic Substances and Diseases Registry, Atlanta, GA.

Anandan S., Vinu A., Venkatachalama N., Arabindoo B., and Murugesan V. 2006. Photocatalytic activity of ZnO impregnated Hb and mechanical mix of ZnO/Hb in the degradation of monocrotophos in aqueous solution, *Journal of Molecular Catalysis A: Chemical*, 256: 312–320.

Barreca D., Gasparotto A., Maccato C., Tondello E., Štangarc U.L., and Suhas R.P. 2009. Photoinduced superhydrophilicity and photocatalytic properties of ZnO nanoplatelets, *Surface and Coatings Technology*, 203(14): 2041–2045.

Brazlauskas M., and Kitrys S. 2008. Synthesis and properties of CuO/Zeolite sandwich type adsorbent-catalysts, *Chinese Journal of Catalysis*, 29(1): 25–30.

Brown S.G., Frankel A., and Hafner H.R. 2007. Source apportionment of VOCs in the Los Angeles area using positive matrix factorization, *Atmospheric Environment*, 41: 227–237.

Cetin E., Odabasi M., and Seyfioglu R. 2003. Ambient volatile organic compound (VOC) concentrations around a petrochemical complex and a petroleum refinery, *Science of the Total Environment*, 312(1–3): 103–112.

Chang C., Lee C., Wu Y., and Jeng F. 2002. Assessment of the strategies for reducing volatile organic compound emissions in the automotive industry in Taiwan, *Resources, Conservation and Recycling*, 34(2): 117–128.

Chang F., Lai S., and Roselin L.S. 2008. Hydrogen production by partial oxidation of methanol over ZnO-promoted Au/Al2O3 catalysts, *Journal of Molecular Catalysis A: Chemical*, 282(1–2): 129–135.

Chen J., and Tang C. 2007. Preparation and application of granular ZnO/Al_2O_3 catalyst for the removal of hazardous trichloroethylene, *Journal of Hazardous Materials*, 142(1–2): 88–96.

Devinny J.S., Deshusses M.A., and Webster T.S. 1999. *Biofiltration for Air Pollution Control* (3rd ed.), Lewis, Boca Raton, FL.

Hernandez-Alonso M.D., Hungria A.B., Martinez-Arias, A., Fernandez-Garcia M., Coronado J.M., Conesa J.C., and Soria J. 2004. EPR study of the photoassisted formation of radicals on CeO_2 nanoparticles employed for toluene photooxidation, *Applied Catalysis B: Environmental*, 50: 167–175.

Huang X., Yuan J., Shi J., and Shangguan W. 2009. Ozone assisted photocatalytic oxidation of gaseous acetaldehyde on TiO_2/H-ZSM-5 catalysts, *Journal of Hazardous Materials*, 171: 827–832.

Ismail, A.A. 2008. Single-step synthesis of a highly active photocatalyst for oxidation of trichloroethylene, *Applied Catalysis B: Environmental*, 85: 33–39.

Jia C., Batterman S., and Godwin C. 2008. VOCs in industrial, urban and suburban neighborhoods, Part 1: Indoor and outdoor concentrations, variation, and risk drivers, *Atmospheric Environment*, 42: 2083–2100.

Kalpana D., Omkumar K.S., Kumar S.S., and Renganathan N.G. 2006. A novel high power symmetric ZnO/carbon aerogel composite electrode for electrochemical supercapacitor, *Electrochimica Acta*, 52(3): 1309–1315.

Kastner J.R., Ganagavaram R., Kolar P., Teja A., and Xu C. 2008. Catalytic ozonation of propanal using wood fly ash and metal oxide nanoparticle impregnated carbon, *Environmental Science and Technology*, 42(2): 556–562.

Kihlman M., Mellqvist J., Samuelsson J., Tang L., and Chen D. 2009. Monitoring of VOC emissions from refineries in Sweden using the SOF method. http://www.fluxsense.se/reports/paper%202%20final%20lic.pdf, accessed June 2009.

Law T.S.C., Chao C., Chan G.Y.W., and Law A.K.Y. 2003. Confined catalytic oxidation of volatile organic compounds by transition metal containing zeolites and ionizer, *Atmospheric Environment*, 37(38): 5433–5437.

Li L., Fan M., Brown R.C., Van Leeuwen J.H., Wang J., Wang W., Song Y., and Zhang P. 2006. Synthesis, properties, and environmental applications of nanoscale iron-based materials: a review, *Critical Reviews in Environmental Science and Technology*, 36(5): 405–431.

Liu Y., Shao M., Lu S., Chang C., Wang J., and Fu L. 2008. Source apportionment of ambient volatile organic compounds in the Pearl River Delta, China: Part II, *Atmospheric Environment*, 42: 6261–6274.

Menesi J., Korosi L., Bazso E., Zollmer V., Richard A., and Dekany I. 2008. Photocatalytic oxidation of organic pollutants on titania-clay composites, *Chemosphere*, 70: 538–542.

Na K., Kim Y.P., Moon K., Moon I., and Fung K. 2001. Concentrations of volatile organic compounds in an industrial area of Korea, *Atmospheric Environment*, 35(15): 2747–2756.

Noll K.E. 1999. *Fundamentals of Air Quality Systems: Design of Air Pollution Control Devices*, American Academy of Environmental Engineers, Annapolis, MD.

Pollution Control Department (PCD). 2011a. Ambient air quality. http://www.pcd.go.th/Info_serv/pol_maptapoot_airVOCs.html, accessed June 2011.

Pollution Control Department (PCD). 2011b. The implementation of pollution in Map Ta Put (in Thai). http://www.pcd.go.th/info_serv/pol_maptapoot_water.html, accessed August 2011.

Pengyi Z., Fuyan L., Gang Y., Qing C., and Wanpeng Z. 2003. A comparative study on decomposition of gaseous toluene by O_3/UV, TiO_2/UV and O_3/TiO_2/UV, *Journal of Photochemistry and Photobiology A: Chemistry*, 156(1–3): 189–194.

Rafson H.J. 1998. *Odor and VOC Control Handbook*, McGraw-Hill, New York.

Rohit M. 2008. *Nanotechnology: Basic Science and Emerging Technologies*, Cyber Tech, New Delhi, India.

Shen Y.S., and Ku Y. 2002. Decomposition of gas phase trichloroethane by UV/TiO_2 process in the presence of ozone, *Chemosphere*, 46: 101–107.

Sobana N., and Swaminathan M. 2007. Combination effect of ZnO and activated carbon for solar assisted photocatalytic degradation of Direct Blue 53, *Solar Energy Materials and Solar Cells*, 91: 727–734.

Thanacharoenchanaphas K., Changsuphan A., Nimnual R., Thongsri T., Phetkasem S., and Lertkanawanit C. 2007. Investigation of BTEX and ozone concentrations in a printing facility in Bangkok, Thailand, *International Journal of Applied Environmental Sciences*, 2: 41–48.

U.S. Environmental Protection Agency (USEPA). 1981. Emission factor and AP 42, can coating. http://www.epa.gov/ttn/chief/ap42/ch04/final/c4s02_2b.pdf, accessed July 2011.

U.S. Environmental Protection Agency (USEPA). 1995. *Compilation of Air Pollutant Emission Factors Volume I: Stationary Point and Area Sources (AP 42)* (5th ed.), U.S. Environmental Protection Agency, Washington, DC.

U.S. Environmental Protection Agency (USEPA). 2002a. Preferred and alternative methods for estimating air emission from the printing, packaging, and graphic art industry. Emission Inventory Improvement Program, Technical Report Series, Vol. 2: Point Sources, Chapter 15, U.S. Environmental Protection Agency, Washington, DC. http://www.epa.gov/ttnchie1/eiip/techreport/volume02/index.html

U.S. Environmental Protection Agency (USEPA). 2002b. Refrigerated condensers. *EPA air pollution cost control manual* (EPA/452/B-02-001) (6th ed), Section 3.1, Chapter 2. U.S. Environmental Protection Agency, Washington, DC.

U.S. Environmental Protection Agency (USEPA) 2002c. Flares. *EPA air pollution cost control manual* (EPA/452/B-02-001) (6th ed), Section 3.2, Chapter 1. U.S. Environmental Protection Agency, Washington, DC.

U.S. Environmental Protection Agency (USEPA). 2008. Basic concept in environmental science, module 6 air pollution and control techniques: VOCs characteristic. http://www.epa.gov/eogapti1/bces/module6/voc/character/character.htm, accessed July 2011.

U.S. Environmental Protection Agency (USEPA). 2011. Waste and cleanup risk assessment glossary. http://www.epa.gov/oswer/riskassessment/glossary.htm, accessed August 2011.

Wang H.C., Chang S.H., Hung P.C., Hwang J.F., and Chang M.B. 2008. Catalytic oxidation of gaseous PCDD/Fs with ozone over iron oxide catalysts, *Chemosphere*, 71(2): 388–397.

Wang S., Ang H.M., and Tade M.O. 2007. Volatile organic compounds in indoor environment and photocatalytic oxidation: state of the art, *Environment International*, 33(5): 694–705.

Wark M., Kessler H., and Schulz-Ekloff G. 1997. Growth and reactivity of zinc and cadmium oxide nano-particles in zeolites, *Microporous Materials*, 8(5–6): 241–253.

Yansheng L., Zhenhua J., Nan Y., and Hong G. 2009. Regeneration of nano-ZnO photocatalyst by the means of soft-mechanochemical ion exchange method, *Journal of Environmental Sciences Supplement*, 21: S69–S72.

Zerbonia A.R., Spivey J.J., Agarwal K.S., Damle S.K., and Sanford W.C. 1995. Survey of control technologies for low concentration organic vapor gas streams, Research Triangle Institute, Office of Air Quality Planning and Standards, U.S. Environmental Protection Agency. http://www.epa.gov/ttn/catc/dir1/low_vo.pdf, accessed July 2011.

Zhang S., Zheng Z., Wang J., and Chen J. 2006. Heterogeneous photocatalytic decomposition of benzene on lanthanum-doped TiO_2 film at ambient temperature, *Chemosphere*, 65(11): 2282–2288.

10

Development of VOC Control Technology for the Paint and Coating Sector in China

Hailin Wang, Wei Zhao, Wei Zhang, and Xiaoke Wang

CONTENTS

10.1 Introduction

10.1.1 Background

Volatile organic compounds (VOCs) belong to a large class of air pollutants that contribute to many environmental problems, either directly as toxic or malodorous substances or indirectly as ozone and smog precursors.

Generally, VOCs are released from natural and anthropogenic sources. The former include vegetation and wild animal emissions, bacterial activities, forest fires, and so on. Vehicle exhaust emission, oil gas evaporation, industrial process emissions (including from use of organic solvents), oil refining, storage and transportation, waste material disposal, and others are among the anthropogenic sources. Worldwide, natural sources release more VOCs than anthropogenic ones; however, in China, especially in a metropolis like Beijing, anthropogenic sources play a more important role than the natural sources (Klimont et al., 2002; Klinger et al., 2002). Among those anthropogenic sources, industrial emissions (associated with all kinds of industrial processes or activities) are the largest anthropogenic source of VOCs, followed by stationary fuel combustion in China (Wei et al., 2008), characterized by high emission intensity and concentrations, long emission duration, and complex compositions (Xi et al., 2010).

Generally, industrial VOC emission sources could be divided into two types. One is pipe emissions, such as sealed pipeline emission and exhaust port and incinerator stack emissions. The other is evaporation emissions, including effusion from operating and treating processes, equipment units, storage tanks, and so on. Due to VOC toxicity, photochemical activity, and large-scale emission, on May 11, 2010, the General Office of the State Council (GOSC) of China issued an important notice, which first considered VOCs as important air pollutants as SO_2, NO_x (nitrogen oxides), and particulate matter (PM), and that some key industries with VOC emissions should be prioritized for control to achieve better regional air quality (GOSC, 2010). Thus, technological control of VOCs is important to achieve the target.

10.1.2 Fundamentals of Catalytic Deep Oxidation Technology for VOC Control

VOCs can be controlled using various techniques, such as thermal incineration, catalytic oxidation, adsorption, and biofiltration. Among these methods, catalytic deep oxidation is one of the most efficient; it provides the potential to oxidize the pollutants generally into carbon dioxide and water without secondary pollution.

High catalytic activity can be achieved two ways. First, the compositions of the active phases can be optimized toward the target VOC molecules. For example, catalysts containing noble metals (mainly Pt and Pd) are quite popular in the catalytic oxidation of nonhalogenated VOCs due to their high intrinsic activity (Bendahou et al., 2008). Second, using a porous support with large surface areas allows the active metals to be highly dispersed as nanoparticles, thus forming many active sites. Also, the pore sizes of the support should be properly designed. Those materials with pore sizes that are too small run the risk of blocking the porosity after the loading of active metals, which makes it difficult for the large VOC molecules to reach the inner active sites, whereas support materials with large pore diameters could solve such problems.

In 1992, a new type of porous silica molecular sieve, M41S, which possesses uniform mesopores with diameters larger than 2 nm, were first synthesized by Mobil Corporation researchers (Beck et al., 1992), who used micelles (aggregate of surfactant molecules dispersed in a liquid colloid) to direct the assembly of silica gel. Since then, there has been much work to synthesize various types of mesostructured support materials. Postsynthesis grafting procedures involve modifying the silica host with some types of organosiloxane-containing organic functional groups, such as thiol, amine, EDTA (ethylenediaminetetraacetic acid), and others (Kresge et al., 1992). The resulting support materials are dispersed in solutions containing metal ions; subsequently, the metal ions are absorbed by the functional groups. On calcination, the metal ions are transformed into oxides and ultimately into metal nanoparticles (after reduction) located within the inner pore channel of the silica host. This is an attractive way to obtain well-defined metallic nanoparticles as the silica host could confine the size and the shape of the guest materials in nanoscales. Furthermore, the size of the nanoparticles could be controlled by tuning the pore diameters of the host.

10.2 Methodology

10.2.1 Methods of Synthetic Catalytic Material Preparation

10.2.1.1 Hexagonal Mesoporous Silica Materials

Hexagonal mesoporous silica (HMS) mesoporous molecular sieve materials were prepared by the procedure described by Zhao et al. (2007). Tetraethyl orthosilicate/water-soluble glass, dodecylamine, ethanol, and deionized water were mixed in the molar ratios of $SiO_2 : C_{12}H_{25}NH_2 : C_2H_5OH : H_2O$ = 1:0.25–0.4:9–11:25–40. Then, a definite amount of lanthanum (La), cerium (Ce), or neodymium (Nd) nitrate was added to the mixture and stirred for 24 h at room temperature. The resultant solid was filtered out, washed, and dried at 70°C for 24 h. After being calcinated at 550°C for 5 h, the mesoporous materials were obtained. The materials were called HMS, La-HMS, Ce-HMS, and Nd-HMS, respectively.

10.2.1.2 Supported Co and Pd Catalysts

The cobalt-loaded HMS and cobalt-loaded Ln-HMS (La-HMS, Ce-HMS, and Nd-HMS) were prepared by the wetness impregnation technique. Various amounts of cobalt nitrate hydrate [$Co(NO_3)_2 \times 6H_2O$] were dissolved in 10 mL of ethanol, mixed with 0.5 g of Ln-HMS, and evaporated at 60°C. Those products were finally calcinated at 500°C for 4 h in static air for decomposition of the impregnated salt to cobalt oxide. The materials were Co-x/y-HMS, Co-x/y-Ln-HMS, where x/y is the cobalt loading (wt%) in HMS and Ln-HMS.

10.2.1.3 Nanosize Pd-Al-HMS Catalyst

Nanosize Al-HMS with an average size of about 50–100 nm was prepared by the following steps: First, a solution mixture of TEOS (tetraethoxysilane), EtOH (ethanol), and $Al(OBu^S)_3$ (aluminum butoxide) was added, under vigorous stirring, to a solution of DDA (dodecylamine) in ethanol and deionized water (1.0 SiO_2, x $Al(OBu^S)_3$, 0.2–0.4 DDA, 7–10 EtOH, 142 H_2O), where the value of x depends on various Si/Al ratios. Second, the mixture was aged at ambient temperature for 24 h. Further, it was filtered, washed, air dried at 70°C, and calcinated in air at 550°C for 5 h to obtain Al-HMS. Pd was loaded onto the Al-HMS via the aqueous impregnated method with palladium chlorides as a precursor. The impregnated solids were calcinated at 500°C for 3 h and then reduced in hydrogen at 400°C for 1 h (Zhao et al., 2010).

10.2.2 Characterization Techniques and Instruments

The porous support materials and catalysts were characterized by the techniques shown in Table 10.1.

10.2.3 Bench-Scale Apparatus for Catalyst Evaluation

The bench-scale experimental system shown in Figure 10.1 was set up for the evaluation of oxidation catalysts for VOC abatement. Benzene and chlorobenzene (C_6H_5Cl) were chosen as the representative VOCs. To create an airstream containing certain concentrations of benzene or chlorobenzene, the VOC vapors flows (generated by air passing through the guaranteed organic reagent) were controlled by valves to maintain a constant flow rate. The resulting mixed VOC flow was then diluted with another stream of pure air before reaching the catalyst bed at the desired VOC concentrations (300–800 ppm).

TABLE 10.1

Techniques and Instruments Used for Material Characterizations

Characterization Method	Instrument Model	Function
BET	NOVA-1200	Porosity determination
TEM	Hitachi H-800	Morphology
XRD	Siemens D5005	Crystalline-phase analysis
Temperature-programmed GC	Varian model 2700	Temperature-programmed reduction
Spectrophotometry	Bruker TENSOR 27 with LN-CMT detector	Diffusive reflectance spectrum
X-ray photoelectron spectrometry	PHI Quantera SXM	Surface elemental composition and valence state

Note: BET, Brunauer-Emmett-Teller; GC, gas chromatography; TEM, transmission electron microscopy; XRD, x-ray diffraction.

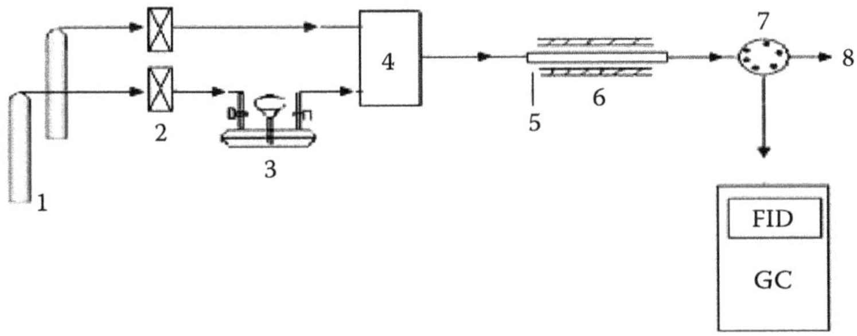

FIGURE 10.1
Scheme of the experimental setup. *Note:* 1, air; 2, gas mass flowmeter; 3, VOC generator; 4, mixing tank; 5, quartz reactor; 6, tubular furnace; 7, six-way valve; 8, exhaust. GC, gas chromatograph; FID, flame ionization detector.

An online gas chromatograph equipped with a flame ionization detector (FID) was used to analyze the benzene, C_6H_5Cl, or CH_2Cl_2 concentration in the inlet and outlet gas streams. The operation parameters were as follows: temperature of the inlet gas varied within 20–80°C; the space velocity for the catalyst varied within 10,000–20,000 h^{-1}.

10.3 Results and Discussion

10.3.1 HMS Catalytic Systems for Removal of Chlorine-Containing VOCs

To obtain cheaper and more effective catalysts instead of noble metal catalysts, pure HMS and M-HMS mesoporous molecular sieves (M denotes the metal atom doped into the framework of HMS) were prepared and employed as the supports of metal oxide catalysts for the complete oxidation of chlorobenzene (C_6H_5Cl).

Chlorine-substituted hydrocarbons are more difficult to oxidize than benzene. Here, the cobalt catalysts were used for C_6H_5Cl oxidation (Table 10.2). The Co/Ce-HMS catalyst had the best catalytic performance, with the conversion of C_6H_5Cl being 98.9% at 480°C. It was also shown that introducing rare earth elements into the framework of the support (HMS) can improve the low-temperature activity of the Co catalysts effectively. For example, comparing 30% Co/HMS with 30% Co/Ce-HMS, there is an abrupt decrease in the conversion temperature of chlorobenzene from 300°C (30% Co/HMS) to 260°C (30% Co/Ce-HMS). In contrast, introducing rare earth elements could decrease the high-temperature activity (Table 10.2).

TABLE 10.2

Catalytic Performance of Various HMS-Supported Catalysts for C_6H_5Cl Oxidation

Catalysts	T_{10}, °C[a]	T_{90}, °C[b]
10% Co/HMS	320	470
30% Co/HMS	300	440
30% Co/La-HMS	280	460
30% Co/Ce-HMS	260	450
30% Co/Nd-HMS	290	460

Note: T_{10} means the temperature at which 10% chlorobenzene is converted, and T_{90} means the temperature at which 90% chlorobenzene is converted.

10.3.2 Pd-Supported Al-HMS for Low-Temperature Oxidation of Benzene

Metallic Pd could be well dispersed on this support, with Pd particle sizes up to 12 nm (Figure 10.2), whereas on traditional support materials (submicrometer HMS) it is impossible to prepare such highly dispersed nanosize Pd. However, the increase in the Pd loading of Pd/Al-HMS catalyst does not enhance the catalytic activity correspondingly (Table 10.3). Keeping the Pd loading constant at 0.9%, the benzene conversion efficiency changed, with the Si/Al ratio of Al-HMS catalyst reaching the maximum catalytic activity (99.6% conversion) at the optimum value of Si/Al = 50. The activity of Pd/Al-HMS at Si/Al = 100 was almost the same as that of Pd/Al-HMS at Si/Al = 50, and the activity of Pd/Al-HMS at Si/Al=20 was about 12%, lower than the maximum efficiency produced by the Pd/Al-HMS (Si/Al = 50). Although with different Si/Al ratios between Pd/Al-HMS (Si/Al = 100) and Pd/Al-HMS (Si/Al = 50), their structures were not obviously different, as shown in Figure 10.3; as a result, the catalytic activity of these two catalysts was very close. The catalyst sample of Pd/Al-HMS (Si/Al = 50; 0.9 wt% Pd), which performed the best in these tests, was selected for further studies. Detailed information can be found elsewhere (Zhao et al., 2010).

For studying the stability of the obtained catalyst, the sample of Pd/Al-HMS (Si/Al = 50; 0.9 wt% Pd) was reduced in situ (the catalyst was not taken out of the reactor) in hydrogen at 400°C for 1 h, and then the next test was performed. It was found that the catalyst kept its activity even after 10 catalytic tests (Zhao et al., 2011).

10.3.3 Performance of the Pilot-Scale Control Device in the Jiayuan Company

The final goal was to develop the deep oxidation catalytic technology with a VOC conversion rate above 95% for application for industrial VOC treatment. A new catalyst coating technique on ceramic honeycomb was developed,

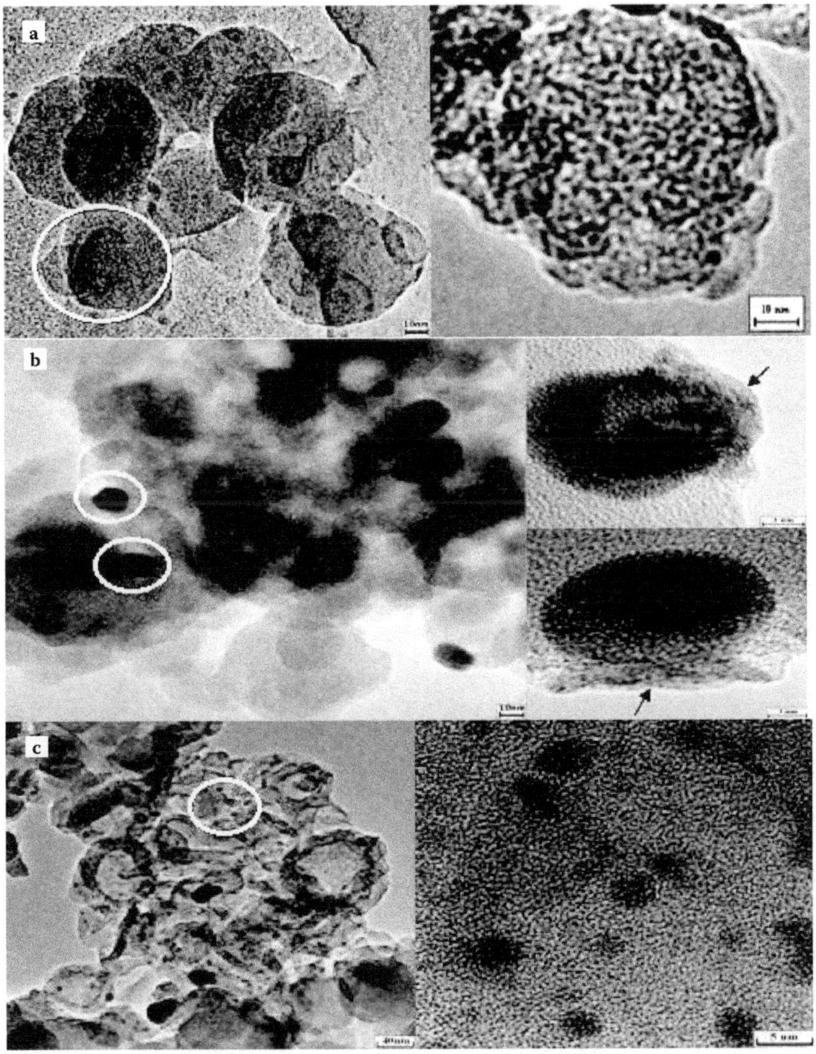

FIGURE 10.2
The TEM images of different samples (the right-side images are the enlarged images of the circled parts appearing in the left-side images): (a) Al-HMS (Si/Al = 50); (b) Pd/Al-HMS (Si/Al = 50; 0.9 wt% Pd); (c) Pd-HMS (0.9 wt% Pd). (*Source:* Zhao W., Li J.J., He C., Wang L.N., Chu J.L., Qu J.K., Qi T., and Hao Z.P. 2010. Synthesis of nanosized Al-HMS and its application in deep oxidation of benzene, *Catalysis Today*, 158: 427–431.)

and an application for a Chinese invention patent has been made (Hao et al., 2008). The invention provides a method for integral-type catalyst preparation; the catalyst has the advantage of a simple preparation process and can be prepared into the integral-type catalyst with a certain shape, size, and strength. The pilot-scale study was done to investigate the influences of

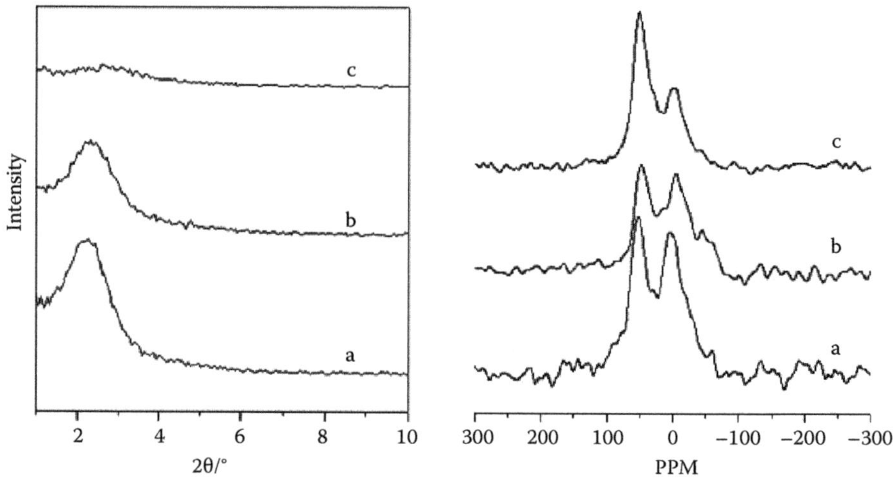

FIGURE 10.3
The XRD (left) and ^{27}Al MAS NMR (right) spectrum of different catalysts. (a) Pd/Al-HMS (Si/Al = 100; 0.9 wt% Pd); (b) Pd/Al-HMS (Si/Al = 50; 0.9 wt% Pd); (c) Pd/Al-HMS (Si/Al = 20; 0.9 wt% Pd). (*Source:* Zhao W., Li J.J., He C., Wang L.N., Chu J.L., Qu J.K., Qi T., and Hao Z.P. 2010. Synthesis of nanosized Al-HMS and its application in deep oxidation of benzene, *Catalysis Today*, 158: 427–431.)

TABLE 10.3
Catalytic Performances of Different Catalysts in Oxidation of Benzene

Catalyst	Pd Loading, wt%	Conversion of Benzene at 200°C, %
HMS	—	0
HMS	0.9	33.8
HMS	1.5	78.1
Al_2O_3	0.9	17.6
Al-HMS (Si/Al = 50)	—	6.1
Al-HMS (Si/Al = 50)	0.6	65.4
Al-HMS (Si/Al = 50)	0.9	99.6
Al-HMS (Si/Al = 50)	1.2	89.3
Al-HMS (Si/Al = 100)	0.9	97.2
Al-HMS (Si/Al = 20)	0.9	87.8

Source: Zhao W., Li J.J., He C., Wang L.N., Chu J.L., Qu J.K., Qi T., and Hao Z.P. 2010. Synthesis of nanosized Al-HMS and its application in deep oxidation of benzene, *Catalysis Today*, 158: 427–431.

pretreatment, surfactants, and loading on the catalyst activity. To evaluate the performance of the newly developed catalytic systems with actual emissions, 120 L of this honeycomb monolith catalyst coated with a mixture of (a) Pd supported on mesoporous γ-Al_2O_3 (70–80%), (b) rare earth oxides (10–25%), and (c) HMS (5–10%) were prepared according to the coating technique on ceramic honeycomb mentioned. The materials were sent to the Jiayuan Company in Fujian, a manufacturer of pollution control devices, for application. The catalytic VOC abatement facility was tested at an emission source containing a mixture of benzene, toluene, and xylenes under a space velocity of 2,000 h^{-1}. This VOC catalytic oxidation system mainly consisted of a preheating chamber, a catalyst chamber, and a heat exchanger. The pilot-scale experiment was conducted for 3 months. The results indicated that the catalytic system was effective (benzene removal efficiency above 95%) and could remove VOCs to meet the domestic regulation of industrial VOC emissions.

10.3.4 Demonstration of Pilot-Scale Control Device and Cost-Benefit Consideration

A demonstration of the adsorption-catalytic combustion installation (Figure 10.4) was made in a stationery manufacturing enterprise in Beijing. The system was applied to control VOCs from emission from a paint-spraying production line in this enterprise. The major VOCs were benzene and toluene. The treatment capacity of this absorption-catalytic oxidation apparatus is 60,000–80,000 m^3/h of exhaust gas. The corresponding operational parameters include inlet gas temperature of 20–80°C, catalytic conversion rates for VOCs greater than 95%, and a space velocity for the catalytic bed of 20,000 h^{-1}.

The information on preliminary costs and benefits based on this pilot-scale demonstration unit are listed in Table 10.4. Note that this is not a full cost-benefit analysis as not all benefits can be assigned a monetary value.

The control device can help the industry meet the national emission standards for benzene and toluene. At a total annual cost of US$16,250, the device can eliminate over 600 kg benzene and 1,400 kg toluene per year. A total reduced levy thanks to the eliminated amounts of these two VOCs would be $US1,440.

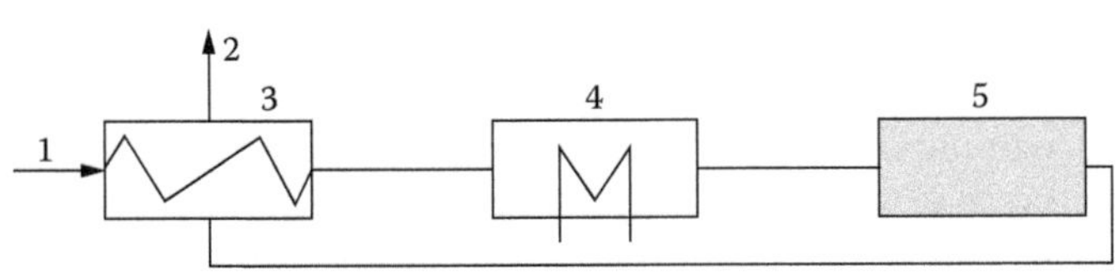

FIGURE 10.4
Scheme of the adsorption-catalytic combustion device. *Note:* 1, exhaust containing VOCs; 2, cleaned air; 3, heat exchange chamber; 4, heating chamber; 5, catalyst chamber.

TABLE 10.4

Information on Costs and Potential Benefits of the Pilot-Scale Demonstration System

Item	Cost, US$	Benefits
Annual catalyst depreciation cost	2,133	
Annual installation depreciation cost	2,375	
Annual operational and maintenance cost	11,740	
Total annual costs	**16,250**	
Amount of benzene removed per year		635 kg
Amount of toluene removed per year		1,424 kg
Benzene removal efficiency		96.7%
Toluene removal efficiency		93%
Reduced levy for less benzene emitted per year		US$895
Reduced levy for less toluene emitted per year		US$548
Meet the national emission standards for benzene and toluene		Yes

10.4 Summary and Conclusions

Mesoporous materials are good supports of oxidation catalysts for VOC control. Their large specific surface area and porous structure are beneficial for the dispersion of the active phase in them. Various Co- and Pd-based mesoporous catalyst materials were prepared and tested on VOC (benzene, chlorobenzene) removal. Co- and Pd-based catalysts were generally prepared by an impregnation method, and benzene and chlorobenzene were treated on a constant-flow, fix-bed reactor at a certain temperature, which showed high removal efficiency for VOCs. Introducing rare earth elements into the mesoporous materials could obviously improve the low-temperature catalytic activity but was not effective for high-temperature activity.

For large-scale demonstration, a monolith catalyst was prepared for the catalytic combustion of VOCs. A pilot-scale device was installed and operated in an industry (Jiayuan Company); a benzene removal efficiency of above 95% was shown. Further, the pilot-scale device was installed at a stationery manufacturing enterprise in Beijing. A preliminary cost-benefit analysis was carried out based on the data obtained from this control device; it showed that implementation of the control device can help the industry be in compliance with the national emission standards for benzene and toluene. The benefits were shown as the amount of VOCs eliminated by the control device, which will bring in multiple benefits for the environment as well as a reduction in levy (tax) associated with VOC elimination.

References

Beck J.S., Vartuli C., Roth W.J., Leonowicz M.E., Kresge C.T., Schmitt K.D., Chu C.T.W., Olson D.H., Sheppard E.W., McCullen S.B., Higgins J.B., and Schlenker J.L. 1992. A new family of mesoporous molecular sieves prepared with liquid crystal templates, *Journal of American Chemical Society*, 114: 10834–10843.

Bendahou K., Cherif L., Siffert S., Tidahy H.L., Benaissa H., and Aboukais A. 2008. The effect of the use of lanthanum-doped mesoporous SBA-15 on the performance of Pt/SBA-15 and Pd/SBA-15 catalysts for total oxidation of toluene, *Applied Catalysis A—General*, 351: 82–87.

General Office of the State Council (GOSC). 2010. General Office of the State Council. Circular on forwarding the guidance of promoting joint prevention and control of air pollution and improving the regional air quality. Issued by Ministry of Environmental Protection and other government agencies. http:// www.gov.cn/gongbao/content/2010/content_1612364.htm.

Hao Z.P., Chen J., Li L.D., and Li J.J. 2008. A monolithic catalyst preparation and application in nature gas and coal bed methane combustion. China Patent No. 20081012196X. Chinese invention patent.

Klimont Z., Streets D.G., Gupta S., Cofala J., Fu L.X., and Ichikawa Y. 2002. Anthropogenic emissions of non-methane volatile organic compounds in China, *Atmospheric Environment*, 36: 1309–1322.

Klinger L.F., Li Q.J., Guenther A.B., Greenberg J.P., Baker B., and Bai J.H. 2002. Assessment of volatile organic compound emissions from ecosystems of China, *Journal of Geophysical Research—Atmosphere*, 107: 4603–4624.

Kresge C.T., Leonowicz M.E., Roth W.J., Vartuli J.C., and Beck J.S. 1992. Ordered mesoporous molecular sieves synthesized by a liquid-crystal template mechanism, *Nature*, 359: 710–712.

Wei W., Wang S.X., Chatani S., Klimont Z., Cofala J., and Hao J.M. 2008. Emission and speciation of non-methane volatile organic compounds from anthropogenic sources in China, *Atmospheric Environment*, 42: 4976–4988.

Xi J.Y., Wu J.L., Hu H.Y., and Wang C. 2010. Investigation of industrial VOCs emission sources and analysis for their emitting characteristics, *China Environmental Science*, 30: 1558–1562.

Zhao W., Li J.J., He C., Wang L.N., Chu J.L., Qu J.K., Qi T., and Hao Z.P. 2010. Synthesis of nanosized Al-HMS and its application in deep oxidation of benzene, *Catalysis Today*, 158: 427–431.

Zhao W., Li J.J., Mu Z., and Hao Z.P. 2007. Synthesis of Co/HMS modified by rare earths and its performance on catalytic combustion of chlorobenzene, *Journal of Chinese Rare Earth Society*, 25: 22–26.

Zhao W., Liu Y.H., Wang L.N., Chu J.L., Qu J.K., Hao Z.P., and Qi T. 2011. Catalytic combustion of benzene on the Pd/nanosize Al-HMS, *Microporous and Mesoporous Materials*, 138: 215–220.

11

Removal of VOCs from Contaminated Gas Streams by Biofiltration

Tyagarajan Swaminathan and Ligy Philip

CONTENTS

11.1 Introduction

Widespread emissions of hazardous pollutants such as benzene, toluene, and xylenes (commonly called BTX) in chemical and petrochemical industries have caused considerable concern over their potential adverse effects on human health and the environment. BTX compounds are among the hazardous air pollutants (HAPs) identified by the U.S. Environmental Protection Agency (USEPA). The presence of these compounds at even low concentrations in the ambient air has adverse impacts on human health. These substances are toxic to the liver, kidneys, and central nervous system as they enter the body by skin contact or by breathing (Akers et al., 1999; Fishbein, 1985; Olaniran et al., 2001). These compounds have been the subject of extreme importance in recent environmental regulations. Although industries emitting volatile organic compounds (VOCs) are required to install control devices to reduce their emissions, it is not widely practiced due to lack of cost-effective techniques, resulting in increasing concentrations of VOCs in ambient air. This need for cost-effective techniques prompted this study on biofiltration of BTX compounds; the results are presented in this chapter.

The common treatment technologies applied so far for treating these VOCs are based on absorption and adsorption processes (Kolkar et al., 2006; Liuzzo et al., 2007). However, existing technologies, such as absorption, carbon adsorption, incineration, and catalytic combustion, are costly, less effective, and more wasteful in terms of energy consumption. Moreover, these processes involve mere transfer of pollutants from the gaseous phase to either a liquid or a solid phase (absorption or adsorption) and may generate new pollutants (combustion). When compared with these conventional technologies, biological waste gas treatment emerges as a promising, cost-effective, and eco-friendly process because they do not produce additional pollutants to the water, air, and soil environments (Devinny et al., 1999; Rula and Cohen, 1999; Shim et al., 2002).

Biofiltration is a cost-effective process that degrades both organic and inorganic pollutants to innocuous end products such as CO_2 and H_2O. It uses microorganisms attached on a porous support media to degrade VOCs and can be potentially applied to treat large volumes of waste gases with low concentrations of biodegradable pollutants (Devinny et al., 1999). In addition, these units can be designed to fit any industrial setting physically, thereby minimizing space requirements. These advantages encouraged employment of biofiltration as a potential control device in this study.

The principal objective of this research work was to evaluate the biodegradation of BTX in batch (liquid phase) and continuous systems (gaseous phase) at laboratory scale (lab scale). In the final stage of this work, the performance of the biofilter was examined with actual waste gas collected from a pharmaceutical industry.

11.2 Methodology

11.2.1 Batch Experiments

The mixed microbial consortia were precultured in 100 mL of the mineral salt medium (MSM) containing 150 mg/L of benzene, toluene, or *o*-xylene as the carbon source for about 40–50 h. MSM had the following composition: 2.0 g/L $(NH_4)_2SO_4$, 2.0 g/L KH_2PO_4, 0.5 g/L NaCl, and 0.025 g/L $MgSO_4{\cdot}7H_2O$ in distilled water. The pH of the MSM was adjusted to 6.8. The batch culture biodegradation of BTX compounds was studied over a concentration range of about 50–500 mg/L (MSM) of BTX individually. Samples collected at regular intervals were analyzed for pH, biomass, and residual benzene, toluene, and xylene concentration.

11.2.2 Continuous Experimental Setup

Garden compost was chosen as the packing material owing to its easy availability, versatility, and chemical composition. Compost contains nutrients such as N, P, and trace elements that are advantageous for microbial growth and sustainability in biofilters. It was mixed with 2-mm diameter ceramic beads in 6:4 volume ratios to enhance the mechanical strength of the packing. The ceramic beads also increased the bed porosity and ensured more homogeneous gas distribution across the filter bed.

A schematic of the experimental setup is given in Figure 11.1. Humidified BTX vapors at constant flow and concentration, controlled through valves, were passed through the packed bed in an upflow mode. The bed was initially inoculated with a mixed microbial culture augmented with individually precultured cells. The bed moisture was maintained by periodic addition of fresh MSM from the top. As compost had most of the micro-/macronutrients required for microbial growth, MSM also can be replaced by water. Experiments were carried out by varying the flow rates of the individual BTX vapors and humidified air independently to get different initial concentrations and residence times in the biofilter. Gas samples were collected from different ports and analyzed for residual BTX concentrations.

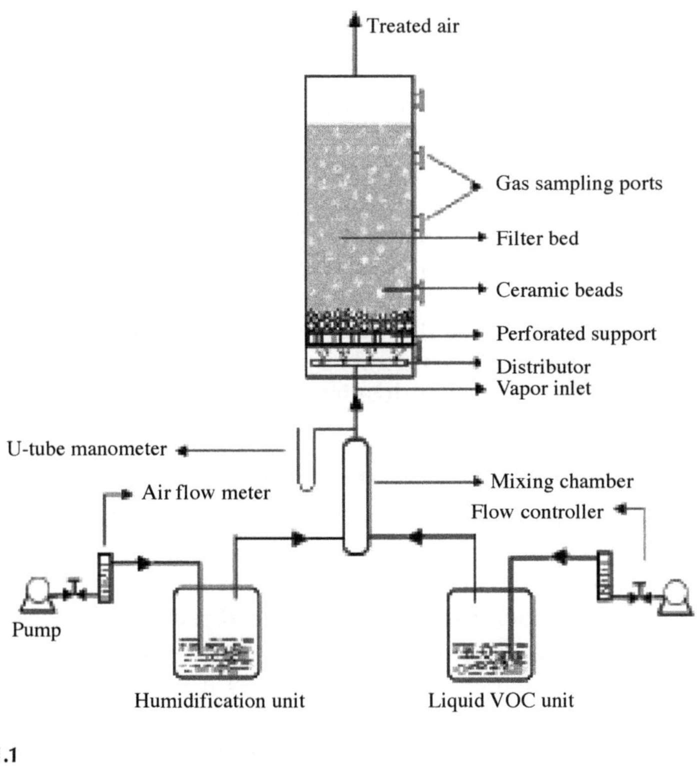

FIGURE 11.1
Schematic of an experimental biofilter setup (continuous) used for BTX removal.

11.2.3 Experimental Materials

All chemicals used in this study were laboratory grade—benzene (>99%), toluene, and xylene (>99%, sulfur free)—and were purchased from Ranbaxy Fine Chemicals Limited, India. The microbial mixed consortia were obtained from the Nesapakkam municipal sewage treatment plant in Chennai. The final inoculum was obtained by a series of repeated inoculations in an MSM.

11.2.4 Analytical Methods

Cell concentration was determined by spectrophotometric method (optical $density_{540}$) using an ultraviolet-visible (UV-Vis) spectrophotometer (Shimadzu UV-1601 PC, Japan). Residual concentrations of benzene, toluene, and xylene in the aqueous and gaseous phases were measured by gas chromatography (Nucon gas chromatograph, model 5765, Aimil Engg, India) fitted with a flame ionization detector using a Chromatopak packed column.

11.3 Results and Discussion

11.3.1 Batch Experiment: Liquid Phase

Substrate versatility studies in batch systems have been carried out to assess qualitatively the possibility of BTX degradation, using microbial cultures precultured in different BTX substrates. Experiments were carried out at various initial BTX concentrations, ranging from about 50 to 500 mg/L.

Figure 11.2 shows that the degradation pattern for all the substrates (benzene, toluene, and xylene) was similar with inhibition occurring at high concentrations (~200 mg/L for benzene and xylene, ~250 mg/L for toluene). Toluene was degraded faster than benzene and xylene. It was also quite apparent that adaptation of the microbes to a more toxic compound like benzene and xylene improved the utilization of toluene (Figure 11.2[b]).

11.3.2 Continuous Experiments: Synthetic Gas Stream

After acclimatizing the individual biofilters for about 40 days with low concentrations of benzene, toluene, and xylene, continuous experiments were carried out at three different flow rates (0.024, 0.036, and 0.072 m^3/h [0.4, 0.6, and 1.2 L/min]) corresponding to the empty bed retention times (EBRTs) of 2.45, 1.63, and 0.81 min, respectively, and at various inlet concentrations (a total of nine experiments for each biofilter) over a period of 4 months. The results of this study demonstrated maximum elimination capacities (ECs) of 52.1, 84.2, and 77.4 $g/m^3.h$, corresponding to inlet loading rates of 110.6, 167.1, and 182.9 $g/m^3.h$ for biofilters treating benzene, toluene, and xylene,

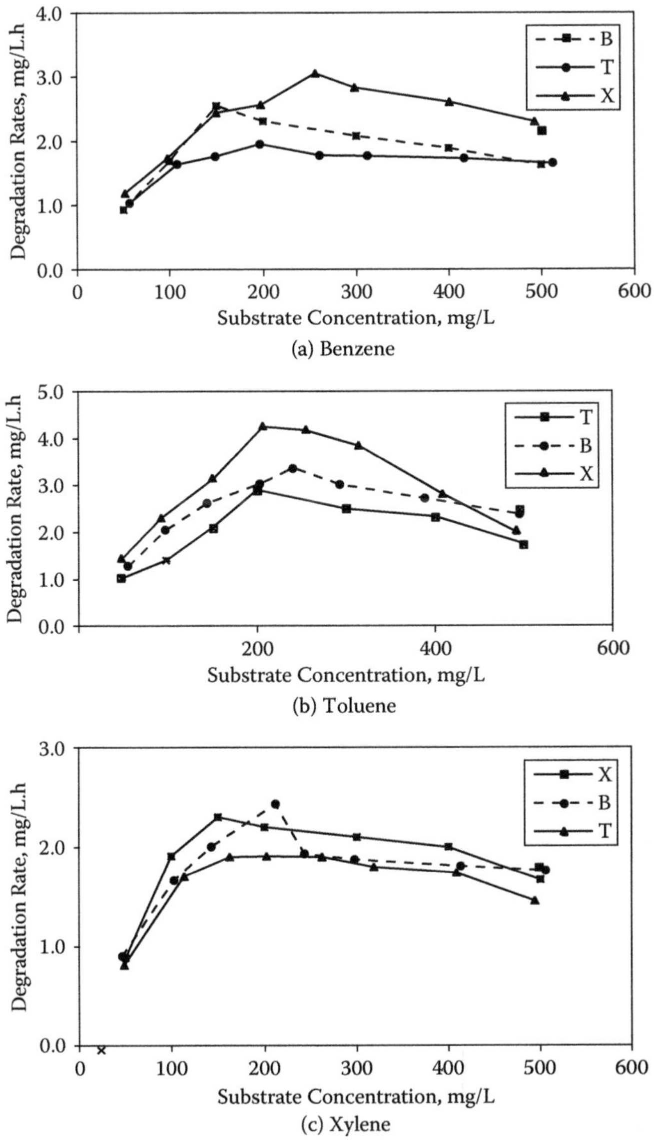

FIGURE 11.2
Degradation rate profiles of degradation of benzene (a), toluene (b), and xylene (c). B, T, and X indicate benzene, toluene, and xylene, respectively, acclimatized culture used for this study.

respectively. The performances of the biofilters for various pollutants under different operating conditions are presented in Figures 11.3 to 11.5.

Removal efficiencies for all three VOCs remained high (80–90%), except at higher VOC loading. It was also observed that the biofilters recovered their

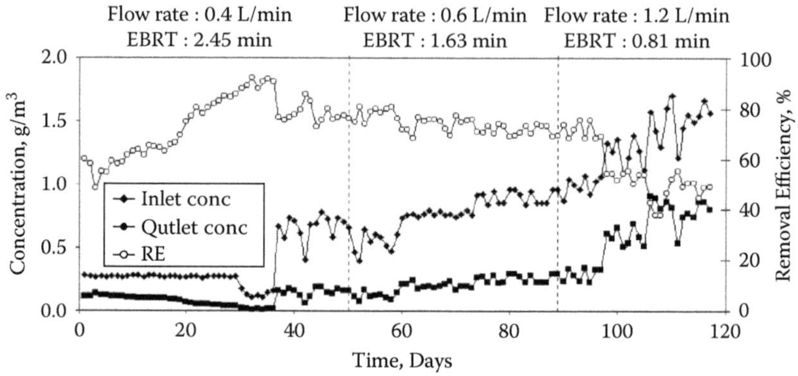

FIGURE 11.3
Effect of gas flow rate and concentration on the removal efficiency of biofilter treating benzene vapors.

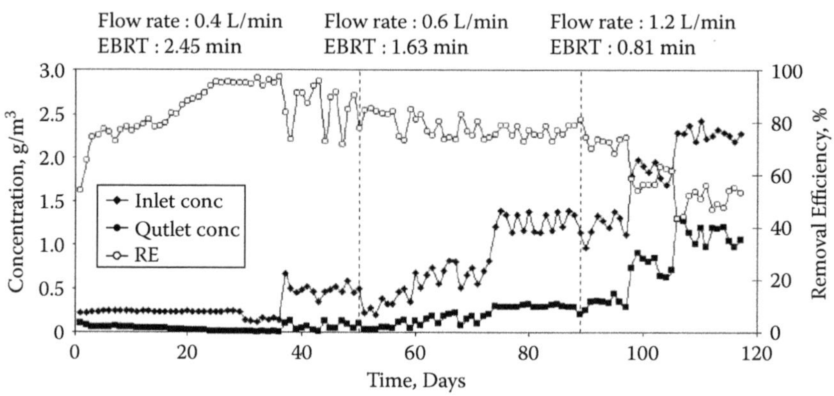

FIGURE 11.4
Effect of gas flow rate and concentration on the removal efficiency of biofilter treating toluene vapors.

efficiencies quickly even when subjected to fluctuations in inlet concentration and flow rate. This indicates the stability of the biofilters and the resilience of the biomass in the compost bed.

11.3.3 Continuous Experiments: Industrial Gas Stream

To study the practical feasibility of biofiltration, the performance of the toluene-acclimated biofilter was studied with a waste gas stream from a pharmaceutical industry (Orchid Chemical and Pharmaceuticals Ltd., India) containing both benzene and toluene at two flow rates, 0.036 and 0.06 m^3/h. The results are presented in Figures 11.6 and 11.7, respectively.

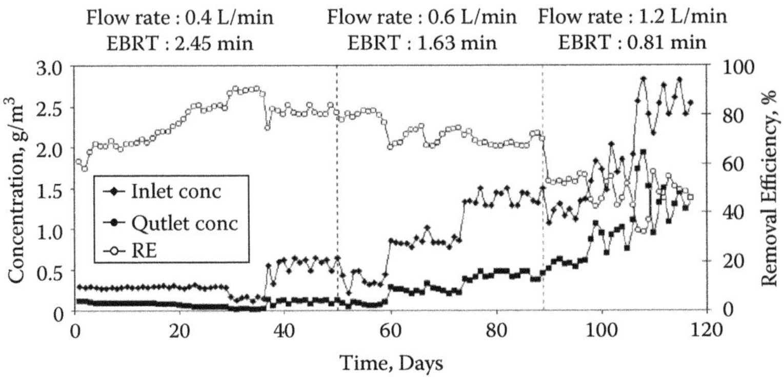

FIGURE 11.5
Effect of gas flow rate and concentration on the removal efficiency of biofilter treating xylene vapors.

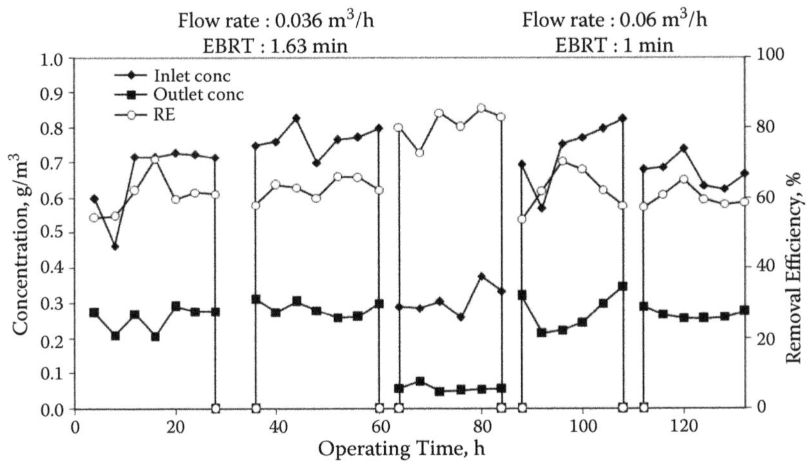

FIGURE 11.6
Effect of gas flow rate and concentration on the removal efficiency of benzene in the biofilter treating industrial gas sample.

The ranges of benzene and toluene concentrations in the flue gas were 0.3–0.85 g/m^3 and 0.4–1.6 g/m^3, respectively. The performance of this biofilter was examined for a period of 140 h at two different flow rates (i.e., 0.036 and 0.06 m^3/h) and by varying the inlet concentrations. Moreover, to evaluate intermittent operation, experiments were stopped for about 4–8 h periodically, and elimination capacity was evaluated. Maximum removal efficiencies of 95.2% and 85.3% for toluene and benzene vapors, respectively, were achieved at a residence time of 1.63 min. This was comparable with the results obtained from biofilters for individual synthetic gas streams. It is interesting to note that such high removal could be obtained for both toluene

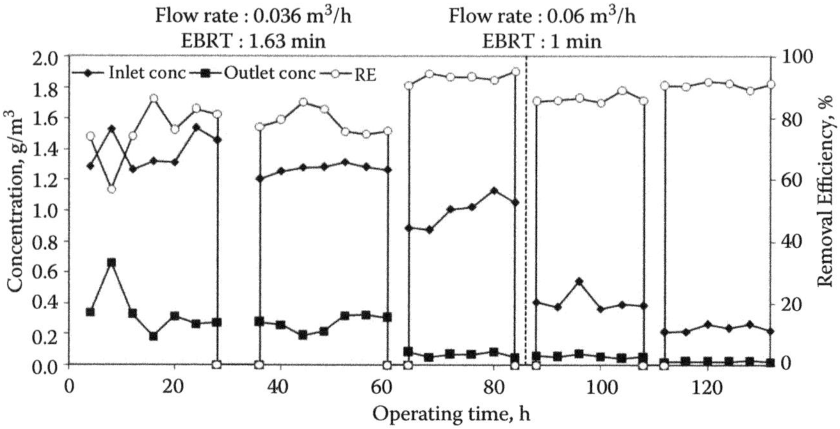

FIGURE 11.7
Effect of gas flow rate and concentration on the removal efficiency of toluene in the biofilter treating industrial gas sample.

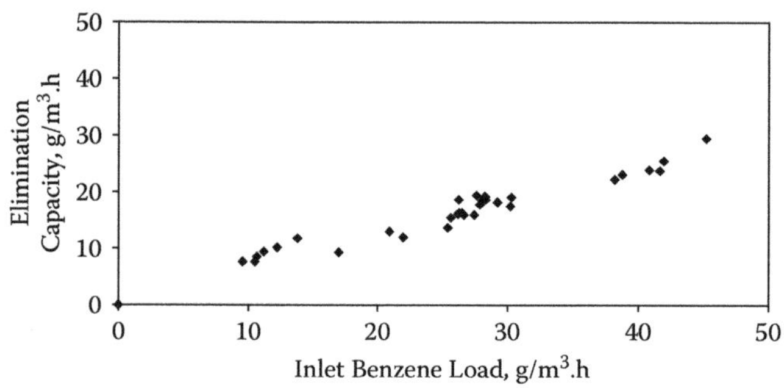

FIGURE 11.8
Effect of inlet loading rate on the elimination capacity for benzene vapors in a biofilter treating industrial gas sample.

and benzene in a mixture. It is generally expected that removal of VOCs would be affected by the presence of other pollutants in the gas stream.

The elimination capacities of the filter bed for benzene and toluene are shown in Figures 11.8 and 11.9, respectively. It could be seen that EC increased with increased VOC loading. However, for benzene the increase in EC above 15 g/m^3 was relatively less. Since the experiments were done in a toluene-acclimated biofilter, the EC for toluene was much higher and kept increasing with increasing toluene load up to 50 g/m^3.

11.3.4 Cost Effectiveness

The most attractive feature of biofiltration is the low operating and maintenance costs relative to other air pollution control technologies. The recirculating

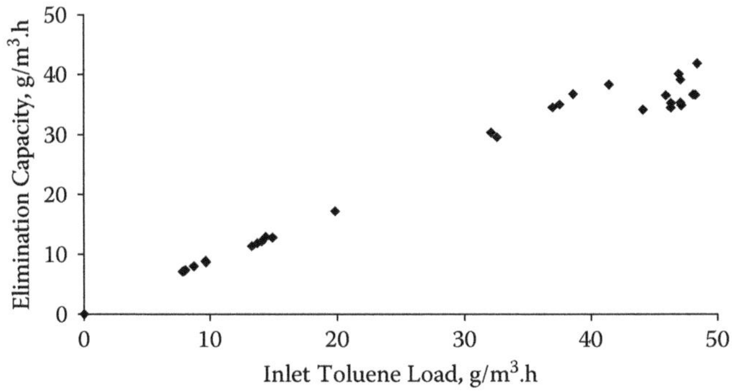

FIGURE 11.9
Effect of inlet loading rate on the elimination capacity for toluene vapors in a biofilter treating industrial gas sample.

pump for the humidifier and the blower that moves the air through the system are the only two energy sinks in the system. A preliminary economic comparison of biofiltration and catalytic oxidation unit has been done assuming a flowrate $Q = 1{,}500$ m^3/h, 90% contaminant removal, and a residence time of 98 s for the biofilter unit and 15 s for the catalytic oxidation unit. Table 11.1 clearly describes the associated costs involved during the operation of these two systems. The annual biofilter operating cost is one-fifth of the photocatalytic reactor operating cost for BTX removal. The annualized treatment cost/1,000 m^3 of waste gas in a biofiltration unit is one-fourth of the cost expended using catalytic oxidation. Hence, the economics indicate that biofiltration should be seriously considered as an alternative to current control technologies.

11.4 Conclusion and Recommendation

The biodegradation potential of BTX compounds and growth kinetics was studied with individually acclimated microbial cultures. Toluene was observed to be the most readily biodegradable, followed by benzene and xylene. Continuous gas phase experiments were also conducted in compost-based biofilters, individually acclimated to each of the BTX compounds, over a period of 4 months at different flow rates (0.024–0.072 m^3/h) and at different inlet VOC concentrations. High (85–95%) removal efficiencies were achieved for all the VOCs when the biofilters were operated at optimal gas flow rate conditions (0.024 m^3/h) and inlet VOC concentrations (<0.5 g/m^3). The performance of the toluene-acclimated biofilter was evaluated with an industrial waste gas stream containing both toluene and benzene. The removal efficiencies for both VOCs in the gas mixture were comparable to

TABLE 11.1

Comparison of Cost Effectiveness between Biofilter (15-Year Life Span) and Photocatalytic Oxidation Unit (20-Year Life Span)

1	**Biofilter:** Dimension: H = 3.4 m, r = 3 m, V = 96 m^3 Construction material: PVC; packing material: garden compost	
1.1	Capital cost (blower, humidification unit, piping, electrical, and site preparation)	Rs 220,000
1.2	Annual operating cost (utilities, labor, chemicals)	Rs 75,000
1.3	Operating cost/1,000 m^3 of gas treated	Rs 6.94
1.4	Annualized capital cost (assuming a life span of 15 years)	Rs 14,600
1.5	Annualized capital cost/1,000 m^3 of gas treated	Rs 1.35
	Total cost of treatment/1,000 m^3 of gas treated (1.3 + 1.5)	Rs 8.3
2	**Photocatalytic unit:** Dimension: H = 2.5 m, r = 0.9m, V = 6.25 m^3 Construction material: reinforced glass; catalyst: TiO_2	
2.1	Capital cost (material, covering chamber, catalyst, lamps, piping, instruments)	Rs 1,052,675
2.2	Operating cost (electrical and catalyst replacement)	Rs 360,000
2.3	Operating cost/1,000 m^3 of gas treated	Rs 33.3
2.4	Annualized capital cost (assuming a life span of 20 years)	Rs 52,635
2.5	Annualized capital cost/1,000 m^3 of gas treated	Rs 4.9
	Cost of treatment for 1,000 m^3 of gas (2.3 + 2.5)	Rs 38.2

Note: Exchange rate of US$1 = Rs 45.50.

their removal in individually acclimated biofilters. This work reflected the ability of mixed microbial culture biofilters to treat BTX vapors effectively. The operation for nearly 7 months without any major operational problem proved their stability in fluctuating conditions.

Pilot plant studies are required to apply this technique in the field. An elaborate evaluation of the microbial activity and the dominant microorganism responsible for contaminant removal has to be carried out. Genetically engineered systems with high elimination capacities can be developed using molecular biology tools. Other challenges include novel configurations of bioreactors, new packing materials, and scale-up for large industrial emissions.

Biofiltration is versatile enough to treat odors, toxic compounds, and VOCs from industries and other commercial establishments. The treatment efficiencies of these contaminants were above 90% for low concentrations of contaminants (<1,000 parts per million by volume). In general, the temperate climatic conditions prevailing in most Asian countries make biofilter operation effective as most diverse microbial communities are active in the mesophilic range (20–45°C). Moreover, the recent understanding about eco-friendly biological systems to treat malodorous emissions from industries and the associated economics would pave the way to implement this technology at the industrial level.

References

Akers K.S., Sinks G.D., and Schultz T.W. 1999. Structure-toxicity relationships for selected halogenated aliphatic chemicals, *Environmental Toxicology and Pharmacology*, 7(1): 33–39.

Devinny J.S., Deshusses M.A., and Webster T.S. 1999. *Biofiltration for air pollution control* (3rd ed.), Lewis, Boca Raton, FL.

Fishbein L. 1985. An overview of environmental and toxicological aspects of aromatic hydrocarbons, *Science of the Total Environment*, 42: 267–268.

Kolkar A., Constance L.S., and Jeffery C.Q. 2006. Mercury in coal and the impact of coal quality on mercury emission from combustion systems, *Applied Geochemistry*, 21: 1821–1836.

Liuzzo G., Verdone N., and Bravi M. 2007. The benefits of flue gas recirculation in waste incineration, *Waste Management*, 27: 106–116.

Olaniran A.O., Babalola G.O., and Okoh A.I. 2001. Aerobic dehalogenation potentials of four bacterial species isolated from soil and sewage sludge, *Chemosphere*, 45: 45–50.

Rula A.D., and Cohen L.A. 1999. Temperature effects and substrate interactions during the aerobic biotransformation of BTEX mixtures by toluene enriched consortia and *Rhodococcus rhodochrous*, *Biotechnology and Bioengineering*, 62: 526–535.

Shim H., Shin E.B., and Yang S.T. 2002. A continuous fibrous-bed bioreactor for BTEX biodegradation by co-culture of *Pseudomonas putida* and *Pseudomonas fluorescens*, *Advances in Environmental Research*, 7(1): 203–216.

12

Novel Bioscrubbing Process for Complete Treatment of NO_x from Flue Gases

Ligy Philip and Sanjeev Arjunagi

CONTENTS

12.1 Introduction

Existing nitrogen oxide (NO_x) control systems cannot be used exclusively for all purposes. Commonly used dry systems such as selective catalytic reduction (SCR), selective noncatalytic reduction (SNCR), and combustion modifications require specified conditions. In particular, the right temperature is required for SNCR (900–1,000°C) and SCR (270–400°C). Wet systems have the drawbacks of expensive chemical addition, high water usage, and excessive hazardous sludge production (Rhoads et al., 1990).

Biological removal of NO_x from contaminated gas streams is emerging as a novel treatment method. Biofiltration and biotrickling filtration are used as attractive alternatives for both organic and inorganic air pollutant treatment. It is reported that NO_2 and SO_2 could be removed effectively using a biotrickling filter/scrubber (Flanagan et al., 2002; Philip and Deshusses, 2003) due to their high solubility in water. However, not many studies have been carried out on the treatment of NO_x (NO + NO_2). Moreover, biological technologies reported so far have many disadvantages, like inhibition in the presence of oxygen and a requirement for high detention times to achieve the necessary degree of treatment.

As the solubility of NO in water is very low (62 mg in 1 kg of water at 20°C and 760 mmHg) (Perry et al., 1984), biological processes such as biofiltering and biotrickling filtering are not successful in removing NO effectively after a short contact time. Thus, conversion of NO to NO_2 or any other soluble form using a suitable technical method and subsequent scrubbing followed by biological denitrification seems to be a viable alternative.

Barman and Philip (2006) reported a novel and effective system for the complete treatment of NO_x from flue gases. The system consisted of three units: (1) photocatalytic or ozone oxidation of NO, followed by (2) scrubbing, and (3) biological denitrification. Maximum photocatalytic oxidation of NO was achieved using powdered TiO_2 at a catalytic loading rate of 10 g/h, relative humidity of 50%, and a space time of 10 s. The used catalyst was regenerated and reused. A total of 72% of oxidized NO was recovered as HNO_3/HNO_2 in the regeneration process. Stochiometrically, 10% excess ozone was able to achieve 100% oxidation of NO to NO_2. Presence of SO_2 adversely influenced the oxidation of NO by ozone. The scrubbing of NO_2, formed by the oxidation of NO by ozone, was effective with distilled water in the scrubber. The scrubber was placed after the oxidation unit. In the biological denitrification, heterotrophic denitrifiers were able to denitrify the leachate with an efficiency of 90%, using sewage (chemical oxygen demand [COD] 450 mg/L) as an electron donor. Thus, in this system, for complete treatment three units are required. It is always desirable to have a minimum number of units for ease in operation and maintenance. Hence, it is essential to try whether the scrubbing and biological denitrification can be achieved in one system, such as a bioscrubber.

Anaerobic ammonium oxidation (ANAMMOX) is a lithoautotrophic biological conversion process mediated by a group of *Planctomycetes* bacteria. These bacteria have a highly unusual physiology in that they live by consuming ammonia in the absence of oxygen. This process involves the oxidation of ammonia with nitrite/nitrate as the electron acceptor to yield gaseous nitrogen. Also, the ANAMMOX organisms grow with CO_2 as the sole carbon source and use nitrite as the electron acceptor to produce cell material (Arts et al., 1995; Yu et al., 2007). The overall process of ANAMMOX is the central energy-generating pathway for these microbes. Even though the *Nitrosomonas* species can also oxidize ammonium anaerobically as well as aerobically (Jetten et al., 1999; Lee and Rittmann, 2000), ANAMMOX is used exclusively for nitrogen removal by these *Planctomycetes*-like bacteria (Jetten et al., 1999).

In this chapter, performance of a bioscrubber employing anaerobic ammonia-oxidizing bacteria for the removal of NO_x from flue gas was evaluated.

The effect of operational parameters such as empty bed contact time and the presence of oxygen on the performance of the system was also studied. The main objective of the present study was to develop and evaluate the performance of a biological system for the complete treatment of NO_x from flue gases.

12.2 Methodology

The chemicals used in this study were procured from Ranbaxy Fine Chemicals Limited (New Delhi, India). Nitric oxide gas was procured from Bhoruka Gas Limited (Bangalore, India). The initial seed sludge was collected from the sewage treatment plant of IIT Madras Chennai. Cow dung sludge acclimatized for the ANAMMOX process in the Environmental Engineering Laboratory, Department of Civil Engineering, IIT Madras, Chennai, was also used for the study. Settled sludge from the sewage treatment plant at IIT Madras or acclimatized ANAMMOX sludge developed from cow dung was added to the reactor (50% of reactor volume with mixed liquid suspended solids [MLSS] of 1,150 mg/L) for acclimatization. Acclimatization was carried out in 12 separate batch reactors (each having a volume of 100 mL, fitted with airtight septum) under anaerobic environments for 30 days. During acclimatization, nutrients were added to the reactors for the growth of bacteria. The composition of minimal salt media was 0.0572 g KH_2PO_4; 2.1 g $NaHCO_3$; 0.5915 g $NaNO_2$; 0.3 g $CaCl_2.2H_2O$; 0.2 g $MgSO_4.7H_2O$; 0.00625 g $FeSO_4$; 0.00625 g EDTA (ethylenediaminetetraacetic acid); 0.455 g $(NH_4)_2SO_4$; and 1 mL trace element solution in 1 L distilled water (Philip and Deshusses, 2003).

12.2.1 Development of ANAMMOX Culture for NO_x Removal

Experiments were carried out in batch systems using serum bottles. Each serum bottle contained 100 mL of minimal salt medium with 140 mg/L of ammonium-nitrogen and 84.5 mg/L nitrate-nitrogen. The study was carried out at 25°C in an orbital shaker at 150 rpm under anaerobic environment. Samples were collected from the serum bottles at different time intervals and were analyzed for NH_4^+, NO_3^-, COD, and biomass concentrations. In the second set of experiments, all other conditions were kept the same except the electron acceptor. Instead of NO_3^-, 152.1 mg/L NO_2^--N was supplied to the system.

12.2.2 Batch System

ANAMMOX reactors were operated until the removal efficiencies of nitrite, nitrate, and ammonia reached about 80%. The concentrations were brought back to initial concentrations, by adding sufficient quantities of nitrite, nitrate, and ammonia, to study the performance of the system in the subsequent removal of nitrite, nitrate, and ammonia. In this study, cow dung sludge performed better in the presence of nitrite as an electron acceptor. Hence, cow dung sludge with nitrite as an electron acceptor was used for the continuous system.

12.2.3 Continuous Reactor System

The reactor was made from transparent Plexiglass pipe (4 cm inner diameter). The schematic of the experimental setup is shown in Figure 12.1. The total length of the reactor was 60 cm, and the bed height was 50 cm. The gas inlet and outlet ports were located at the bottom and top lids, respectively, of the reactor. The liquid inside the column consisted of mineral medium with the composition specified previously. There was no external carbon source supplied to the column except for CO_2 and carbonate. CO_2 was present in the simulated flue gas, whereas carbonate was present in the nutrient medium. The reactor was seeded with about 300 mL of acclimatized sludge containing a MLSS concentration of 1,000 mg/L, and the remaining volume contained mineral media. During the startup of the continuous experiment, 150 mg/L of NO_2^--N and 230 mg/L of NH_4^+-N were maintained in the system throughout since 150–200 ppm of NO_x gas was supposed to enter the system daily, with a gas flow rate of 1 L/min. Based on the stoichiometry, the concentrations of nitrite nitrogen and ammonia nitrogen discussed were chosen. The system was operated with a hydraulic retention time (HRT) of 24 h. Once the system was stabilized, liquid nitrite feed was replaced by gaseous NO in the feed.

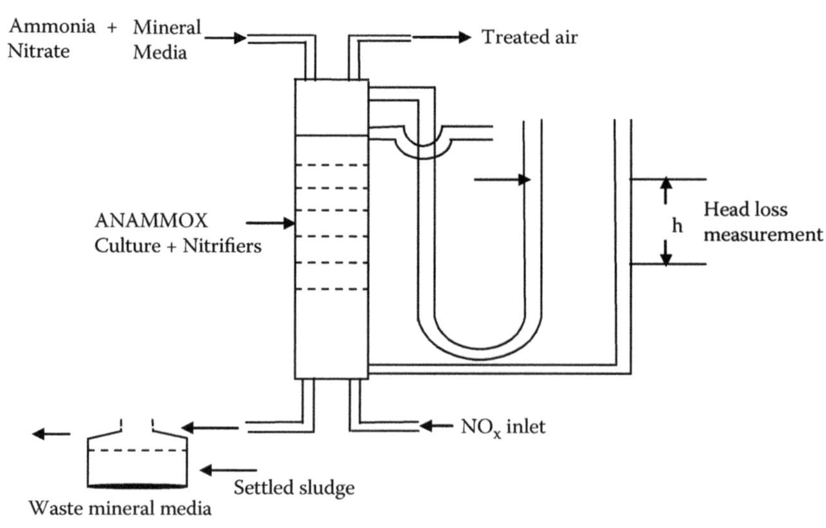

FIGURE 12.1
Schematic of the continuous biological system for NO_x removal.

Simulated flue gas was prepared by mixing a metered flow of approximately 1% NO_x (5,000 ppm), 10% CO_2, and 89% N_2. The gases were homogenized in a mixing chamber and fed to the bioreactor at a flow rate of 1,000 mL/min with the help of a rotameter. Initial studies were carried out in the absence of oxygen in the system with an empty bed retention time (EBRT) of 60 s. The EBRT was gradually reduced up to 10 s. Once satisfactory performance was observed in the system, 5% oxygen was introduced in the simulated gas stream, and the reactor was operated with a retention time of 60 s. The inlet and outlet NO and NO_2 concentrations were measured; ammonia, nitrite, and nitrate concentrations were also measured from the liquid effluent. The mineral medium was added to the system, and effluent was removed from the bottom of the system and allowed to settle in a closed reactor. The supernatant was thrown away, and the settled biomass was circulated to the system. Further, to increase the efficiency of the system, NO was oxidized to NO_2 with the help of ozone.

For the oxidation of NO by ozone, the experiment was conducted in a 5-L borosil bottle at room temperature (27°C). The inlet NO_x concentration was maintained at 150 ppm (145 ppm NO and 5 ppm NO_2), and space time (10 s) was adjusted by controlling the inlet stream flow rate. Ozone was supplied from an ozone generator (Vortex, India), and flow was controlled by a built-in controller.

12.3 Results and Discussion

12.3.1 Performance of Reactors during Batch Study

One of the objectives of the study was to check whether NO can be used as an electron acceptor by ANAMMOX bacteria for the generation of N_2 gas. If this route is possible, control of NO_x from flue gases through biological systems can be achieved easily. To check this concept, reactors were developed for the removal of different concentrations of nitrite and nitrate in the presence of ammonia under anaerobic conditions. Once a strong ANAMMOX culture is developed, it may be easy to replace nitrite and nitrate with NO_x. The performances of these reactors were monitored in terms of COD, nitrite, nitrate, and ammonia removal. The MLSS concentrations in the reactors were also monitored periodically.

Ammonia was one of the critical parameters to be monitored to determine the efficiency of the ANAMMOX process. In this study, ammonia removal in the presence of both nitrite and nitrate as electron acceptors was monitored. Two reactors were operated with different seed sludges (sludge collected from the sewage treatment plant [aerated lagoon] of IIT Madras Chennai and cow dung sludge acclimatized for the ANAMMOX process).

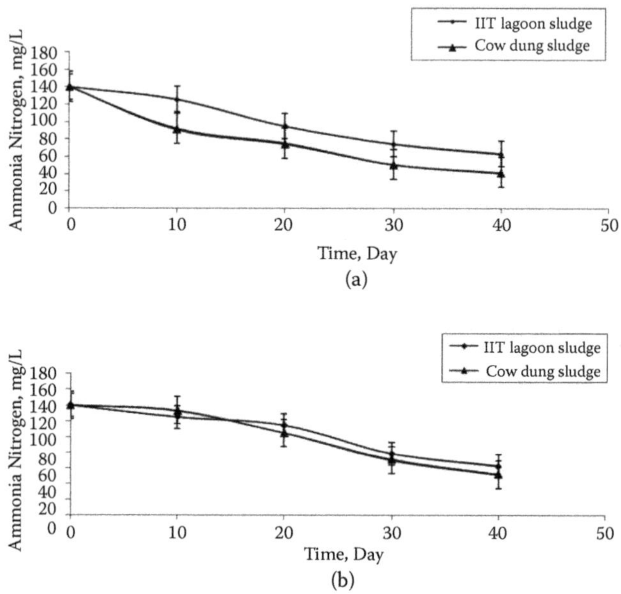

FIGURE 12.2
Ammonia oxidation with (a) nitrite as electron acceptor and (b) nitrate as electron acceptor.

Figures 12.2(a) and 12.2(b) show the result of ammonia oxidation in the presence of nitrite and nitrate as electron acceptors by both the cultures. Ammonia removal was increased in the presence of cow dung sludge with nitrite as the electron acceptor. This may be due to the prior acclimatization of the sludge for the ANAMMOX process. With NO_3^- as the electron acceptor, both the cultures performed in almost a similar manner. The sludge collected from IIT Madras sewage treatment plant may already have contained a considerable population of nitrifiers and denitrifiers. The initial concentration of NH_4^+-N was 140 mg/L, and after the test run of 40 days, the concentration was reduced to 42 mg/L. The NO_2^--N concentration changed from 182 to 38 mg/L. The stoichiometry of the reaction in this study is shown in Equation 12.1:

$$NH_4^+\text{-N}/NO_2^-\text{-N} = 0.69 \quad (12.1)$$

As per reported ANAMMOX studies (Strous et al., 1998), the ratio of NH_4^+-N/NO_2^--N was 1.3. Stoichiometrically, the reported value and the one obtained in the present study did not match well. This may be due to the increased presence of nitrifiers and denitrifiers in the system. A significant quantity of ammonia might be oxidized into nitrite/nitrate in the system. It is reported that anoxic nitrification is possible by a certain group of microbes under oxygen stress conditions.

When nitrate was employed as the electron acceptor, the initial concentration of NH_4^+-N was 140 mg/L, and after the test run of 40 days, the concentration was reduced to 52 mg/L. NO_3^--N concentration changed from 88 to 21 mg/L. The stoichiometry of the reaction is shown in Equation 12.2.

$$NH_4^+\text{-N}/NO_3^-\text{-N} = 1.31 \tag{12.2}$$

As per reported ANAMMOX studies (Mulder, 2003), the ratio of NH_4^+-N/NO_3^--N was 0.6, which is different from the value obtained in this study. It may be inferred that the developed sludge used in this study had more denitrifiers. They might have reduced the nitrate to nitrogen gas utilizing the available COD in the system. There was a significant quantity of COD reduction in the system. The initial MLSS concentration of reactor with cow dung sludge was 1,500 mg/L, which was gradually increased and after the end of batch study (40 days) reached 1,760 mg/L. The reactor with the IIT lagoon sludge was operated with an initial MLSS concentration of 2,100 mg/L, and at the end of the batch study the MLSS concentration in the system was increased to 2,250 mg/L. Although the reactor with IIT lagoon sludge had more MLSS, the performance was better in the reactor with the cow dung sludge. This might be due to the prior acclimatization of this sludge for ammonia oxidation using nitrite/nitrate as electron acceptors.

12.3.2 Performance of Reactors during Continuous Study

The batch study presented clearly showed that cow dung sludge was effective in the removal of nitrite and ammonia (89% of nitrite and 69% of ammonia); hence, cow dung sludge was used for the continuous studies. The system was run continuously in a sequential batch mode with a detention time of 24 h and with initial concentration of 150 mg/L of NO_2^- N. Initially, the study was carried out without introducing NO_x gas. The significance of this study was to make the system become acclimatized to nitrite removal with ammonia oxidation. The amount of nitrite in the system was equivalent to nitrite generated in the system by the scrubbing of NO_x from flue gas for a flow rate of 1 L/min with a concentration of 100 ppmv (parts per million by volume) for 24 h. The residual concentrations of nitrite and ammonia after 24 h was monitored. Every 24-h interval, the concentration of nitrite and ammonia was made up to the initial concentration.

The study of ammonia oxidation with nitrite as an electron acceptor was carried out in batch systems but with a longer HRT, as presented in Section 12.3.1. To assess the ability of developed ANAMMOX culture to remove nitrite, a continuous system was operated with high loading. The results are presented in Figures 12.3 and 12.4.

After 25 days of acclimatization, the continuous system showed 71% ammonia removal and 76% nitrite removal with a detention time of 24 h. The initial concentration of NH_4^+-N was 233 mg/L, and after 23 days of continuous

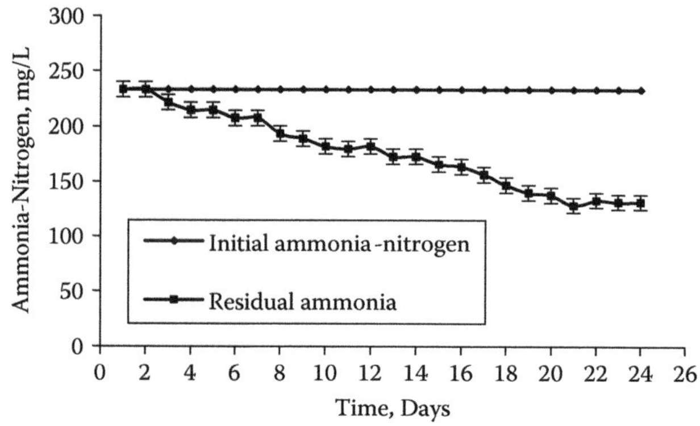

FIGURE 12.3
Kinetics of ammonia-N removal in the system.

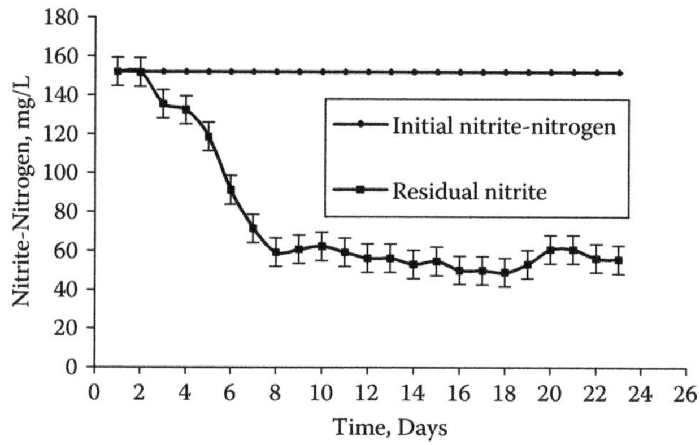

FIGURE 12.4
Kinetics of nitrite-N removal in the system.

operation, the concentration was reduced to 131.4 mg/L. The corresponding nitrite removal was from 152 to 35 mg/L. The stoichiometry was NH_4^+-N/NO_2^--N = 0.87. From the results of this study, it can be inferred that ammonia oxidation and nitrite removal were due to the presence of nitrifiers and denitrifiers. Similar studies were also carried out using nitrate as an electron acceptor (results not presented).

12.3.2.1 Performance of Reactors during Continuous Study without Ozone Oxidation

12.3.2.1.1 NO_x Removal in the Absence of Oxygen

Initially, the study was carried out in the absence of oxygen. During the entire study period, an inlet NO_x concentration of 100 ppmv was maintained in the system, and the outlet concentration was measured regularly. The results of the study are presented in Figure 12.5(a). After the system was operated

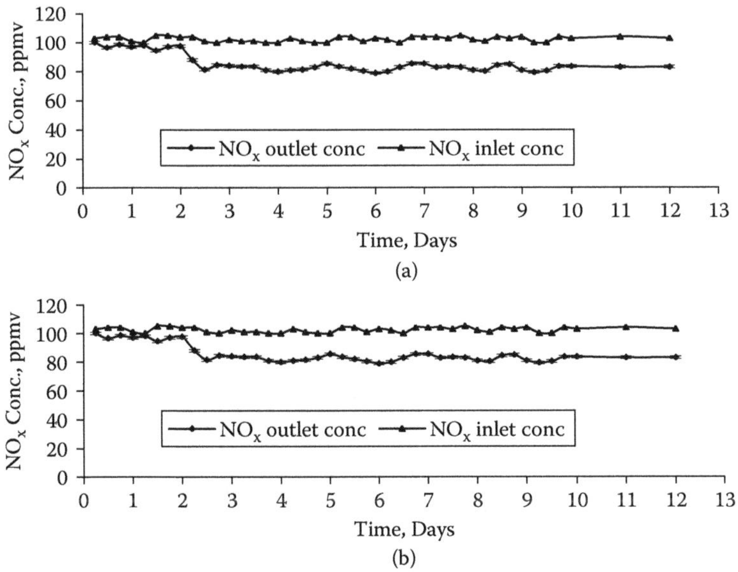

FIGURE 12.5
NO_x removal (a) in absence of oxygen and (b) in presence of oxygen.

continuously for 11 days, the outlet NO_x concentration was reduced to 78 ppmv. The results indicated that the system performance for NO_x removal was not high, with NO_x removal of 22%. This might have been due to the low solubility of NO_x gas, as mentioned.

12.3.2.1.2 NO_x Removal in the Presence of Oxygen

Flue gas normally has 3–8% oxygen. The performance of the system was evaluated in the presence of 5% oxygen by volume. In the presence of oxygen, the following mechanism was possible: Ammonium becomes oxidized to nitrite with the help of autotrophic nitrifiers, and ammonium reacts with nitrite to produce nitrogen gas. A study was carried out with an inlet NO_x concentration of 100 ppm along with 5% oxygen, keeping all other parameters the same as in the previous case (in absence of oxygen). Outlet NO_x concentration was monitored regularly. After 11 days of continuous operation, the outlet NO_x concentration was 80 ppm, and NO_x removal was 20% (Figure 12.5(b)). Although the presence of oxygen did not hinder the activity of the biological system, the overall removal efficiency was still very low, which might have been due to the low solubility of NO_x.

12.3.2.2 Performance of Reactors during Continuous Study after Ozone Oxidation

From the study results presented, it is clear that, using the developed biosystem, it was not possible to remove NO_x in the form of NO due to its low solubility. Hence, to increase the efficiency of NO_x removal, first NO was oxidized with the help of ozone to form NO_2, which was allowed to enter the biosystem. Inlet and outlet NO and NO_2 concentrations were measured along with nitrite, nitrate, and ammonia removal in the aqueous phase of the biosystem. The optimization of NO oxidation by ozone was presented in Barman and Philip (2006). Similar conditions were maintained for the present study. A NO_x concentration of 200 ppmv (175 ppmv NO and 25 ppmv NO_2) was used. Ozone flow was gradually increased, and the oxidation of NO to NO_2 was monitored. It was found that almost all (99%) NO was converted to NO_2 when a slight excess (10%) of a stoichiometric amount of ozone was passed.

12.3.2.2.1 NO_x Removal after Ozone Oxidation

NO_x removal studies were carried out with an inlet NO_x concentration of 150 ppm (145 ppm NO_2; 5 ppm NO) and an EBRT of 60 s. The results are presented in Figure 12.6(a), which clearly indicates that the efficiency of NO_x removal increased significantly after the partial oxidation of NO to NO_2. After 15 days of operation, a NO_x removal efficiency of 79% was achieved in the biosystem. Gradually, the contact time was reduced to 30 s and then to 10 s to understand system behavior at varying detention times. When the detention time was reduced to 30 s, the efficiency of the system was reduced to 60%. Further reduction in EBRT to 10 s deteriorated the performance of the system. This might have been due to the decreased contact of the pollutants with the bioscrubbing medium. The addition of packing material or reducing the bubble size by modifying the diffuser may marginally improve the efficiency of the system.

12.3.2.2.2 Nitrate Removal in the System

Figure 12.6(b) shows the nitrate removal from the system after ozone oxidation. Initially, the nitrate concentration in the system was increased during the study. But after 3 days, the nitrate concentration gradually decreased and reached a steady state in 12 days. After reaching a steady-state nitrate removal of 75%, there was no further improvement. The residual nitrate concentration may be the result of ammonia oxidation. However, as the nitrate loading to the system was increased (by reducing the EBRT and increasing the gas flow rate), the residual nitrate concentration at steady state also was increased marginally. At the retention time of 30 s, 55% nitrate removal was observed, whereas the 10-s retention time (gas) gave a nitrate removal efficiency of 45%.

12.3.2.3 Performance of the Control Reactor

A control reactor was operated in identical conditions, except for the absence of the microbes in the system, to assess the abiotic removal of NO_x in the system. The system was operated at a detention time of 60 s.

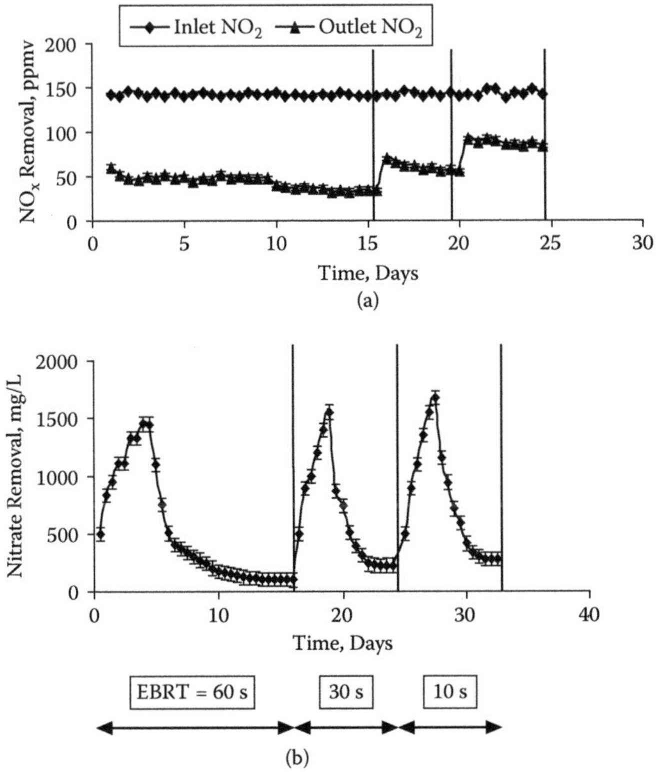

FIGURE 12.6
(a) NO_x removal in the continuous system after ozone oxidation. (b) Nitrate removal in the system. (*Note:* The reactor was operated with different EBRTs. The different sections represent EBRTs of 60, 30, and 10 s.)

12.3.2.3.1 NO_x Removal in the Control Reactor

The results are presented in Figure 12.7(a). There was about 74% NO_x removal initially, and gradually the NO_x removal efficiency was reduced to 68%. From the result, it is clear that the major mechanism of NO_2 removal was scrubbing. When microbes are present in the system, the accumulated nitrate is utilized. This gives a favorable concentration gradient to enhance the NO_x removal, which may be the reason for better performance with microbes present in the system.

12.3.2.3.2 Nitrate Removal in the Control Reactor

Figure 12.7(b) shows abiotic nitrate removal from the system. As time progressed, the nitrate concentration kept increasing. This was due to the absence of autotrophic ANAMMOX culture in the control system. In the biological system presented, ANAMMOX culture was responsible for

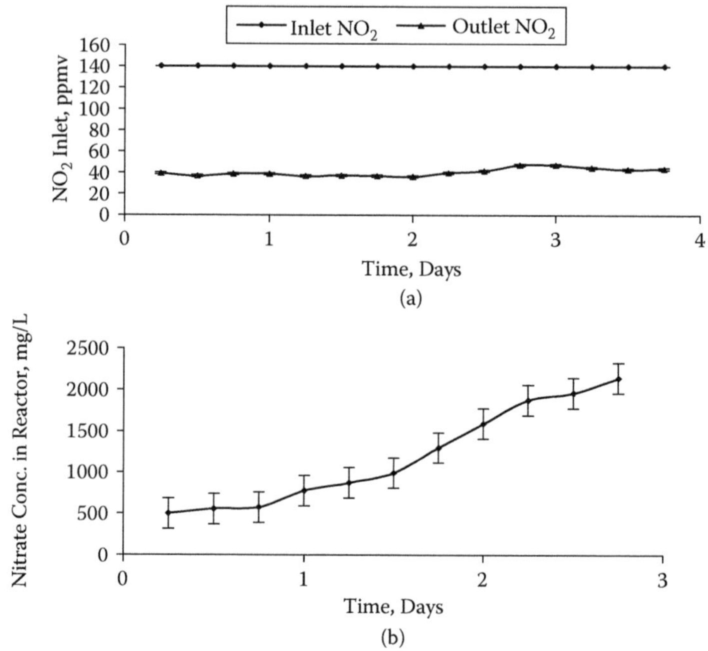

FIGURE 12.7
(a) NO_x removal and (b) Nitrate accumulation in the system with identical reactor but without microbes.

nitrate removal from the system. When 150 ppmv of NO_2 gas was allowed to enter the system for a detention time of 1 min, NO_2 in the gas phase was converted to nitrate in the presence of water in the system. The presence of autotrophic culture was responsible for removal of up to 75% of nitrate in the biological system. However, in the absence of microbes in the abiotic system (i.e., in this control reactor), nitrate was not removed; instead, it accumulated in the system.

12.4 Conclusion

In this study, a novel bioscrubber was developed for the complete treatment of NO_x from flue gases. As a first step, an autotrophic ANAMMOX system was developed using ammonia as the electron donor and nitrite/nitrate as the electron acceptor. Once an efficient ANAMMOX culture was developed, nitrate or nitrite was replaced by NO_x. Batch experiments indicated that ammonia oxidation and simultaneous nitrite/nitrate removal was possible in the presence of acclimatized cow dung sludge. Maximum removal efficiency of 79% of NO_x gas at a detention time of 60 s, 60.1% at 30 s, and 40.3% at 10 s was achieved after ozone oxidation. Less removal of NO_x was observed before ozone oxidation due to the low solubility of NO gas. The biological system containing ANAMMOX cultures seems to be a feasible option for treating nitrogen oxides from flue gas.

References

Arts A.M.P., Robertson L.A., and Kuenen J.G. 1995. Nitrification and denitrification by *Thiosphaera pantotropha* in aerobic chemostat cultures, *FEMS Microbiology and Ecology*, 8: 305–316.

Barman S., and Philip L. 2006. Integrated system for the treatment of oxides of nitrogen from flue gases, *Environmental Science and Technology*, 40: 1035–1041.

Flanagan W.P., Apel W.A., Barnes J.M., and Lee B.D. 2002. Development of gas phase bioreactors for the removal of nitrogen oxides from synthetic flue gas stream, *Fuel*, 81: 1953–1961.

Jetten M.S.M., Strous M., van de Pas-Schoonen K.T., Schalk J., van Dongan U.G.J.M., van de Graff A.A., Logemann S., Muyzer G., van Loosdrecht M.C.M., and Kuenen J.G. 1999. The anaerobic oxidation of ammonium, *FEMS Microbiology Review*, 22: 421–437.

Lee K.C., and Rittmann B.E. 2000. A novel hollow-fibre membrane biofilm reactor for autohydrogenotrophic denitrification of drinking water, *Water Science and Technology*, 41: 219–226.

Mulder A. 2003. The quest for sustainable nitrogen removal technologies, *Water Science and Technology*, 48(1): 67–75.

Perry R.H., Green D.W., and Maloney J.O. 1984. *Perry's Chemical Engineers Handbook* (6th ed.), McGraw-Hill International Editions, New York, pp. 3–98.

Philip L., and Deshusses M.A. 2003. Sulfur dioxide treatment from flue gases using a biotrickling filter-bioreactor system, *Environmental Science and Technology*, 37: 1978–1982.

Rhoads T.W., Marks J.R., and Siebert P.C. 1990. Overview of industrial source control for nitrogen oxides, *Environmental Progress*, 9(2): 126–130.

Strous M., Heijnen J.J., Kuenen J.G., and Jetten M.S.M. 1998. The sequencing batch reactor as a powerful tool for the study of slowly growing anaerobic ammonium-oxidizing microorganisms, *Applied Microbiology and Biotechnology*, 50: 589–596.

Yu X., Qi Z., Zhang X., Yu P., Liu B., Zhang L., and Fu L. 2007. Nitrogen loss and oxygen paradox in full-scale biofiltration for drinking water treatment, *Water Research*, 41: 1455–1464.

Section IV

Integrated Management Case Studies

13

Integrated Approach to Rice Straw Management for Reduction of Field-Burning Activity

Nguyen Thi Kim Oanh

CONTENTS

13.1 Current Practice of Rice Straw Field Burning and Emission

Rice (*Oryza sativa*) is a globally cultivated crop and is by far the most popular crop in Asia (Kadam et al., 2000). At least 114 countries in the world grow rice, and more than 50 have an annual production of 100,000 metric tons (tonnes) or above. Asian farmers produce about 90% of the total, with two countries, China and India, growing more than half of the crop. Annual world rice production, in million tonnes of milled rice, increased from 432 in 2007/08 to 449 in 2010/11 and is projected to go up by 2% to 459 in 2011/12 (International Grains Council [IGC], 2012). The annual increase is expected to concentrate in Asian countries to meet the domestic and export demand. In Southeast Asian countries, rice is the staple food that provides 50–80% of the total calories consumed. The region also housed the world's top two rice exporters in 2009, Thailand and Vietnam (Rice-Trade, 2010).

Field burning of rice straw (RS) after harvesting, an inexpensive method of crop residue disposal, is practiced in many parts of the world. This is done to clear the surface biomass from land for faster crop rotation; to control undesirable weeds, pests, and diseases; and to return some nutrients to the soil. As farmers become wealthier, the amount of crop residues used as cooking fuel or animal feedstock is projected to decrease. Cao et al. (2006) cited a confirmed relationship between the fraction of crop straw field burned and the peasants' income level in China, with a correlation coefficient of 0.81. Future increase in the rice production and concomitant reduction in RS off-site uses would increase field burning if no control measures are taken.

This RS field-burning activity, uncontrolled and with incomplete combustion by nature, emits a large amount of toxic air pollutants (fine particles, inorganic and organic gases) and climate forcers (Andreae and Merlet, 2001; Reid et al., 2004). Agricultural field-burning activities are linked to a deterioration of air quality in many locations: the United States (Jiminez et al., 2007), Europe (Viana et al., 2008), and Asia, such as Taiwan (Yang et al., 2006) and Thailand (Tipayarom and Kim Oanh, 2007). In Asia, intensive field-burning activities commonly take place during the dry season when stagnant atmospheric conditions normally exist, which consequently leads to even higher air pollution levels in rural and adjacent urban areas (Tipayarom and Kim Oanh, 2007). Overall, emissions from field-burning activities deteriorate local air quality and have been reported to cause high personal exposure (Wu et al., 2006) and result in adverse health effects (Torigoe et al., 2000).

Annually, a huge amount of crop residue is field burned, which emits a large amount of toxic pollutants and climate forcers. In Asia, the crop residue field burning is an important type of biomass open burning. This activity has been reported to contribute substantially to national emissions. In China, for example, about 19.4–24.4% of crop residue is field burned (Cao et al., 2006; Yan et al., 2006), contributing about 100 gigagram (Gg) of black carbon (BC) and about 400 Gg of organic carbon (OC), which corresponds to a share of 6.6% and 9.3% of the national total emission (all sources) of the respective pollutants in 2000 (Cao et al., 2006). Yan et al. (2006) also reported emissions of other species from the crop residue field burning in China for 2000 (rounded values in Gg) of particles with an aerodynamic diameter less than 2.5 μm ($PM_{2.5}$) 476; CO_2 184,951; CO 11,231; CH_4 330; and nonmethane hydrocarbons (NMHCs) 1,917. In Thailand, in 2007 the crop residue field burning was estimated to contribute 10 Gg BC (7% of the country's total BC

TABLE 13.1

Field Burning of Rice Straw (Tg) in Selected Countries and Emission (Gg)

Pollutants	India (1999–2005)[a]	Philippines (2002–2006)[a]	Thailand (2002–2006)[a]	Thailand (2007)[b]	China (2004)[c]
RS field burned, Tg (% of total RS)	13.92 (14%)	10.15 (95%)	10.45 (48%)	20.34 (68%)	24.09 (24%)
CO_2	16,253	11,850	12,207	15,360	18,952 ± 299
$PM_{2.5}$	144	105	108	108	
BC				6.5	
OC				39	
CH_4	13.4	9.7	10.0	125	
NMHC	44.5	32.5	33.4	91	
CO	386	282	290	1,210	1,538 ± 117
N_2O	0.78	0.57	0.59		
NO_x	35	25	26	29.7	43.5 ± 2.2
PAH	0.21	0.15	0.16		

Sources: [a]Gadde et al. (2009); [b]Kanabkaew and Kim Oanh (2010); [c]presented values were estimated from Zhang et al. (2008) based on the shares of RS field burning and fuel use.

emission) and 54 Gg OC (17%); 143 Gg PM_{10} (particles with an aerodynamic diameter less than 10 μm) (11%); 1,453 Gg CO (15%); 42 Gg nitrogen oxides (NO_x) (5%), and 108 Gg (4%) NMHCs (Kanabkaew and Kim Oanh, 2010).

In many Asian countries, RS by far is the largest agricultural residue subjected to field burning. The portion of RS burned in the field varies among the countries, and it is estimated to range from as high as 95% in the Philippines and 48% in Thailand to 14% in India (Gadde et al., 2009). A survey conducted in the central part of Thailand showed even a larger portion of RS field burning, above 90%; hence the country average portion of RS field burned may be as high as 68% (Kanabkaew and Kim Oanh, 2010). Accordingly, RS field burning has a high emission share in the country's total crop residue field burning, that is, about 80% on average, and ranges from 65% for BC to 95% for CH_4 (Kanabkaew and Kim Oanh, 2010). A summary of the RS field-burning activities and emission generated in selected countries is given in Table 13.1.

Despite its high contribution to air pollution, the RS field-burning emission is often overlooked in air quality management in Asia, as highlighted in Chapter 1. It is important to demonstrate the significance of this source in terms of emissions of toxic air pollutants and climate forcers and consequent effects. The source-impact relationship associated with this field-burning activity should be analyzed to generate information pertinent to policy makers and the public. Such information can be disseminated to minimize RS field burning effectively and to promote nonburning alternatives. This chapter presents and analyzes integrated RS management measures based on the results of a multiyear study conducted at the Asian Institute of Technology (AIT).

13.2 Characterization of Rice Straw Burning Emission

A comprehensive size and compositional characterization of air pollution emission from RS burning was conducted by AIT during 2003–2006 using both in situ experiments and laboratory hood experiments. Emission factors (EFs) of mass and composition for different particle size ranges, and gaseous pollutants were determined for both spread- and pile-burning conditions.

13.2.1 Pile versus Spread Field Burning of RS and Emission

Spread burning of RS refers to burning on harvested paddies after combine harvesters have been used (e.g., in Thailand). The equipment cuts the upper part of RS and spreads it in windrows while leaving the lower (standing) part virtually untouched. The spread RS is mostly burned, while the standing parts in many cases are only partially burned, especially when the RS moisture is high. In places where a manual harvesting technique is still used (e.g., many regions of Vietnam or Indonesia), the RS after harvest is piled up at a paddy corner and burned, largely under smoldering conditions, with a visible dense smoke plume.

The pile-burning emission characterization was conducted in a laboratory fume hood at AIT, whereas the spread-burning experiments were done both in situ in paddy fields in Pathumthani, Thailand, and in the laboratory hood. The results showed that EFs varied significantly, reflecting a number of factors that affect the combustion conditions (e.g., biomass loads, weather conditions for in situ field burning, moisture content of RS and paddy soil) (Kim Oanh et al., 2011). In particular, the emission strongly depended on the burning types (i.e., spread or pile burning), with pile burning emitting significantly higher amounts of air pollution. The pile-burning experiments with biomass loads of 1.0–1.3 kg/m^2 (i.e., about two times the average biomass load in paddy fields in central Thailand, 0.58 ± 0.14 kg/m^2) produced particulate matter (PM) emission of 20 ± 8 g/kg dry RS, which is about two to three times the amount of $PM_{2.5}$ (8 ± 3 g/kg RS) or PM_{10} (9 ± 4 g/kg RS) emitted from spread burning.

13.2.2 Size of Particles in RS Field-Burning Smoke

The majority of particles in the RS-burning smoke in spread field burning experiments were in the submicron size range. Small particles of the size 0.4–1.1 μm contribute the largest fraction of PM mass (Figure 13.1). These fine particles can enter deep into the lung and hence are more health hazardous than coarser particles (Pope et al., 2009).

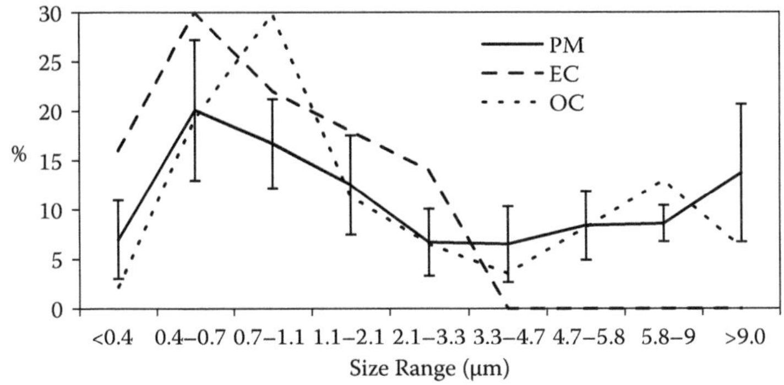

FIGURE 13.1
Percentage of PM mass, organic carbon (OC), and elemental carbon (EC) in different particle size ranges of spread RS field burning smoke. (*Source:* Adapted from Kim Oanh N.T., Thuy L.B., Tipayarom D., Manadhar B.R., Pongkiatkul P., Simpson C.D., and Liu L.S. 2011. Source characterization of aerosol emission from field burning of rice straw, *Atmospheric Environment*, 45: 493–502.)

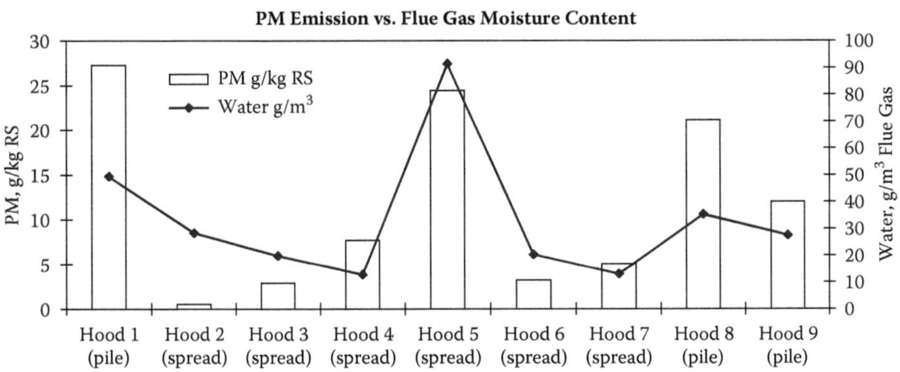

FIGURE 13.2
Particulate matter emission versus moisture content in flue gas based on hood experiments. (*Source:* Adapted from Kim Oanh N.T., Thuy L.B., Tipayarom D., Manadhar B.R., Pongkiatkul P., Simpson C.D., and Liu L.S. 2011. Source characterization of aerosol emission from field burning of rice straw, *Atmospheric Environment*, 45: 493–502.)

13.2.3 Moisture Content and PM in RS Hood Burning Smoke

The relationship between PM EFs (g/kg dry RS burned) and the flue gas moisture content (g/m^3 flue gas) was analyzed based on results obtained in the hood burning experiments and is shown in Figure 13.2. The results showed that in general the higher the flue gas moisture content was, the higher PM emission was. The moisture content in flue gas reflects not only the moisture content of RS but also that of the surface paddy soil; both were included in the RS samples used for the hood burning experiments. This implies that burning dry RS on dry paddy soil would generate less PM air pollution. It is therefore important to avoid burning when RS or paddy soil is not dry enough to minimize emission.

13.2.4 Particulate Matter Composition

The most significant components of PM (both $PM_{2.5}$ and PM_{10}) include OC, water-soluble ionic species, elemental carbon (EC), and levoglucosan. The levels of the major components (in mg/g of $PM_{2.5}$) are 330 for OC, 160 for water-soluble ions (collectively), 57 for EC, and 56 for levoglucosan (Kim Oanh et al., 2011). A ratio between EC and TC (total carbon) of 0.15 was obtained for $PM_{2.5}$ emitted from RS burning, which is significantly lower than that (above 0.70) obtained for $PM_{2.5}$ emitted from diesel vehicles (Kim Oanh et al., 2010). This feature serves as an important marker to distinguish between these two source categories for PM source apportionment studies by the receptor modeling tool as discussed in Chapter 3.

A range of toxic organic compounds, such as polycyclic aromatic hydrocarbons (PAHs), organochlorine pesticides, and polychlorinated biphenyls (PCBs), has also been identified in PM of RS-burning smoke (Kim Oanh et al., 2011), which further confirmed the health hazards of the emission. Relative abundance of certain compounds among PAHs (e.g., fluoranthene and pyrene) and methoxyphenols (e.g., acetylsyringone) can also serve as markers for RS-burning PM in receptor modeling studies.

13.2.5 Estimation of Emission from Crop Residue Field Burning in Thailand

The total rice cultivation area in Thailand is 108,283 km^2 (Office of Agricultural Economics [OAE], 2007), which produces a large amount of RS annually. Even though only a portion of this huge amount of RS is field burned, a large quantity of air pollution emission is expected. The results of a multipollutant emission inventory for crop residue field burning in Thailand (Kanabkaew and Kim Oanh, 2010) showed the most likely (best) emission estimate (in Gg) from this activity in 2007 was 128 of $PM_{2.5}$; 143 of PM_{10} ; 4 of SO_2 ; 21,400 of CO_2; 1,453 of CO; 42 of NO_x; 59 of NH_3; 132 of CH_4; 108 of NMHCs; 10 of EC; and 54 of OC. RS burning provided by far the largest contribution (i.e., around 80% on average) to the country's total crop residue field-burning emissions. The emission was highest during the dry season and in the central part of the country. Figure 13.3 shows the spatial distribution of PM_{10} and CO emission strength from the crop residue open burning in Thailand.

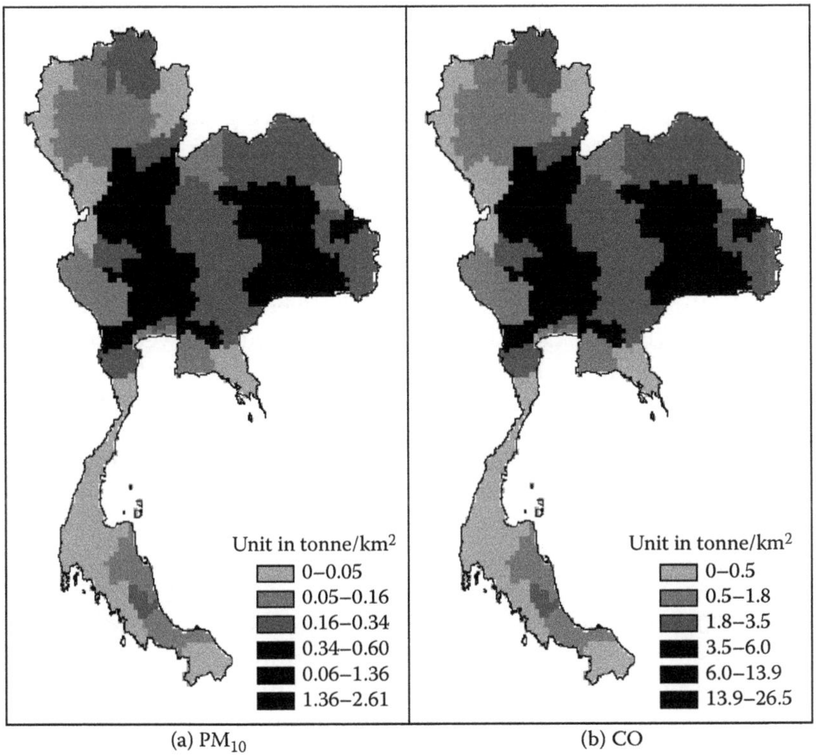

FIGURE 13.3
Spatial distribution (0.1° × 0.1°) of the annual strength of PM_{10} and CO emission from crop residue field burning in Thailand (RS burning contributed about 80% of the total crop residue field burning emission). (*Source:* Adapted from Kanabkaew T. 2011. Assessment of effects from crop residue field burning on particulate matter air quality, doctoral dissertation, Asian Institute of Technology, Pathumthani, Thailand.)

13.3 Potential Impacts of Rice Straw Field-Burning Emission

13.3.1 Relationship between RS Field Burning and Ambient Air Quality

In Pathumthani Province, located in the central part of Thailand, farmers grow on average five rice crop cycles over 2 years (i.e., two or three crop cycles annually). The RS of the major crop (harvested in November–December) and secondary crop (harvested mainly in March–April) is always burned in the field. The burning takes place 2–7 days following the harvesting of a paddy and generally in the afternoon, between 1100 and 1700. Only the RS from the third crop, which is harvested during the wet season, is not always burned in the field due to weather conditions. Overall, 90% of RS produced is burned in this part of the country (Tipayarom and Kim Oanh, 2007).

Tipayarom and Kim Oanh (2007) analyzed the casual link between the RS field burning and ambient air pollution in the Pathumthani Province using satellite data (MODIS [Moderate Resolution Imaging Spectroradiometer] hotspots that represent open fires) and ambient air-monitoring data. Each hotspot on a satellite image may contain one or more active burning fires within a 1-km pixel. The hotspot data were obtained from Web sites (http://earthobservatory.nasa.gov/NaturalHazards/ and http://maps.geog.umd.edu/). The ambient monitoring data for CO and PM_{10} at the Rangsit automatic monitoring station were collected from the Pollution Control Department (PCD) of Thailand.

Generally there were good correlations between hotspot counts and the pollutant levels (Figure 13.4). The linear relationship between the monthly hotspot counts and monthly levels of air pollutants had an R^2 of 0.56 for CO and 0.77 for PM_{10}. Higher numbers of hotspots were observed during the RS-burning period (from November to April the next year) and lower numbers of hotspots observed during the rainy months (May to October). Note that there is actually no forest in Pathumthani, so the detected hotspots most likely indicate crop residues or grassland open fires. RS burning mostly covers small paddy plots (much less than 1 km each side) and for a short period of time (1–2 h). The MODIS satellite passes over Thailand twice per day, at 1030 and 1330 local time (see Chapter 4), so it cannot capture short-lived fires that happen at other times. Thus, hotspot counting based on the available satellite imagery may underestimate short-lived fire spots in an area.

In the dry season when the northeast monsoon is prevalent, the rice-growing area of Pathumthani is upwind of Bangkok City; hence, emission

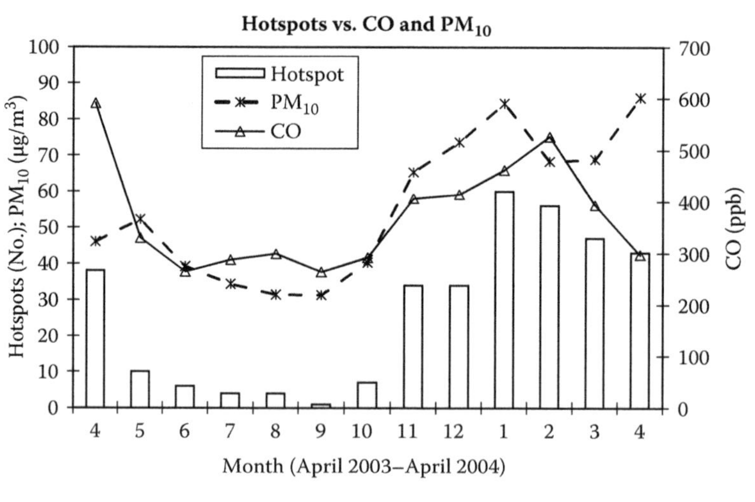

FIGURE 13.4
Hotspots versus monthly average CO and PM_{10}. (*Source:* Adapted from Tipayarom D., and Kim Oanh N.T. 2007. Effects from open rice straw burning emission on air quality in the Bangkok metropolitan region, *ScienceAsia*, 33(3): 339–345.)

from intensive RS field burning affects not only local air quality but also air quality in downwind regions, including Bangkok. Results of the HYSPLIT (hybrid single-particle Lagrangian integrated trajectory) model (http://www.arl.noaa.gov/ready/hysplit4.html) confirmed the main directions of air mass movement (trajectories) toward the city, and the travel time from the source region to the city center was estimated as less than 10 h. Thus, the emission from RS burning can contribute directly to air pollution in Bangkok, adding to other intensive emission sources already existing in the city.

13.3.2 Polycyclic Aromatic Hydrocarbons in Ambient Air in Intensive Burning Areas

Many PAHs are toxic and carcinogenic air pollutants. They are emitted from an incomplete combustion process. Ambient air quality monitoring was conducted during 2006–2007, focusing on PAHs in two areas in Thailand: the rural area of Klong Luang, Pathumthani, where intensive RS burning is common, and a remote area of Khaoyai National Park with less RS burning (Klinmalee, 2008). The monitoring was done in both dry (more burning) and wet (less burning) seasons for 16 U.S. Environmental Protection Agency (USEPA) priority PAHs, ranging from naphthalene to ideno(1,2,3-c,d)pyrene following the USEPA TO-13A method (USEPA, 1999). A calibrated high-volume sampler was used to collect PAH samples in PM phase (on a quartz fiber filter) and gaseous phase (on XAD-2 adsorbent).

The monitoring results (Figure 13.5), averaged over 10–20 samples for each site, clearly showed that PAH levels were significantly higher in the RS-burning area during the dry season, that is, it totaled about 400 ng/m^3 (excluding naphthalene), as compared to about 10–40 ng/m^3 during the wet season. The levels of PAHs measured in the remote area with less open burning were low in both seasons (i.e., 1–2 ng/m^3). PAHs were mostly found in the gas phase. The detected PAH compounds were mainly of low molecular weight (from acenaphthene to pyrene), typical markers of biomass burning (Kim Oanh et al., 2011).

13.3.3 Exposure to Burning Smoke and Health Effects

Emission from intensive RS field-burning activities results in higher exposure to air pollution of people living around RS plantation areas during the burning season (dry season), which may result in measurable health effects. Klinmalee (2008) compared the exposure of schoolchildren in the RS growing area of Klong Luang, Pathumthani, and schoolchildren living in the remote area of Khao Yai National Park during the 2006 dry season. The personal exposure of two selected students from each area was monitored. The monitoring equipment was with each selected student for 24 hours per monitoring day (i.e., at bedtime, the equipment was kept near the bed). The

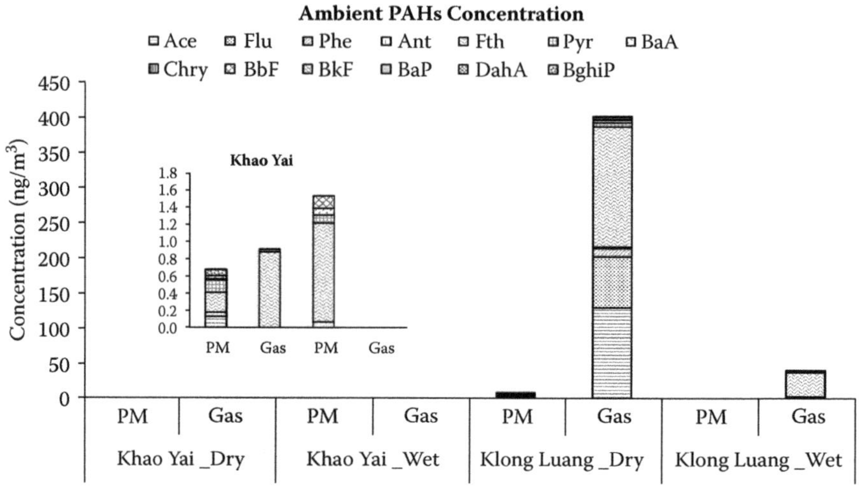

FIGURE 13.5
Ambient levels of PAHs at Khao Yai and Pathumthani during dry and wet seasons. acenaphthylene (Acy); acenaphthene (Ace); fluorene (Flu); phenanthrene (Phe); anthracene (Ant); fluoranthene (Fth); pyrene (Pyr); benz(a)anthracene (BaA); chrysene (Chry); benz(b)fluoranthence (BbF); benz(k)fluoranthence (BkF); benz(a)pyrene (BaP); dibenz(a,h)anthracene (DahA); and benz(g,h,i)perylene (BghiP). (Adapted from Klinmalee A. 2008. The development of assessment procedure for exposure to open rice straw burning emission: a case study of school children in Thailand, doctoral dissertation, Asian Institute of Technology, Pathumthani, Thailand.)

results showed that during both weekdays and weekend, the schoolchildren in Klong Luang area were exposed to higher $PM_{2.5}$ levels than those in the Khao Yai National Park (Figure 13.6).

Measurements of lung functions, including the forced expiratory volume in 1 s (FEV_1) and peak expiratory flow (PEF), were collected for 50 schoolchildren between the ages of 14 and 15 in each area. The ratio between FEV_1 and body mass index (BMI; body mass in kilograms divided by the square of height in meters, kg/m^2) was calculated and utilized as an indicator together with PEF. In the dry season with intensive RS burning, the average and the range of FEV_1/BMI ratios, expressed in liters per kilogram per square meter, for children in the Khao Yai National Park were 0.1 (0.06–0.16), higher than that of 0.07 (0.03–0.11) for the Klong Luang children. The *t*-test results showed that the FEV_1/BMI ratios between the two areas were statistically different, at a significance level of 0.02. However, the PEF average values and range (in parentheses) in liters per minute of 364 (183–583) for children in Khao Yai and 345 (183–709) for children in Klong Luang were not statistically different. Further studies for longer time periods and including other relevant biomarkers are necessary to better reveal the relationships between the exposure to RS-burning smoke and health effects of schoolchildren and people living in intensive burning areas.

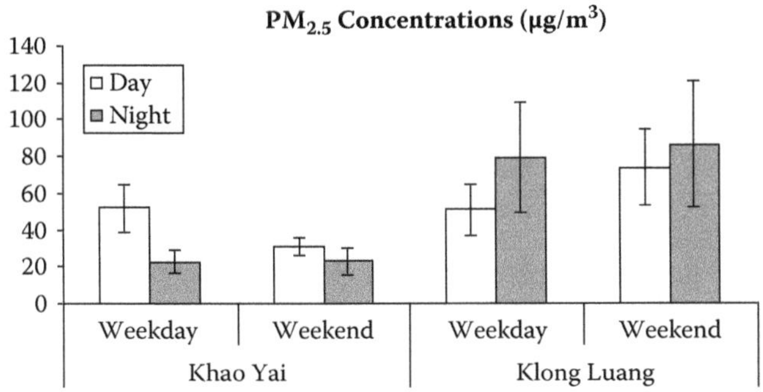

FIGURE 13.6
Schoolchildren's personal exposure to $PM_{2.5}$ in dry season during daytime (0800–1530) and nighttime (rest of the day) in two monitoring areas. (Adapted from Klinmalee A. 2008. The development of assessment procedure for exposure to open rice straw burning emission: a case study of school children in Thailand, doctoral dissertation, Asian Institute of Technology, Pathumthani, Thailand.)

13.4 Nonburning Alternatives

13.4.1 Current and Potential Nonburning Alternatives for RS Management in Asia

To introduce relevant sustainable RS management practices to local farmers, investigation and promotion of nonburning alternatives were conducted through the Sustainable Rice Straw Management for Urban Air Pollution Reduction in Bangbuathong, Nonthaburi Province, Thailand project (Kanokkanjana and Bridhikitti, 2007). Nonthaburi Province is located about 20 kilometers from Bangkok City. Approximately 50% (or 305 km^2) of the province area is utilized for agricultural purposes, of which paddy fields are the majority. The traditional rice-growing method in this province is similar to other central provinces of Thailand (i.e., about five crop cycles per 2 years). There are two main types of rice species cultivated in this area: the 90-day rice crop and the 150-day rice crop. In Thailand, rice-harvesting machines (combine harvesters) are commonly used, and RS spread over paddy fields after harvesting is left in the field for 7–10 days to dry before burning (Tipayarom and Kim Oanh, 2007).

Thailand ratified the Association of Southeast Asian Nations (ASEAN) Agreement on the Transboundary Haze Pollution in July 2003. The country also has initiated a master plan for controlling open burning of agricultural residues. The government has policies to promote nonburning alternatives, and the Department of Agricultural Extension (DAE) and PCD are responsible for implementation. The DAE and PCD promoted a field composting method that involves using a powerful plow machine on harvested paddy to bury RS in the soil. Water, bioextracted liquid, and biofertilizer are applied to accelerate the residual degradation. However, a survey conducted by Kanokkanjana and Bridhikitti (2007) showed that generally the field composting method was not perceived as suitable by local farmers, primarily because of the long period it requires, which delays crop rotation. Cost of such a plow machine was also a problem.

Although RS field burning is still the choice, local farmers are also aware of negative effects of field RS burning on paddy soil quality that may affect the productivity of the next crops (Kanokkanjana and Bridhikitti, 2007). For example, after burning the soil structure becomes tight and hard. Besides, some organic nutrient substances on the topsoil are also burned to ash. The impacts of smoke on human health are not yet well perceived, but a number of local residents complained about the decrease in visibility for road transportation due to RS-burning smoke. The survey also showed that there is a large demand for RS for cattle feedstock, gardening, and mushroom growing in the area. Especially during dry and hot periods, the supply of grass is insufficient; hence, RS is an important substitute for cattle feedstock. Baled or packed RS is preferable for this purpose; unfortunately, these types are not locally available in Nonthaburi, so people need to purchase the RS from other provinces. Local gardeners in the province traditionally use RS (loosely packed) for mulching to prevent moisture evaporation and to control weeds on gardens growing vegetation and orchards. RS can also be used as a medium to culture mushrooms, which can be done at the household level, thus bringing in additional income for the rural poor. Currently, RS utilization as fuel for domestic combustion, industry, and power plants has only a small share in the country. For example, only 3 of 77 biomass power plants in Thailand used RS for power generation, while the rest mainly used bagasse and rice husks, which in principle are not subject to field burning (Department of Alternative Energy Development and Efficiency [DEDE], 2007). Promotion of off-site uses of RS can help minimize field burning and implement nonburning alternatives successfully.

A foremost prerequisite for off-site RS uses is its collection from paddy fields and transportation to the use sites. RS collection, however, is both time and labor intensive due to a large RS volume spread thinly over vast paddy fields after combine harvesters are used. The bulky RS is also difficult to transport and store. Kanokkanjana and Bridhikitti (2007) demonstrated the application of mechanical equipment, a RS baler (Figure 13.7), which efficiently removed most spreading parts of RS and produced compressed RS bales. Such an RS baler can be used on paddy fields, normally 2–3 days after harvesting, hence quickly and properly clearing land for the next crop.

This RS collection method appears to provide a win-win solution for all involved. Farmers can remove RS in time for the next cultivation while generating additional income from selling the RS bales. However, such baler equipment with the operator's service should be available at a reasonable

FIGURE 13.7
Rice straw baler. (Photo courtesy of K. Kanokkanjana.)

cost. Good coordination with potential RS buyers is necessary to ensure there is a market for RS bales.

Other uses of collected RS are also possible with a high potential to be successfully promoted in Asia. Those include traditional uses in handicrafts (hats, mats, decoration items) or as construction material. The latter is initiated in a World Bank project in Vietnam, "Low-Cost Housing: Waste Rice Straw Construction Panels" (World Bank [WB], 2010). RS can also be used as fuel in local industries and power plants and for domestic cooking. In addition, RS can be processed to produce biofuel. A prerequisite of all the off-site uses, however, is the efficient collection and transportation of RS.

13.4.2 Prevented Emission of Air Pollution and Climate Forcers

Nonburning alternatives for RS management will help to relieve a huge emission burden of air pollution and climate forcers. The emission from burning of 1 km^2 of rice paddy was estimated and is presented in Table 13.2. The EFs were obtained from the in situ field experiments conducted in Thailand by Kim Oanh et al. (2011) for typical paddies harvested by combine harvesters. Note that for locations where manual harvesting is still practiced and the pile RS burning is conducted, even higher amounts of air pollution will be released.

Nonburning alternatives thus would help prevent an enormous amount of emissions generated by RS field-burning activities in Asia, where the rice cultivation area is about 3 million km^2 (Food and Agricultural Organization [FAO], 2010). Assuming the average dry biomass load of 0.58 kg/m^2 of paddy (Kim Oanh et al., 2011), the total amount of RS produced from the cultivation area per crop cycle would be about 1,700 million tonnes (Tg). If the current

TABLE 13.2

Estimates of Air Pollution and Climate Forcers Emission from Field Burning 1 km^2 of Rice Paddy per Crop Cycle

Species	EF, g/kg RS[a]	Emission, tonnes/km^2	Primary Effects
$PM_{2.5}$	8.3	4.8	Toxic and visibility/climate effects
PM_{10}	9.4	5.5	Toxic and visibility/climate effects
CO	97	56	Toxic and ozone pollution precursor
BC	0.51	0.3	Toxic and strong climate warming forcer
OC	2.99	1.7	Toxic and climate cooling forcer
CO_2	1147	665	GHG (greenhouse gas) but assumed to be adsorbed by next crop

[a] EF were taken from Kim Oanh N.T., Thuy L.B., Tipayarom D., Manadhar B.R., Pongkiatkul P., Simpson C.D., and Liu L.S. 2011. Source characterization of aerosol emission from field burning of rice straw, *Atmospheric Environment*, 45: 493–502, assuming the dry RS load is 0.58 kg/m^2 of paddy.

practice of field burning (about 20% of the RS produced is burned in the field) can be stopped, then a huge amount of air pollutants and climate forcer emissions could be eliminated. Application of nonburning alternatives for RS management would bring in significant cobenefits for air quality and climate-forcing mitigation.

13.5 Summary and Recommendations

Rice is a globally cultivated crop and is by far the most popular crop in Asia, where the annual production is rapidly increasing to meet domestic and export demand. Field burning of RS after harvesting is commonly practiced by farmers in the region. This activity generates large amounts of toxic air pollutants, such as fine particles; CO; volatile organic compounds (VOCs); semi-VOCs, including carcinogenic PAHs; and climate forcers and contributes significantly to the national and Asian regional total emission. A comprehensive RS-burning emission characterization conducted at AIT showed that air pollution emission varied significantly depending on a number of factors that affect the combustion conditions (e.g., biomass loads, weather conditions, moisture content of RS and paddy soil). Pile burning of RS emits substantially higher amounts of toxic pollutants than spread RS burning. Likewise, emission from RS burning increases with the moisture content of RS and of paddy soil. This suggests that if field burning is unavoidable, in certain cases the activity should be conducted only when RS and paddy topsoil are dry enough to reduce the emission.

There were linear relationships between satellite hotspot (representing open-burning fires) counts and ambient air pollution levels (CO and PM_{10})

measured in Pathumthani. Higher levels of toxic and carcinogenic PAHs were observed in the rural area of Klong Luang, Pathumthani, during the dry season with intensive RS-burning activities, as compared to those measured in the Khao Yai National Park. The results of the personal exposure monitoring showed that schoolchildren in Klong Luang were exposed to higher levels of toxic $PM_{2.5}$ than those in the Khao Yai National Park.

Implementations of nonburning alternatives for RS management can eliminate enormous emission of air pollution and climate forcers, hence providing significant cobenefit. Collection of RS for subsequent off-site utilizations for cattle feedstock, garden mulching, mushroom growing, construction materials, fuels, and so on would also bring in additional income to farmers from selling RS bales. The use of RS for mushroom growing or handicraft making will also create jobs and additional income for the rural poor. Mechanical equipment such as RS balers can be used for RS collection and can quickly clear the land for fast crop rotation but should be made available and at a reasonable cost for local farmers.

Models of good practices should be created and demonstrated to Asian farmers as showcases of sustainable RS management. Technology, information sharing, and educational measures should be integrated in such models to help farmers switch from RS field burning to nonburning methods of RS disposal. Subsidy for the appropriate technology, such as RS balers, may be considered. The community participatory approach is a key to promote nonburning alternatives; this should be done through public hearing, dialogue, dissemination of information on the cause-effect relationship between RS-burning air pollution emission and health effects.

References

Andreae M.O., and Merlet P. 2001. Emission of trace gases and aerosols from biomass burning, *Global Biogeochemical Cycles*, 15: 955–966.

Cao G., Zhang X., and Zheng F. 2006. Inventory of black carbon and organic carbon emissions from China, *Atmospheric Environment*, 40: 6516–6527.

Department of Alternative Energy Development and Efficiency (DEDE). 2007. Biomass Energy, Department of Alternative Energy Development and Efficiency. http://www.alternative.in.th/, accessed 2011.

Food and Agricultural Organization (FAO). 2010. Agro-maps: global spatial database of agricultural land use statistics, Food and Agricultural Organization. http://www.fao.org/landandwater/agll/agromaps/, accessed 2011.

Gadde B., Bonnet S., Menke C., and Garivait S. 2009. Air pollutant emissions from rice straw open field burning in India, Thailand and the Philippines, *Environmental Pollution*, 157: 1544–1558.

International Grains Council (IGC). Grain Market Report No. 418. 2012. http://www.igc.int/downloads/gmrsummary/gmrsumme.pdf/, accessed 2012.

Jiminez J., Claiborn C., Dhammapala R., and Simpson C.D. 2007. Developing a source fingerprint for burning of wheat and Kentucky bluegrass stubble in eastern Washington and northern Idaho, *Environmental Science and Technology*, 41(22): 7824–7829.

Kadam K.L., Forrest L.H., and Jacobson W.A. 2000. Rice straw as a lignocellulosic resource: collection, processing, transportation and environmental aspects, *Biomass and Bioenergy*, 18: 369–389.

Kanabkaew T. 2011. Assessment of effects from crop residue field burning on particulate matter air quality, doctoral dissertation, Asian Institute of Technology, Pathumthani, Thailand.

Kanabkaew T., and Kim Oanh N.T. 2010. Development of spatial and temporal emission inventory for crop residue field burning, *Environmental Modeling and Assessment*, doi: 10.1007/s10666-010-9244-0.

Kanokkanjana K., and Bridhikitti A. 2007. *Sustainable Rice Straw Management for Urban Air Pollution Reduction in Bangbuathong, Nonthaburi Province, Thailand,* final report of the CIDA-AIT PARTNERSHIP, Alumni demonstration project, Asian Institute of Technology, Pathumthani, Thailand.

Kim Oanh N.T., Thiansathit W., Bond T.C., Subramanian R., Ekbordin W., and Ittipol P. 2010. Compositional characterization of $PM_{2.5}$ emitted from in-use diesel vehicles in Asia, *Atmospheric Environment*, 44(1): 15–22.

Kim Oanh N.T., Thuy L.B., Tipayarom D., Manadhar B.R., Pongkiatkul P., Simpson C.D., and Liu L.S. 2011. Source characterization of aerosol emission from field burning of rice straw, *Atmospheric Environment*, 45: 493–502.

Klinmalee A. 2008. The development of assessment procedure for exposure to open rice straw burning emission: a case study of school children in Thailand, doctoral dissertation, Asian Institute of Technology, Pathumthani, Thailand.

Office of Agricultural Economics (OAE). 2007. Statistic year book 2005, Office of Agricultural Economics. http://www.oae.go.th/statistic/yearbook48/, accessed 2007.

Pope C.A., III, Ezzati M., and Dockery D.W. 2009. Fine particulate air pollution and life expectancy in the United States, *The New England Journal of Medicine*, 360: 376–386.

Reid J.S., Koppmann R., Eck T.F., and Eleuterio D.P. 2004. A review of biomass burning emissions, part II: Intensive physical properties of biomass burning particles, *Atmospheric Chemistry and Physics Discussions*, 4: 5135–5200.

Rice-Trade. 2010. Major rice producing countries, Rice Trade B2B Marketplace. http://www.rice-trade.com/major-rice-producing-countries.html, accessed 2010.

Tipayarom D. 2004. Source characterization for air pollution emission from open rice-straw burning in Thailand, master's thesis, Asian Institute of Technology, Pathumthani, Thailand.

Tipayarom D., and Kim Oanh N.T. 2007. Effects from open rice straw burning emission on air quality in the Bangkok metropolitan region, *ScienceAsia*, 33(3): 339–345.

Torigoe K., Hasegawa S., Numata O., Yazaki S., Matsunaga M., Boku N., Hiura M., and Ino H. 2000. Influence of emission from rice straw burning on bronchial asthma in children, *Pediatrics*, 42: 143–150.

U.S. Environmental Protection Agency (USEPA). 1999. Compendium method TO-13A: Determination of polycyclic aeromatic hydrocarbons (PAHs) in ambient air using gas chromatography/mass spectrometry (GC/MS), Office of Research and Development, Center for Environmental Research Information, U.S. Environmental Protection Agency. http://www.docstoc.com/docs/7827128/EPA-R, accessed 2011.

Viana M., López J.M., Querol X., Alastuey A., García-Gacio D., Blanco-Heras G., López-Mahía P., Piñeiro-Iglesias M., Sanz M.J., Sanz F., Chi X., and Maenhaut W. 2008. Tracers and impact of open burning of rice straw residues on PM in Eastern Spain, *Atmospheric Environment*, 42: 1941–1957.

World Bank (WB). 2010. Low-cost housing: waste rice straw construction panels, World Bank. http://web.worldbank.org/external/projects/main?%20Projectid=P114893, accessed 2010.

Wu C.F., Jimenez J., Claiborn C., Gould T., Simpson C.D., Larson T., and Sally Liu L.J. 2006. Agricultural burning smoke in eastern Washington: part ii: exposure assessment, *Atmospheric Environment*, 40: 639–650.

Yan X., Ohara T., and Akimoto H. 2006. Bottom-up estimate of biomass burning in mainland China, *Atmospheric Environment*, 40: 5262–5273.

Yang H.H., Tsai C.H., Chao M.R., Su Y.L., and Chien S.M. 2006. Source identification and size distribution of atmospheric polycyclic aromatic hydrocarbons during rice straw burning period, *Atmospheric Environment*, 40: 1266–1274.

Zhang H., Ye X., Cheng J., Yang X., Wang L., and Zhang R. 2008. A laboratory study of agricultural crop residue combustion in China: emission factors and emission inventory, *Atmospheric Environment*, 42: 8432–8441.

14

Integrated Management Strategies for VOC Emission Control in the Paint and Coatings Sector in China

Yahui Zhuang, Hailin Wang, Wei Zhang, Zhenping Hao, and Xiaoke Wang

CONTENTS

14.1 Introduction

Volatile organic compounds (VOCs) are the precursors of secondary particulate and ground-level ozone, which are formed in the atmosphere by reactions of VOCs and oxides of nitrogen (NO_x) in the presence of sunlight. Most of the VOCs themselves are harmful to human health. Their photochemical products, namely smog and ozone, have a wide variety of human health effects and cause agricultural crop loss and damage to forests and ecosystems, as discussed in Chapters 1 and 5 of this book.

According to our survey, in China the paint and coating sector now becomes the second-largest anthropogenic VOC emission source, after VOCs from the mobile source. There have already been many studies devoted to vehicular emission control in the country, but few studies have been carried out for VOC emission management systems in the paint and coatings sector. This led to the choice of VOC control in the paint and coatings sector for this case study.

Paints and coatings can be generally classified, according to their application fields, into four main types: (1) architectural coatings used in homes and buildings, such as interior and exterior paints, primers, sealers, and varnishes; (2) industrial coatings that are factory applied to decorate and protect various manufactured goods (automotives, appliances, machinery, containers, etc.) as a part of the production process; (3) special-purpose coatings, such as aerosol paints, marine paints, and high-performance maintenance coatings; and (4) allied paint products, including putties, paint/varnish removers, paint thinners, pigment dispersers, and paintbrush cleaners (U.S. Environmental Protection Agency [USEPA], 2006). In Chinese literature, all four types of coatings are collectively called "coatings," whereas the old term "paint" is still widely used, generally referring to the industrial coatings.

Paints and coatings are manufactured in plants through the following basic steps: addition of raw materials (resins, dry pigments, water or organic solvents, depending on the type of paints); mixing/dispersion; filtration; and packaging the paint or coating for sale. Hence, the in-plant management to prevent or control VOC emissions is a part of the environmental management in the paint and coatings sector. Paints and coatings are used in virtually all sectors for corrosion protection, decoration, insulation, and many other purposes and are applied on the surfaces of metals, wood, concrete, fabrics, plastics, films, and papers. From the integrated management viewpoint, VOC prevention and abatement should be taken into account in various sectors during paint or coating application as the concept from start to finish demands such an approach.

From the viewpoint of control technologies, the integrated control strategies include pollution prevention by cleaner technologies, end-of-the-pipe pollution control, and pollution trading. Both end-of-the-pipe VOC control technologies and cleaner production technologies are available in China.

From the viewpoint of management systems, the integrated management strategies cover institutional development, economic means, legislative measures, technical demonstration and training, and public education. In China, virtually all these aspects of integrated management have been implemented. Environmental protection agencies at the national, provincial, and municipal levels have been set up. Environmental laws and regulations, air quality standards, emission standards, and production norms have been promulgated. A levy system for violation of environmental quality has also been adopted (National Development and Reform Commission [NDRC], 2007b).

Nevertheless, there are serious problems in the whole management system because it was estimated in 2004 that about 90% of the total anthropogenic VOC emissions from this sector in China remained uncontrolled according to our survey. Apparently, the control measures that have been taken to abate VOC emissions seem ineffective.

Consequently, the first task of development of integrated management strategies for this sector presented in this case study is to find out the causes of management failure and to make the current integrated management system more effective. The second task is to demonstrate practical VOC abatement technologies to ease dissemination in this sector.

14.2 Methodology

First, a situation analysis of the existing management system in this sector was made. Up-to-date industry data were collected from publications of the light industry and chemical industry bureaus of the national and local governments, the Chinese Coating Industry Association, Chinese furniture associations, Chinese newspapers and magazines, and our in-house databases at the Research Center for Eco-Environmental Sciences, Chinese Academy of Sciences (RCEES). In view that the environmental management system (EMS) in the paint and coatings sector consists of (a) cleaner production (waste minimization), (b) reduction of air emissions (end of the pipe), and (c) promotion of the International Organization for Standardization (ISO) 14,000 system (environmental accounting system), the major activities in these three areas were reviewed. Failures in the existing management system were identified and cut-in points recommended to improve the effectiveness of the management system.

Second, the practical technological options available for VOC abatement were screened. Catalytic technology has been recognized as a promising end-of-the-pipe technique for VOC control. Consequently, catalytic combustion systems for VOC control were developed in RCEES. As discussed in Chapter 10 of this book, pilot-scale demonstrations for the practical control of industrial benzene, toluene, and xylenes emissions were performed several years ago, and a national demonstration and dissemination project is under way as a follow-up step.

14.3 Current Situation of the Paint and Coatings Industry

The paint and coatings industry has grown significantly in the past decade. The total annual production of all sorts of coatings increased dramatically, from 1.8 million metric tons (tonnes) in 2000 to 3.8 million tonnes in 2005, reaching 7.5 million tonnes in 2009 (Lu et al., 2010). The urgent demand from transportation and housing stimulates the soaring production and consumption of architectural and industrial coatings in China, as an enormous consumer market is now emerging. In Beijing, for example, the demand for architectural coatings was even greater because of the construction of the Olympic Village, stadiums, and hotels.

Large quantities of organic solvents are used in the production of solvent-based paints and coatings because they can dissolve or disperse other constituents. Organic solvents are also used in the production of water-based paints and coatings, although in smaller amounts. As the annual production of coatings in China reached 2.6 million tonnes/year in 2004, the total weight of VOCs used in these coatings in China was estimated to be approximately 600,000 tonnes in that year (National Coating Information Center [NCIC], 2005). Our survey showed that the paint and coatings sector now turns out to be the second-largest anthropogenic VOC emission source, next only to the mobile source, as mentioned.

VOC emissions occur not only when coatings are manufactured inside production facilities but also when paint and coating products are applied to building structures, consumer products, and other surfaces. For meeting low-VOC content or emission requirements, there is an increasing trend of the production of environmentally friendly coatings with low-VOC (water-based and high-solid coatings) or no-VOC contents (powder coatings). A gradually diminishing trend of inferior and high-VOC coatings (solvent-based, low-solid coatings) is also observed. The statistical data indicated an unprecedented increment of powder coatings. The annual production of powder coatings (heat-curable coatings), only 12,000 tonnes in 1990, increased to 570,000 tonnes in 2005 and reached 800,000 tonnes in 2009 (Lu et al., 2010). To meet the soaring demand of new application lines in the country, additional high-end powder and high-solid coatings (polyesters and acrylics) have been imported, whereas relatively low-end products, such as phenolic and modified-resin coatings, have been exported to other developing countries.

Environmental management plays an increasingly important role in promoting consolidation of the paint and coatings industry, marked by a large number of mergers, acquisitions, and spin-offs since the beginning of the twenty-first century. Environmentally friendly products have been manufactured in China by many international large companies in the coating industry, such as Akzo Nobel, Nippon Paint Limited, BASF, Sherwin-Williams, ICI Taigu, Bayer, Savcor, Du Pont, Dow, Naret, and many others. These big manufacturers introduced various cleaner production processes in China (NCIC, 2005, 2006).

Nevertheless, the industry is also facing some serious problems. Many township enterprises have joined the paint and coatings sector. They produce and sell low-grade paints and coatings (NCIC, 2005). Fierce competition for market shares and raw materials thus occurred. Nearly 60% of the coatings produced in 2004 belonged to "solvent-based coatings" with high-VOC contents. Hence, the abatement of VOC emissions in the coating sector becomes a major concern. The small manufacturers usually cannot find sufficient investments and advanced technologies to compete with big companies. Some of them stepped out of the business or were merged into big companies, whereas others attempted to use illegal means to sell their inferior products by rebating, bribing, or using fake labels imitating prominent trademarks. Although some illegal manufacturers have been punished, many of them still can survive because some users or customers prefer cheap paints, unaware of the health and environmental impacts of VOCs. Raising public awareness is thus an important component in the integrated management system to phase out inferior products.

The Chinese central government continues to support consolidation in this sector since mergers and acquisitions favor a real and growing need for environmental quality. Product quality and environmentally friendly innovation in coating production are prerequisites for survival. Integrated environmental management is a crucial tool for the healthy development of the paint and coatings industry.

14.4 Overview of the Existing Environmental Management System for VOC Control

The reduction of VOC emissions from the paint and coatings sector can be achieved through upgrading environmental regulations and voluntary efforts of manufacturers. Consumers' preferences for cleaner products can also contribute to VOC reduction, as mentioned. The existing EMS in the paint and coatings sector includes cleaner production (waste minimization), reduction of air emissions, and promotion of the ISO 14,000 system. An overview of the EMS was made. The successful and unsuccessful strategies and measures for the control of VOC emissions from the paint and coatings sector in China are highlighted next.

14.4.1 Managing and Minimizing VOC Emissions (Cleaner Production)

The paint and coatings sector is working to reduce waste generation and promote waste recycling, as well as to address the life-cycle analysis to assess the impact of VOCs in the coatings products. A new and comprehensive Cleaner Production Promotion Law became effective January 1, 2003 (Central People's Government [CPG], 2002). This new legislation measure is an initiative of the Chinese administration to establish cleaner production nationwide as one of China's key strategies for sustainable development. It serves as a foundation for the promotion and implementation of cleaner production in the coating sector.

14.4.1.1 Reduction in Environmental Release

Since 2000, the sector has made progress in reducing releases of toxics, including VOCs. Most of these releases were to air. A part of the wastes produced was managed through recycling. Reduction of VOC emissions can be attained through a shift from solvent-based, low-solid coatings (50–70% organic solvent content) to high-solid coatings (15–35% organic solvent content), waterborne coatings (3–18% organic solvent content), or heat-curable powder coatings and radiation-curable coatings (<5% organic solvent content). Radiation-cured coatings, solidified with the aid of certain sort of radiation (electron beams, ultraviolet [UV] or visible light), contain very small amounts of organic solvents (Sun et al., 2004) and thus are good alternatives for solvent-based coatings. The share of water-based coatings in architectural coatings in the world market has increased from 35% to over 80% in recent years (USEPA, 2006). VOC release could be minimized as water replaces a significant part of organic solvents.

VOC emissions during coating application can also be reduced by employing cleaner deposition processes. For instance, the electrophoretic deposition process is widely used to paint or coat automobile bodies or any complex fabricated parts with both inside cavities and an outer surface to be coated (BASF, 2006). The unique feature of this process is that colloidal coating particles suspended in a liquid medium (usually water) migrate under the influence of an electric field (electrophoresis) and are deposited onto a conductive metallic workpiece that serves as an electrode.

As a low-cost or no-cost option for cleaner production, good housekeeping during the production and application of coatings can also reduce a part of the VOC emissions. For instance, the Sherwin-Williams (Shenzhen) EMS has recycling/rework programs aimed at minimization of cleaning materials, maximization of solvent reuse and recycling, and reworking of mistinted paints into future batches (Sherwin-Williams, 2006).

In recent years, the industry has initiated cleaner production demonstration and dissemination (Ministry of Industry and Information Technology [MIIT], 2010). Installation of closed (airtight) production systems and a shift from solvent-based coatings to water-based or UV-curable coatings are the major options for cleaner production in the near future. These measures can reduce significant amounts of VOC emissions as shown in Table 14.1.

TABLE 14.1

Implementation and Dissemination of Cleaner Production in the Coating Sector in China

Coating Category	Cleaner Production Technology	VOC Reduction Rate, kg VOC/tonnes coating	Annual VOC Reduction Goal in 2014, tonnes
Solvent-based coatings	Closed production system with airtight mixing, grinding, blending, piping, and packaging	20–30	1×10^4
Water-based anticorrosion coatings for bridges	Substitution of solvent-based coatings by water-based ones	200	5×10^3
Water-based coatings for automobiles	Substitution of solvent-based primers, sealers, and topcoats by water-based ones	200	3×10^4
Water-based coatings for containers	Substitution of solvent-based coatings by water-based ones	200	4×10^3
UV-curable coatings	Production of solid coatings with very low VOC contents	300	2×10^4
Water-based coatings for wooden ware	Substitution of solvent-based coatings by water-based ones	500	7×10^4

Source: Data from MIIT. 2010. Implementation and dissemination of clean production in the coating sector. http://jns.miit.gov.cn/n11293472/n11295091/n11299329/n13564959files/n13564939.pdf, accessed 2011.

14.4.1.2 Life-Cycle Management

The paint and coatings sector has reduced a number of harmful constituents, such as VOCs, lead, and mercury, through life-cycle management. Further VOC management is feasible through recycling greater amounts of leftover paint for the substitution of virgin raw materials in the production of new paint and coating products. China's increasing share of radiation-curable coatings is evidence of cleaner production (Table 14.2). Coatings, which harden or "cure" by photoinduced polymerization, offer a way to reduce air pollution and to conserve energy. With such coatings, the change from liquid to solid is accomplished by a cross-linking process rather than by removal of organic solvents (Zhu, 2004).

14.4.2 Reducing VOC Emissions through the End-of-the-Pipe Control

The practical technological options available for end-of-the-pipe VOC abatement are as follows: (a) VOC emissions from conventional spray application can be reduced with high-volume, low-pressure spray guns, which can minimize overspray. Overspray can also be minimized via an electrostatic coating

TABLE 14.2

China's Consumption Share in Radiation-Curable Coatings in the World

Region	Annual Consumption of Radiation-Curable Coatings, tonnes			Average Annual Growth Rate, % 2004–2009
	2001	2004	2009 (estimated)	
North America	67,840	79,160	117,500	8
Europe	83,850	101,500	129,020	5
Japan	42,090	44,780	54,740	4
China	—	32,600	52,500	10
Other Asian	—	42,800	60,000	7
Total	193,780	300,840	413,760	7

Source: Data from Socially Responsible Investment (SRI) Consulting Group. 2005. Specialty chemicals radiation curable coatings. http://www.radtech-europe.com/files_content/rccoatingssri2005.pdf, accessed 2011.

process that employs charged paint particles to paint a workpiece more efficiently (Baidu-Baike, 2010). Paint, in the form of either powdered particles or atomized liquid droplets, is initially projected toward an electrically conductive workpiece using normal spraying methods and is then accelerated toward the workpiece by a powerful electrostatic charge, as mentioned. (b) Robots can replace workers for spray applications to reduce human exposures and achieve closed production (Kawasaki-Robot, 2008). (c) VOCs in the air or exhausts ventilated from the spray workshops can be removed by thermal or catalytic incineration, adsorption, or biofiltration (Xi and Hu, 2006).

A downward trend in VOC emissions can be anticipated in the future due to public awareness and upgrading of regulatory requirements for VOC limits in coatings.

14.4.3 Efforts to Promote the Existing Environmental Management Systems

The adoption of EMSs within the paint and coatings sector is increasing rapidly. The implementation of the ISO 14,000 system is now encouraged but is not compulsory. Implementing an EMS component by a manufacturer is a condition of obtaining an export permit for its products. The efforts discussed next have been made by the administration to promote the existing EMS.

14.4.3.1 Evaluation Indicator System of Cleaner Production

To foster cleaner production in the paint and coatings sector, the NDRC has initiated the implementation of three separate evaluation indicator systems for the solvent-based, water-based, and powder coatings industries, respectively (NDRC, 2007a). Five primary indices and 18 secondary indices have been selected for the system. The indices concerning consumption rates for raw materials and energy resources, pollutant levels, and recycling rates for waste resources are quantitative indicators, whereas those indices related to product characteristics, environmental management, health, and safety are qualitative ones. Environmental performance indicators attempt to provide a numerical basis for assessing the environmental performance of cleaner production or the effectiveness of EMSs of manufacturers.

14.4.3.2 Environmental Management Certificate and Environmental Labeling in China

Since 2002, Beijing has become more aware of environmental protection. One of the contributing factors was China's accession to the World Trade Organization (WTO). The preparation for the 2008 Olympic Games, which was then under way, also raised awareness of environmental protection. To ensure wide adoption of environmental products and technologies reaching international standards and fulfilling all the environmental criteria, Beijing has placed great emphasis on promoting an internationally recognized Environmental Management Certificate and the China Environmental Label (NCIC, 2005). The China Environmental Labeling Scheme was first introduced in early 1993. The labeling scheme is implemented by the China Environmental Labeling Committee formed by the National Environmental Protection Administration, the Center for Quality and Technology Inspection, the National Import/Export Goods Customs, and a group of experts to review the establishment and operation of the scheme. There are a total of 42 categories for the China Environmental Label (Green-Council, 2004). Applications are open to any parties interested in the environmental labels. The coverage of the scheme will expand in response to demand. From 1994 to 2005, environmental labels were awarded to 305 enterprises for 1,274 products. In the paint and coatings sector, however, up to 2005 only 104 manufacturers had been awarded the environmental label (China Environmental Labeling [CELC], 2006). Thousands of paint and coating manufacturers in China did not get the environmental labels. Nevertheless, they could meet the nonstringent VOC emission standard, so they were still allowed to produce and sell their inferior coating products.

Failure to eliminate low-end products with high-VOC contents was associated with the confusing policies and ineffective enforcement. On one hand, the government promulgated a flexible (not stringent) national VOC emission standard in 2002 (Table 14.3), so that many small domestic manufacturers still could survive. On the other hand, it encouraged international and domestic producers to meet the "noncompulsory" ISO 14,000 standards voluntarily by issuing green labels since 2004. The overall regulation on VOC emissions was under dispute (NCIC, 2005, 2006). The industry was lacking systematic stringent standards and sufficient investment in research and development to reduce VOC emissions effectively. Only by the end of the 2000s did the

TABLE 14.3

Limits of Harmful Substances in Interior Architectural Coatings in Standards GB18582-2001 and GB18582-2008

Standard Code	Item	Limit
GB18582-2001 (Effective January 1, 2002)	VOCs	200 g/L
	Free formaldehyde	0.1 g/kg
	Soluble Pb	90 mg/kg
	Soluble Cd	75 mg/kg
	Soluble Cr	60 mg/kg
	Soluble Hg	60 mg/kg
GB18582-2008 (Effective June 1, 2010)	VOCs	120 g/L
	Benzene + toluene + xylene + ethylbenzene	300 mg/kg
	Free formaldehyde	100 mg/kg
	Soluble Pb	90 mg/kg
	Soluble Cd	75 mg/kg
	Soluble Cr	60 mg/kg
	Soluble Hg	60 mg/kg

Sources: Data from Standardization Administration of China (SAC). 2002. *Limits of Volatile Organic Compound Contents in Interior Architectural Coatings (GB18582-2001)*, Standard Press of China, Beijing; SAC. 2009. *Limits of Volatile Organic Compound Contents in Interior Architectural Coatings (GB18582-2008)*, Standard Press of China, Beijing.

administration acknowledge this issue and strengthen the VOC limits in architectural and wooden-ware coatings since 2009 (Table 14.3). New VOC limits for automobile coatings, exterior architectural coatings, as well as coatings for toys also have been promulgated.

14.4.3.3 CCC Certification

In December 2001, the State General Administration of Quality Supervision, Inspection and Quarantine, issued the Regulations for Compulsory Product Certification (China Compulsory Certification Center [CCC], 2003). The Compulsory Product Certification System began to replace the original Quality License System for Commodity Inspection and Safety Certification System for Electrical Equipment. The abbreviation for *China Compulsory Product Certification* is CCC (3C). It is the statutory compulsory safety certification system and the basic approach to safeguard the consumers' rights and interests and protect personal and property safety; it is adopted widely by international organizations. This is a kind of inspection (Waratah, 2006) that controls both product quality (ISO 9000) and environmental performance (ISO 14024). For instance, the VOC-free coatings of Waratah Paint (Shenzhen) Limited have not only won the certificate of environmentally friendly marked products of China but also gained the qualified certificate, "The Standard of Harmful Substances on National Interior Decoration Materials." The company has passed both ISO 9001 and ISO 14001 (Waratah, 2006).

Coating products that cannot pass this certification have not been allowed to enter the market since August 2005 (NCIC, 2005, 2006). In short, low price is no longer the weapon of competition. The crucial points of competition are product quality and environmental quality.

14.5 Cut-in Points for EMS Improvement

Since individual components of integrated management of VOC emissions from the paint and coatings sector have been introduced and implemented for several years in China and in Beijing, it is necessary to identify the existing policy failure and to find out the causes of the ineffective measures as cut-in points for management improvement. The strategic recommendations to the decision makers on policy changes as the outcomes of this case study presented at the end of this chapter are based on the results of the assessment and survey conducted by this study but not on any textbook definition of integrated management.

14.5.1 Innovative Restructuring Mechanism in the Paint and Coatings Sector

The market shares in the paint and coating sector in 2005 were as follows: 50–60 prominent international and joint manufacturers provided two-fifths of the total production, and 2,500 domestic small manufacturers produced the remaining three-fifths (NCIC, 2005). For medium and large manufacturers, modifying the product mix according to market demands, upgrading manufacturing facilities and equipment, as well as strengthening management capacity are their priorities because they have experience and can find financial and human resources to do this work. Regarding new small investigators in this sector, they should avoid duplicated production of inferior or polluting coatings. Some of them can be consolidated into big companies, whereas the others need to choose coating production from natural raw materials or waste materials as a sustainable option, as detailed in Section 14.5.4. Since natural raw materials and waste materials usually occur in a scattering way, small-scale production is thus suitable for avoiding long-distance transportation of raw materials to centralized large plants. Innovative restructuring mechanisms should be encouraged.

14.5.2 Economic and Technological Incentives for Effective VOC Control

The VOC control strategies of the environmental protection agencies up to 2005 were not effective, as pointed out previously. The failure stems from the nonstringent limits for VOCs in various coatings (NCIC, 2005, 2006). The policy makers were afraid of bankruptcy of small and medium Chinese manufacturers due to more stringent standards for VOCs. They also feared that such a measure would favor the development of international large companies in the coating sector. In fact, such moderate standards could not save the small businesses in the long run because users are becoming aware of the hazardous nature of poor coatings and tend to use coatings made by reliable international manufacturers. The only effective strategies are to promulgate more stringent limits for VOC contents in coatings, as well as to offer economical and technological help to the domestic small businesses. Such an approach is now accepted by the administrations as VOC limits in major categories of coatings have been tightened since 2009 (Table 14.3), and cleaner production options are being demonstrated and disseminated, as described in the previous sections. However, only the big companies can afford these changes, and small owners are still facing financial and technical troubles. As a reviewer of United Nations Development Program (UNDP) projects, the first author of this chapter witnessed that the Global Environmental Facility (GEF) and UNDP have set up a successful example in China to eliminate ozone-depleting substances (which also are volatile organic compounds) by offering both technical training for alternative manufacturing processes and financial support for the procurement of necessary equipment to small businesses. Such an approach should be followed as well in the abatement of VOCs in the paint and coatings sector in the country.

14.5.3 Lesson on Setting Up an Ambiguous VOC Content Limit for Waterborne Coatings

There was a crucial ambiguity in the definition of VOC content in the Chinese standard for waterborne coatings (NCIC, 2005, 2006). The Chinese government took the US and EU VOC content limits for granted (Table 14.4) but did not adopt the same definition of VOC content. The definition of *VOC content* made by the USEPA and the Californian Air Resource Board (CARB) clearly requires the exclusion of exempt solvents and water content in the coatings (Equation 14.1). Exempt solvents are a group of compounds or classes of compounds, such as acetone, that have been determined by the USEPA to have negligible contribution to tropospheric ozone formation. When a compound is exempted, it is excluded from the USEPA's definition of VOCs for federal regulations related to attaining the national ambient air quality standards for ozone. The effect of this exemption is that a manufacturer using an exempt solvent in formulating a coating would not have to count that solvent as a VOC in determining whether the coating meets regulatory obligations.

The VOC regulatory formula adopted by the USEPA and CARB is presented in Equation 14.1, which calculates the VOC content in coating materials (USEPA, 1998).

TABLE 14.4

Comparison of VOC Content Limits for Water-Based Interior Architectural Coatings in the Emission Standards of China and the United States

Country	Standard Code	Coating Category	Item	Limit	Effective Date
China	GB18582-2001	Flat	VOC content	200 g/L	Jan 1, 2002
	GB18582-2008	Flat	VOC content	120 g/L	June 1, 2010
United States	Rule 63FR48848	Flat	VOC content	250 g/L	Sept 11, 1998
		Nonflat	VOC content	380 g/L	Sept 11, 1998
		Nonflat high gloss	VOC content	380 g/L	Sept 11, 1998
	SCAQMD rule 1113	Flat	VOC content	60 g/L	July 1, 2008
		Nonflat	VOC content	50 g/L	July 1, 2006
		Nonflat high gloss	VOC content	50 g/L	July 1, 2007

Sources: Data from U.S. Environmental Protection Agency (USEPA). 1998. *Federal Register Environmental Documents,* National Volatile Organic Compound Emission Standard for Architectural Coatings, Vol. 63, pp. 48848–48887; South Coast Air Quality Management District (SCAQMD). 2007. Rule 1113 architectural coatings. http://www.aqmd.gov/rules/reg/reg11/r1113.pdf, accessed 2011; Data from SAC. 2002. *Limits of Volatile Organic Compound Contents in Interior Architectural Coatings (GB18582-2001),* Standard Press of China, Beijing; SAC. 2009. *Limits of Volatile Organic Compound Contents in Interior Architectural Coatings (GB18582-2008),* Standard Press of China, Beijing.

$$\text{VOC Regulatory} = \frac{(\text{weight volatiles}) - (\text{weight water}) - (\text{weight exempt})}{(\text{volume coating}) - (\text{volume water}) - (\text{volume exempt})} \quad (14.1)$$

In Equation 14.1, volatiles are the compounds that evaporate when the coating is applied (i.e., VOC solvents, water, and exempt solvents). The weight of the water and exempt compounds is subtracted from the weight of the volatiles in the numerator. In the denominator, the volume of water and exempt compounds is subtracted from the volume of the coating.

The Chinese VOC content definition in GB18582-2001 (Table 14.3) does not require the exclusion of exempt solvents and water content (SAC, 2002). The inconsistency between the Chinese and USEPA formulas to calculate the VOC concentration lies in this requirement to subtract the volumes of water and exempt compounds in the denominator causes the controversy. Subtracting water or exempt solvents from the volume of the coating in the denominator makes the denominator a smaller number. This results in a larger VOC regulatory concentration than if the water or exempt solvents were not subtracted from the denominator.

However, in the revised Standard GB18582-2009 (SAC, 2009), the formula for VOC content calculation also requires the subtraction of water volume in the denominator, which is now consistent with the international definition. Nevertheless, exempt solvents are still not taken into account. This is a lesson that policy makers should learn for promulgating new emission standards for many other categories of coatings in the future.

14.5.4 Sustainable Coatings from Natural and Recycled Materials

Development of sustainable natural coating materials should be encouraged. China has a long tradition of using natural lacquer, such as Chinese lacquer *Shengqi* and *Tong* oil for protecting wooden furniture and boats or ships. These natural coatings do not contain synthetic solvents and are solidified through an "enzyme-activated" oxidation cross-linking mechanism. Since these coatings are produced from natural plants (Chinese lacquer trees and tong trees), their production is sustainable. Unfortunately, their current shares in the coatings production are less than 1% (Baidu-Baike, 2007a, 2007b). Investigations are ongoing to accelerate the slow solidification process, which might last for months by introducing minute amounts of initiators or promoters.

There are also other options offered by using waste material as raw material for the production of coatings. In Tianjin, a new technology has been developed using orange peel as the raw material for coatings, with a yield of 1 tonne of coating from 4 tonnes of orange peel (Liuhong-Tech, 2007). It is also reported that a self-healing coating can be produced from crab shells (Ghosh and Urban, 2009). The administrations should support small businesses in developing such sustainable coating production by offering low-interest loans and technical training, as mentioned.

14.6 Outcomes from VOC Control Demonstration Projects

All the options mentioned for VOC control (i.e., shift from solvent-based coatings to water-based or high-solid coatings, reduction of overspray, replacement of workers by robots, and removal of VOCs in exhausts) have been implemented in Beijing. The last option, the end of pipe (VOC removal in exhausts), is suitable for the existing small businesses as the manufacturing processes and facilities need not be changed. Only a VOC treatment facility is required to be added at the end of the exhaust pipe. Compared with thermal incineration, stripping, and adsorption in the last option, catalytic technology has been recognized as a promising technique. Consequently, catalytic combustion systems for VOC control were developed in RCEES as presented in Chapter 10. For dilute VOC emissions, preconcentration of dilute VOC emissions by adsorption and subsequent catalytic combustion have been successfully demonstrated as well.

A pilot-scale demonstration was made to control the industrial emissions of benzene, toluene, and xylenes in the Fujian Jiayuan Company. The pilot-scale results indicated that the catalytic system developed in this study was effective under a space velocity of 2,000 m^3/h and met the domestic regulation of industrial VOC emissions.

Finally, the RCEES efforts for VOC emission control have been peer reviewed and recognized by the national government. A new national spin-off demonstration and dissemination project for VOC control in the whole Pearl River Delta region is now being carried out under the coordination of the RCEES research team. According to the RCEES survey on the barriers encountered in the Pearl River Delta region as well as a preliminary cost-benefit analysis based on the catalytic combustion of VOCs emitted from a paint-spraying production line of stationeries in Beijing (detail in Chapter 10), the following management recommendations are presented:

1. For small manufacturers, a catalytic combustion technique is recommended as a good process modification option.
2. The current catalytic oxidation-combustion systems, though effective, are rather expensive for the small entrepreneurs in China. Consequently, small low-interest loans are needed for dissemination of this technology. This economic measure has been proven to be effective in GEF/UNDP projects for phasing out ozone-depleting substances in China. Meanwhile, cheaper and better catalytic devices for VOC removal are being developed by the RCEES team to overcome this economic barrier.
3. Regulations and enforcement of VOC emission control in various sectors should be promulgated, updated, and implemented. There is the "three synchronicities" policy in China that requires the manufacturers to undertake a three-step process: design, construct, and operate pollution control devices to mitigate pollution. However, often it turned out that the devices for pollution control, after construction, were simply shut down to avoid operating costs. The pollution levies are usually so low that they are below the pollution control operating costs. So, a somewhat perverse incentive structure is formed: The local environmental protection bureaus (EPBs) are glad to receive fees, the polluters are also pleased to have lower operating costs, and no higher governmental authority steps in to ensure regulatory compliance. As a result of strengthened enforcement, the increased costs of monitoring, inspection and maintenance, and prosecution should be taken into account (United Nations Development Program [UNDP]/China International Center for Economic and Technical Exchange [CICETE], 2000).
4. Capacity building for the public officials on economic external costs of production is an important task to improve the regulatory system in China (UNDP/CICETE, 2000). Pollution has long been recognized as the external cost of industrial production or a "market imperfection" that is not being properly taken into account in individual market

transactions. Correcting such negative external costs is certainly one of the principal roles of governments. Unfortunately, in many cases local officials and even local EPB officials in China, being less aware of the external costs, are more interested in pursuing gross domestic product (GDP) goals than environmental protection. Since the rule of law remains undeveloped, and since environmental damages are not monetized, production almost always wins. It does not mean that environmental damages are not real. The increasing incidences of bronchial illness, lung cancer, and retarded neurological development in children, as well as the increasing crop and forest damages, are evidence of environmental policy failure. As a result of low awareness of externality, the new accounting system "green GDP" advocated by the Ministry of Environmental Protection for several years still faces strong resistance and cannot be implemented.

14.7 Summary and Recommendations

VOC emission management of the paint and coatings sector was selected for this case study because this is an important pollution source in China, and few systematic studies of it have been carried out. On the basis of industry data and governmental regulations, survey and situation analysis were conducted in this study. The success and failure in the control of VOC emissions from the paint and coatings sector in China were highlighted. Technological, economic, and legislative means for VOC pollution control in China and in Beijing were reviewed. Based on the identified cut-in points in the VOC emission management system, the following recommendations are proposed:

1. Demonstration and technological training on the catalytic combustion for VOC abatement in the paint and coatings sector should be organized. Low-interest loans for the small and medium manufacturers in the paint and coatings sector should be provided by environment-oriented banks or funding agencies so that the small businesses can afford necessary equipment for VOC control. Development of sustainable coatings from natural and waste materials is recommended for the small manufacturers.
2. Enforcement of more stringent emission standards and inspection are necessary to eliminate inferior products with high-VOC emissions. Capacity building for the public officials on economic external costs of production is an important task to improve the regulatory system. An economic compensation and reorientation system for phasing out inferior products from less-developed districts in the

country should be established to eliminate any social unrest. The unemployed can be attracted to other small businesses and obtain some loans to start their new careers.

3. Public awareness of the health damages caused by poor coatings should be raised to reduce the market share of inferior products with high-VOC contents. It is recommended that in the future all pollution control projects should have a "public awareness" component, with electronic, print, and other media outreach programs designed to increase awareness about the damages caused by pollution, measures taken to control pollution and mitigate the damage, as well as steps that common citizens can take to contribute to these control efforts.

References

Baidu-Baike. 2007a. Chinese lacquer (*Rhus verniciflua* Stokes). http://baike.baidu.com/view/91679.htm, accessed 2011.

Baidu-Baike. 2007b. Tong oil (*Aleurites fordii* Hemsi). http://baike.baidu.com/view/196583.htm, accessed 2011.

Baidu-Baike. 2010. Electrostatic spraying. http://baike.baidu.com/view/142785.htm, accessed 2011.

BASF. 2006. Electrophoretic coatings for automobiles. http://www.greater-china.basf.com/apex/GChina/en-GB/content/BASF-China/1,2_Products/, accessed 2011.

China Compulsory Certification Center (CCC). 2003. Circular relevant to the implementation of the compulsory. http://www.ccc-cn.org/cccindex.htm, accessed in 2011.

China Environmental Labeling (CELC). 2006. China environmental labeling—the tenth ring type I sign. http://www.10huan.com, accessed 2011.

Central People's Government (CPG) of PRC. 2002. Cleaner Production Promotion Law of the PRC. http://www.gov.cn/gongbao/content/2002/content_61640.htm, accessed 2011.

Ghosh B., and Urban M.W. 2009. Self-healing oxetane-subsituted chitosan polyurethane networks, *Science*, 323: 1458–1460.

Green-Council. 2004. China Environmental Labeling Program. http://www.green-council.org/eng/greenlabel/china.asp, accessed 2011.

Kawasaki-Robot. 2008. Painting robot and painting set, Unit K. http://www.kawasakirobot.com.cn, accessed 2011.

Liuhong-Tech. 2007. Coating from orange peel, *Sichuan Chemical Engineering*, 10(4): 53.

Lu N., Ye D., and Zhang S. 2010. Annual market research report of Chinese coating industry. http://www.coatings.hc360.com/10dq/SecondMenu/Detail_2009dqyjbg1.html, accessed 2011.

Ministry of Industry and Information Technology (MIIT). 2010. Implementation and dissemination of clean production in the coating sector. http://jns.miit.gov.cn/n11293472/n11295091/n11299329/n13564959files/n13564939.pdf, accessed 2011.

National Coating Information Center (NCIC). 2005. Analysis on China coating industry, China Coating Industry Association, Changzhou. http://www.chinaccm.com/48/4805/480501/news/20050411/143257.asp, accessed 2011.

National Coating Information Center (NCIC). 2006. China coating industry. Report 2006. Research in China, November 1. http://www.researchinchina.com/htmls/Report/2008/2813.html, accessed 2011.

National Development and Reform Commission (NDRC). 2007a. Clean production evaluation indicator system of clean production in the coating sector. http://www.sdpc.gov.cn/zjgx/W020070428556955287922.pdf, accessed 2011.

NDRC. 2007b. The levy system in China. http://www.sdpc.gov.cn/jggl/jgqk/t20070404_126543.htm, accessed 2011.

Standardization Administration of China (SAC). 2002. *Limits of Volatile Organic Compound Contents in Interior Architectural Coatings (GB18582-2001)*, Standard Press of China, Beijing.

Standardization Administration of China (SAC). 2009. *Limits of Volatile Organic Compound Contents in Interior Architectural Coatings (GB18582-2008)*, Standard Press of China, Beijing.

South Coast Air Quality Management District (SCAQMD). 2007. Rule 1113 architectural coatings. http://www.aqmd.gov/rules/reg/reg11/r1113.pdf, accessed 2011.

Sherwin-Williams. 2006. Sherwin-Williams EMS. http://oem.sherwin-williams.com/, accessed 2011.

Socially Responsible Investment (SRI) Consulting Group. 2005. SRI report 2005. Specialty chemicals radiation curable coatings. http://www.radtech-europe.com/files_content/rccoatingssri2005.pdf, accessed 2011.

Sun Z.Y., Wei J., and Qin C.X. 2004. UV-light curable powder coating, *Thermally Solidified Coatings*, 1: 9.

United Nations Development Program/China International Center for Economic and Technical Exchange (UNDP/CICETE). 2000. *Urban Air Pollution Control in China*, China Science and Technology Press, Beijing.

U.S. Environmental Protection Agency (USEPA). 1998. *Federal Register Environmental Documents*, National Volatile Organic Compound Emission Standard for Architectural Coatings, Vol. 63, pp. 48848–48887. http://www.epa.gov/ttn/atw/183e/aim/fr1191.pdf

U.S. Environmental Protection Agency (USEPA). 2006. Paint and coatings. http://www.epa.gov/sectors/sectorinfo/sectorprofiles/paint.html, accessed 2011.

Waratah. 2006. Fresh Air products. http://www.waratahpaint.com, accessed 2011.

Xi J.Y., and Hu H.Y. 2006. Advances in the treatment of VOC in exhausts by biofiltration, *Environmental Science and Technology* (in Chinese), 29(10): 106–108.

Zhu L.D. 2004. *Cleaner Production Technologies of Coatings*, Chemical Industry Publishing House, Beijing.

15

Integrated Air Quality Management to Reduce Traffic Emission in Bandung City

Puji Lestari

CONTENTS

15.1 Introduction

Bandung is a capital city of the West Java Province of Indonesia with unique climate and topography. The particulate matter (PM) pollution levels in Bandung by far exceeded the Indonesian National Ambient Air Quality Standards (NAAQS) for road site monitoring (Lestari and Savitri, 2003; Lestari, 2004b).

The transportation sector and industry have been identified as the major sources of toxic air pollutants in the ambient air of Indonesia. Recent growth in the number of vehicles, urban population, and industry has worsened air pollution problems in most big cities in the country. In Bandung, the increase in the number of vehicles, which reaches more than 10% per year, frequently causes traffic congestion, especially during the rush hours on weekdays. On the weekends, high traffic density is also observed as many visitors travel to the city. The total vehicle populations in Bandung City from 2004, 2005, and 2006 were 461,303, 487,391, and 521,678 units, respectively. The total number of vehicles increased from 732,536 in 2007 to 1,179,800 in 2010 (Lestari and

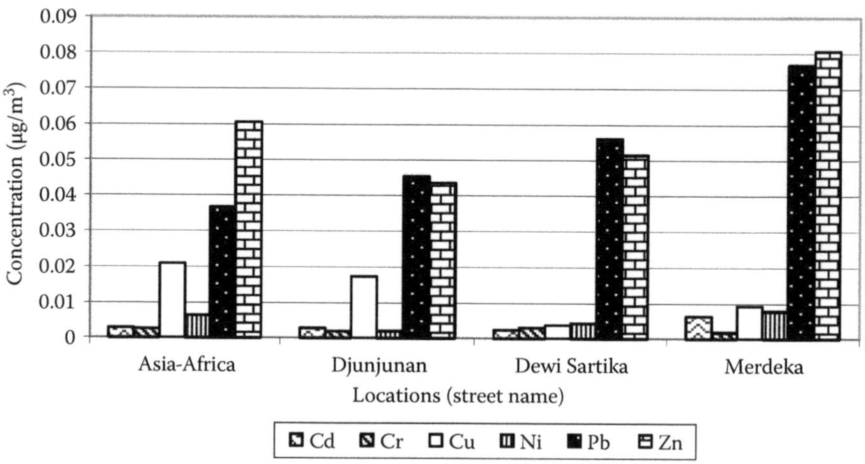

FIGURE 15.1
Average concentrations of elements at road sites in Bandung. (*Source:* Lestari P. 2004a. Lead in the air and lead in the school children's blood: case study in Bandung, paper presented at the KPBB workshop, Borobudur Hotel, Jakarta, December 15.)

Leopold, 2008; Statistical Center Bureau [SCB], 2011). These huge increases over the 4 years worsened traffic congestion in the city. The vehicle fleet composition is dominated by motorcycles (64%); followed by private passenger cars (21%); light duty, including small trucks, buses, and pick-ups (8%); heavy duty, mostly trucks and buses (6%); and public passenger cars (1%).

A study measured the blood lead level (BLL) of schoolchildren in Bandung when leaded gasoline was still used indicated a high blood lead content (Lestari et al., 2006b). High levels of lead in the ambient air at various road sites in the city were also reported for the period by Lestari (2004a), as shown in Figure 15.1. Lead is a neurotoxin or a nerve poison that can reduce the intelligence level and brain function of children. For adults, lead could cause high blood pressure, anemia, infertility, or even death (Centers for Disease Control and Prevention [CDC], 2000). Assessment of the BLL in the urban population was conducted and potential effects were proven to stimulate and speed up phasing out leaded gasoline in many places (Lovei, 1996).

The air pollution problem in Bandung City is mostly caused by transportation (Lestari, 2004b), which needs to be properly managed to reduce the source emission and subsequently the adverse effects. This integrated air quality management (AQM) case study focused on traffic emission in the city and was conducted during the second phase of the AIRPET (Asian Regional Air Pollution Research Network) project (2004–2008) by the Indonesian team. The major tasks of the case study were (a) determination of the BLL for schoolchildren in the city, (b) vehicle emission testing for development of emission factors and emission profiles for motorcycles and passenger cars, and (c) development of vehicle emission inventory for Bandung.

15.2 Methodology

The target source of this integrated AQM case study was emission reduction for the transportation sector, especially vehicles powered by leaded gasoline. The study was initiated in 2004 when leaded gasoline was still used in Bandung and was completed in 2008, that is, 2 years after the phasing out of leaded gasoline in 2006. The main source of lead in the ambient air was from the transport sector; hence, the main target of this case study was to stimulate the process of phasing out leaded gasoline in the city. The activities carried out included the development of source profiles and emission factors from vehicles, measurements of the lead levels in the ambient air, BLL tests for schoolchildren, and public campaigns to promote unleaded gasoline. In addition, an emission inventory was also conducted to determine types of vehicles contributing most to the air pollution in the city. The emission tests were conducted for several vehicle and fuel technologies, including motorcycles (two- and four-stroke gasoline-powered motorcycles) and passenger cars (biodiesel and bioethanol powered) equipped with carburetor and electrical injection systems.

The BLL test study was conducted in cooperation with the Lead Information Center (LIC), Jakarta. The development of source profiles and emission factors for vehicles was also conducted in cooperation with the Institute of Thermodynamic, Combustion Engine, and Propulsion (BTMP), Jakarta.

15.2.1 Development of PM Emission Source Profiles and Emission Factors

The source identification process is an important step in AQM (Hopke, 1991). Data on profiles and emission factors generated for local sources would be more relevant for characterization of local air pollution problems. In particular, local PM source profiles are also certainly required to determine the PM source contributions using receptor modeling. Information on lead levels in vehicle exhaust are also useful for stimulating the phasing out of leaded gasoline for eventual elimination of this toxic emission.

In this study, source profiles and emission factors were developed especially for motorcycles with two- and four-stroke engines powered by leaded gasoline and passenger cars using diesel, gasoline, and biofuel. The vehicle tests were performed using a chassis dynamometer operated under laboratory conditions following Economic Commission for Europe (ECE) method 83-1999 (SM/III/EKB/1). Each vehicle was tested according to the ECE driving cycle, which included the operating cycle for part 1 (urban driving cycle, UDC) and part 2 (extraurban driving cycle, EUDC), as illustrated in Figures 15.2. The UDC has low vehicle speeds, low load, and low temperature. The EUDC has high speeds, reaching 120 km/h. One operating cycle

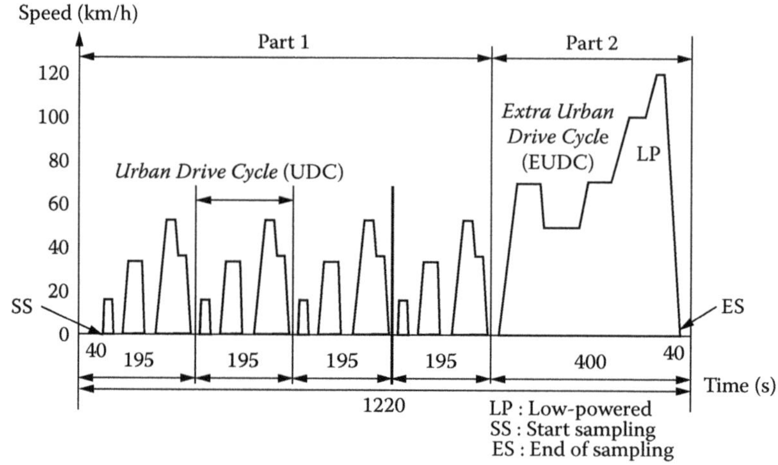

FIGURE 15.2
Driving cycles (United Nations Economic Commission for Europe [UN-ECE] No. 83-04) used in this study. (*Source:* UN-ECE. 2006. Regulation No. 83: Uniform provisions concerning the approval of vehicles with regard to the emission of pollutants according to engine fuel requirements, *Official Journal of the European Union.* http://eur-lex.europa.eu/LexUriServ/LexUriServ.do?uri=OJ:L:2006:375:0223:0495:EN:PDF.)

consists of four operating times with UDC and one with EUDC. For biodiesel-fueled passenger cars, each vehicle was tested three times using different types of blended fuels: B-0 (0% biodiesel by volume), B-5 (5% biodiesel), and B-10 (10% biodiesel). Similarly, for the cars with bioethanol the blended fuels were BE-0 (0% bioethanol), BE-5 (5% bioethanol), and BE-10 (10% bioethanol).

The emission test was conducted using a single-roll chassis dynamometer (48-inch diameter) and an exhaust gas-sampling system with a constant-volume sampling system. A nondispersive infrared analyzer was used for carbon monoxide (CO) and carbon dioxide (CO_2), a chemiluminescent analyzer for nitrogen oxides (NO_x), and a flame ionization detector analyzer for hydrocarbon (HC) measurements.

The PM composition analysis was done using an atomic absorption spectrophotometer with flame ionization and graphite furnace detectors for elements, ion chromatography for ions, and smoke stain reflectometer for black carbon.

15.2.2 Emission Inventory

The emission inventory of conventional pollutants (PM, CO, SO_2, and NO_x) and greenhouse gases (GHGs) for the transportation sector of Bandung City was conducted for the years 2004–2006. Two approaches were used: (a) based on the vehicle kilometers travelled (VKT) and (b) based on the fuel (including gasoline and diesel) consumption (Lestari et al., 2006a). Vehicle classifications were determined based on the available vehicle data in Indonesia. The results of the study were used to identify the contribution of various vehicle types to the emissions of the transportation sector in the city.

15.2.3 Blood Lead Level Tests

BLL tests for schoolchildren were conducted, and the information obtained was disseminated widely for promoting the phaseout of leaded gasoline in the country. The activities were done through cooperation with the LIC Jakarta and the Department of Public Health, University of Indonesia, with strong support from the local government of the city. Four hundred schoolchildren from 40 elementary schools participated the BLL testing campaign, together with policy makers, members of the House of Representatives, as well as the city mayor. Simultaneously, IQ tests were also performed for the schoolchildren. The results from the BLL study served as a driving force for unleaded gasoline implementation in Bandung.

15.3 Results

15.3.1 Vehicle Exhaust PM Emission Profiles

A PM source profile was obtained for each type of vehicle fuel system tested. As an example, the results of the source profile for private passenger cars using bioethanol are presented in Table 15.1. Black carbon was the main component of the exhaust PM, 16–30% by weight. The PM composition also changed with the fuel blending rate, with the percentage of BC in PM generally higher in BE-5. The source profiles of PM emissions obtained from vehicle exhaust were the local specific data, which can be used in PM source apportionment studies by receptor modeling. The local source profile data are expected to improve results for source contributions. Subsequently, information that is more reliable can be provided for policy development to prioritize the emission sources for control efforts.

15.3.2 Vehicle Exhaust Emission Factors

The emission factors developed in this study are presented in Table 15.2. Cars equipped with a carburetor system emitted significantly higher air pollution than those with an electrical fuel injection system. The most significant emission reductions were recorded for CO and HC. When the fuel blending rate increased from BE-0 to BE-10, the emission factors of CO and PM were reduced, whereas the emission factors of HC and NO_x increased. The emission factor results obtained in the vehicle tests were used in the emission inventory for the transportation sector in the city.

TABLE 15.1

Exhaust Source Emission Profiles of PM for Bioethanol Fueled Passenger Cars, in Percents (%) of PM Weight

	Electrical Injection System			Carburetor System		
Parameters	**BE-0**	**BE-5**	**BE-10**	**BE-0**	**BE-5**	**BE-10**
Black carbon	16.64	29.30	23.67	16.78	17.41	17.24
Ca^{+2}	6.50	10.64	10.40	3.90	3.95	3.11
Na^{+}	13.07	5.96	3.03	1.01	4.40	1.39
SO_4^{-2}	3.22	1.66	4.62	1.06	0.90	0.71
K^{+}	2.66	2.73	2.13	1.17	1.88	0.84
Fe	3.40	4.06	4.37	0.94	1.24	1.71
Al	0.00	20.58	22.85	0.00	17.79	15.70
Mg	19.04	11.73	15.52	24.34	22.67	27.12
Pb	0.10	0.86	0.20	0.29	0.30	0.31
Other water soluble ions	1.43	5.81	3.92	1.07	3.18	1.83
Other elements	0.36	2.28	0.64	0.54	0.56	0.45
Unclassified	33.56	4.37	8.64	48.91	25.71	29.59
Total, %	100	100	100	100	100	100

Source: Lestari et al., 2008.
Note: BE is bioethanol and the number showing the fuel blending in percent by volume.

TABLE 15.2

Emission Factors from Cars with Bioethanol Fuel (Operating Cycle)

	Emission Factors (g/km)					
Blended Fuel	**CO**	**HC**	**NOx**	**HC+NOx**	**CO_2**	**PM**
Car with Carburetor System						
BE-0	48.524	5.535	1.508	7.042	202.281	0.0313
BE-5	45.813	5.560	1.695	7.255	196.607	0.0288
BE-10	41.844	6.389	2.137	8.526	195.995	0.0238
Car with Electrical Fuel Injection (EFI) System						
BE-0	3.807	1.154	2.231	3.384	191.856	0.0234
BE-5	3.589	1.163	2.187	3.350	190.473	0.0068
BE-10	3.210	1.305	2.441	3.746	191.358	0.0097

Note: One operating cycle combines 4 UDC and 1 EUDC

15.3.3 Emission Inventory

The vehicles were classified into four groups: passenger cars, light duty, heavy duty, and motorcycles. The passenger cars were divided into two subclasses: private passenger cars (car, minibus, and jeep) and public passenger cars (taxi and mikrolet). Mikrolet is a type of small public transport typical in Indonesia. Detailed fuel consumption by vehicle types in Bandung

TABLE 15.3

Annual Fuel Consumption by Different Types of Vehicles in Bandung

		Annual Fuel Consumption, KL (Thousand Litter)					
		2004		2005		2006	
Type of Vehicles		**Gasoline**	**Diesel**	**Gasoline**	**Diesel**	**Gasoline**	**Diesel**
Passenger car	Private car	54,830	0	60,929	0	53,471	0
	Minibus	105,789	2,767	112,958	4,417	107,601	30,459
	Taxi	8,567	0	10,367	0	8,795	0
	Mikrolet	45,653	0	49,833	0	56,529	0
	Jeep	21,715	854	23,280	1,328	21,734	8,712
Light Duty	Pick Up	51,766	2,149	57,520	2,780	59,301	18,686
	Microbus	0	529	0	805	0	9,122
Heavy Duty	Bus	0	65,010	0	80,244	0	416,815
	Truck	0	12,818	0	17,147	0	117,704
Motorcycle	Motorcycle	82,701	0	90,321	0	116,186	0
Total		371,021	84,127	405,208	106,720	423,618	601,497

Source: Lestari and Leopold, 2008.

is presented in Table 15.3. Accordingly, all heavy-duty vehicles and light-duty microbuses were diesel powered; all private cars, motorcycles, taxis, and mikrolets were gasoline powered. Other vehicle types (pickup, jeep, and minibus) used either diesel or gasoline fuels.

The results of the GHG emission inventory study showed significant increases in the emission of CO_2 and CH_4 from the year 2004 to the year 2006, that is, 18.7% and 24.8%, respectively. The results also indicated that the results of CO_2 and CH_4 emissions calculated based on the VKT approach were higher than those obtained by the simple top-down approach using fuel consumption. Vehicles may have purchased fuel outside the city for driving in the city; hence, estimation using the fuel sale data in Bandung may underestimate emission. The results also indicated that the main sources of CO_2 in Bandung were the heavy-duty vehicles, passenger cars, and motorcycles, while the main sources of CH_4 were motorcycles and passenger cars (Lestari and Leopold, 2008).

The emission inventory for CO, NOx, PM, and HC pollutants was conducted in cooperation with the local West Java Environmental Protection Agency (BPLHD) for 2 years, 2005 and 2006. The obtained gridded emission for the city showed that the emission strengths in different grids were well correlated with traffic density. The results of the emission inventory for the year 2006 revealed that CO and NO_x emissions were mainly from private passenger cars, while the main sources of HC were motorcycles, followed by the private passenger cars. For illustration, Figure 15.3(a) shows the source emission shares for CO in Bandung City. For particulate matter (PM_{10}, particles with an aerodynamic diameter less than 10 μm), the main sources were private passenger cars, heavy-duty

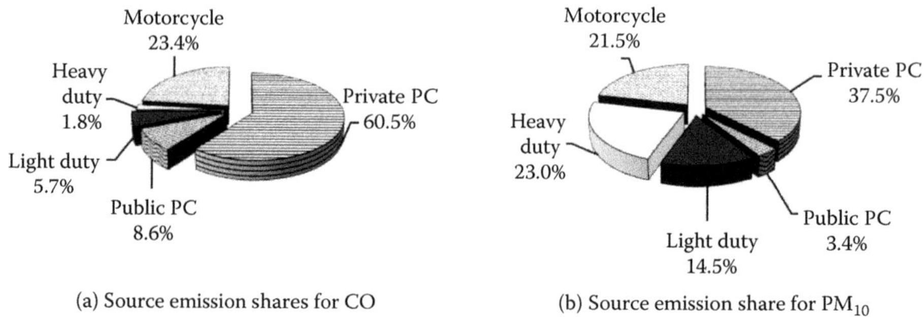

FIGURE 15.3
Source contribution for emission of (a) CO and (b) PM_{10} in Bandung City, 2006. PC: passenger cars. (From Lestari et al., 2006a. Emission inventory of pollutant CO, NO_x and SPM from transportation sector in Bandung basin, paper presented at the Environmental Technology and Management Conference, Bandung, Indonesia, September 6–7.)

vehicles, and motorcycles as shown in Figure 15.3(b). The total CO and PM_{10} emissions in 2006 from the transportation sector in Bandung were estimated at 185,476 tonnes and 1,113 tonnes, respectively. By fuel type, combining the contributions of heavy-duty and light-duty vehicles, the main contributor of PM_{10} was diesel-powered vehicles. Note that the passenger cars also included the minibus and jeep, which partially use diesel fuel (Table 15.3).

The results from this study were reported to BPLHD and served as the scientific basis for development of the new regulation of air pollution control for the West Java Province. The new regulation and new standard were launched in 2006 (Perda no. 11, November 2006).

15.3.4 BLL Results

The BLL test results revealed that more than 65% of schoolchildren (264 of 400 children) had a BLL above the World Health Organization (WHO) guideline of 10 μg/dL (Figure 15.4), and the average value was 14.13 μg/dL. The BLL test conducted for schoolchildren indicated a huge potential impact of lead air pollution on children's health, which may cause developmental problems. The results from the survey also indicated that the main source of lead in the blood was from the transportation sector (Lestari, 2006). The results were disseminated in local as well as national media, such as newspapers, TV stations, and radio stations, with recommendations to implement unleaded gasoline in Bandung City. The information is believed to have contributed to speeding up the phasing out of leaded gasoline in the city. At the end of 2006, Bandung City was free from leaded gasoline.

To reveal the impact of the policy (phasing out of leaded gasoline) on air quality and health, in 2008 the BLL test was conducted for 83 schoolchildren

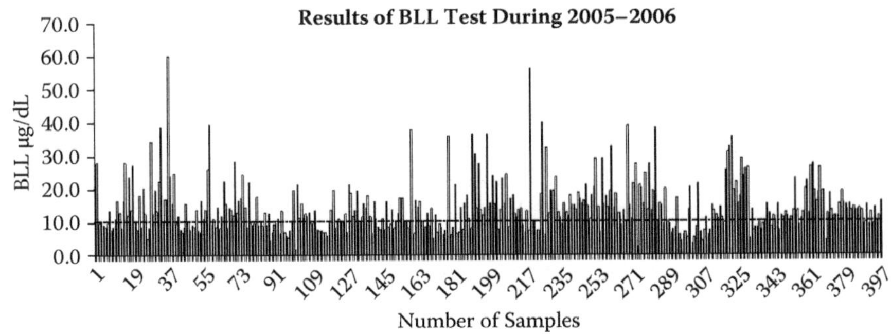

FIGURE 15.4
Blood lead level test results in Bandung (400 schoolchildren) before phasing out leaded gasoline.

from 10 schools in Bandung. In addition, the lead content in the fuel from five gas stations was tested. The results indicated that there was a significant reduction in the BLL in children in Bandung City, with an average BLL value of 9.1 μg/dL in May 2008, as compared to 14.1 μg/dL in 2005. This significant reduction was achieved by the reduction of lead content in the gasoline. The measurements conducted on this research project in 2008 indicated that the current lead content in gasoline distributed in Bandung was 0.01 g/L, as compared to 0.133 g/L obtained by the fuel quality tests conducted by the Ministry of Environment (MOE) and Joint Committee for Leaded Gasoline Phase Out (KPBB) in 2006, before phasing out leaded gasoline (MOE-KPBB, 2006).

15.4 Summary and Conclusions

This integrated AQM case study was conducted focusing on transport sector emissions. The results from the vehicle testing showed that engine technology is an important factor affecting emission. Vehicles with an electrical injection system emitted lower amounts of pollutants than vehicles with a carburetor system per km of traveling distance. The emission inventory results indicated that private passenger cars were the main contributor to PM_{10}, CO, and NO_x emissions in Bandung City, while motorcycles were the main contributors of HC. The results suggested that private passenger cars need more attention for emission control efforts to reduce overall air pollution emissions from the transportation sector in Bandung. The obtained source profiles indicated that black carbon was the major component in the PM emitted from vehicle exhaust. The fuel blending rate also affected the PM emission profiles and air pollution emission factors. CO and PM emission was reduced with increasing blending rate from 0% to 10%, while that of NO_x and HC increased.

The results of the BLL study for schoolchildren were considered a driving force to speed up the phasing out of leaded gasoline in Bandung, which finally occurred by the end of 2006. The phasing out of leaded gasoline had a significant impact on the BLL levels in schoolchildren in Bandung City. Significant reductions in the BLL in schoolchildren were recorded in 2008, which was 2 years after leaded gasoline was phased out, with the average value of BLL reduced from 14.13 µg/dL (2006) to 9.14 µg/dL (2008). In the meantime, the lead content in gasoline was reduced from 0.133 g/L in 2006 to 0.01 g/L in 2008.

References

Centers for Disease Control and Prevention (CDC). 2000. Blood lead levels in young children and selected sites, 1996–1999, *Morbidity and Mortality Weekly Report (MMWR)*, 49: 1133–1137.

Hopke P.K. 1991. *Receptor Modeling for Air Quality Management. Data Handling in Science and Technology*, Vol. 7, Elsevier Science, Amsterdam.

Lestari P. 2004a. Lead in the air and lead in the school children's blood: case study in Bandung, paper presented at the KPBB workshop, Borobudur Hotel, Jakarta, December 15.

Lestari P. 2004b. Study of air quality and its impact in Bandung City, paper presented at the AIRPET National workshop, Bandung, Indonesia, April 12.

Lestari P. 2006. Factors affecting BLL on school children in Bandung, *Journal Purifikasi*, 6(2).

Lestari P., and Leopold A. 2008. Emission inventory of GHGs (CO_2 and CH_4) from transportation sector using VKT and fuel consumption approaches in Bandung city, paper presented at the BAQ conference, Bangkok, Thailand, November.

Lestari P., Kurniawan E., and Dewi K. 2006a. Emission inventory of pollutant CO, NOx and SPM from transportation sector in Bandung basin, paper presented at the Environmental Technology and Management conference, Bandung, Indonesia, September 6–7.

Lestari P., Minarni D., and Mahalana A. 2006b. Study of blood lead level test for school children in Bandung, paper presented at the BAQ conference, Yogyakarta, Indonesia, December.

Lestari P., Reksowardojo I., Prawoto, Simanjuntak N., and Marshal. 2008. The effect of using bio-ethanol fuel to vehicle's emission on gasoline vehicles during urban and extra urban cycles, *Purifikasi Journal, 7*(2).

Lestari P., and Savitri. 2003. Atmospheric particulate concentration measured in an urban area Bandung, *Journal of PAGEOPH*, 160: 107–116.

Lovei M. 1996. *The World Bank's Approach to Removing Lead from Gasoline*, Environment Department, World Bank.

Ministry of Environment and Joint Committee for Leaded Gasoline Phase Out (MOE-KPBB). 2006. *Indonesia Fuel Quality Report, Clean Fuel: A Requirement for Air Quality Improvement*, Ministry of Environment and Joint Committee for Leaded Gasoline Phase Out (KPBB). Jakarta, Indonesia.

Statistical Center Bureau (SCB). 20011. *Biro Pusat Statistik Kodya Bandung, BPS, 1995 to 2006. Kotamadya Bandung Dalam Angka* (annual report), Statistical Center Bureau. Bandung, Indonesia.

United Nations Economic Commission for Europe (UN-ECE). 2006. Regulation No. 83: uniform provisions concerning the approval of vehicles with regard to the emission of pollutants according to engine fuel requirements, *Official Journal of the European Union*. http://eur-lex.europa.eu/LexUriServ/LexUriServ.do?uri=OJ:L:2006:375:0223:0495:EN:PDF.

16

Integrated Air Quality Management for Brick-Manufacturing Village in Vietnam

Hoang Xuan Co, Nghiem Trung Dung, and Nguyen Thi Kim Oanh

CONTENTS

16.1 Introduction: Brick Making and Environmental Impacts

Brick manufacturing in many Asian countries is predominantly a cottage industry using traditional kilns, which are energy inefficient and highly polluting (Gomes and Hossain, 2003). The brick industry in developing countries grows rapidly thanks to the increasing demand from intensive construction activities (Regional Energy Resources Information Center [RERIC], 2003).

Generally, brick manufacturing brings in overall development to their locations, with profits for kiln owners from selling bricks, profits for farmers from leasing land, and employment for local people (Singh and Asgher, 2005). This industry often belongs to the informal sector and is largely unlicensed and virtually unregulated (Brick Industry Association [BIA], 2004; Blackman, 2000; Food and Agriculture Organization [FAO], 1993). Rapid development of this industry coupled with weak environmental control often leads to severe adverse effects on the environment that should not be overlooked. For example, loss of land, removal of nutrients and humus from soil, increased erosion, and so on can adversely affect agricultural production in areas where brick kilns are located (RERIC, 2003).

Air pollution emission from brick kilns in developing countries has recently gained local and international attention in Asia (Clean Air Initiative for Asia Cities [CAI-Asia], 2008; Ferdausi et al., 2008). Backward kiln technologies used for brick firing and lack of emission control devices often result in large emission of a wide range of incomplete and complete combustion products. These include CO, VOCs (volatile organic compounds), PM (particulate matter), NO_x (nitrogen oxides), and CO_2. Sulfur oxides (SO_x) emission largely depends on the sulfur content of the fuel used for brick firing. Emission from brick kilns also includes those pollutants that originated from the raw materials of the bricks that are subsequently released during the firing stage. These include SO_x (from S in soil), VOCs (volatilized from brick raw materials), and phytotoxic fluoride compounds (RERIC, 2003; Zhang, 1997).

A brick-manufacturing village normally concentrates a large number of kilns that are densely distributed. These villages are clustered in rural and semiurban areas and are surrounded by green agricultural fields and populated communities. Air pollution emission from the brick-making villages could cause severe adverse effects to human health and agricultural crops in the surrounding areas (Singh and Asgher, 2005).

16.1.1 Brick Making in Vietnam

The brick industry is one of the largest industries in Vietnam, generating much employment and income. The brick-manufacturing sector is traditionally a small- and family-based industry and commonly located in rural areas. The brick industry in Vietnam has rapidly developed, with production increasing from 11.8 billion pieces in 2001 to 18.4 billion pieces in 2009, that is, an average growth of about 6% per year. In Vietnam, similar to other developing countries, brick manufacturing also largely belongs to the informal sector. The majority of the brick production in the country is by the nonstate sector, including the medium-, small-, and family-based enterprises, with a share that has increased steadily from 61% in 2001 to over 91% in 2009 (Table 16.1). The large amount of air pollutants released from the brick-firing process is of particular concern (Le and Kim Oanh, 2010).

TABLE 16.1

Brick Production of the Country of Vietnam and Bac Ninh Province (Million Pieces)

Year	Area	2001	2002	2003	2004	2005	2006	2007	2008	2009
State	Country	2,555	2,837	3,187	2,934	2,822	2,007	1,397	1,475	1,502
	Bac Ninh	131	161	138	86	0	0	0	0	0
Nonstate	Country	7,236	8,508	9,597	11,705	13,692	15,975	13,709	16,803	16,996
	Bac Ninh	477	696	905	1,146	1,535	1,445	1,287	925	1,113
Foreign Invested	Country	20	20	26	21	0	0	0	0	0
Country Total		11,812	13,367	14,813	16,664	16,530	18,005	15,106	18,278	18,498
Bac Ninh Total		608	857	1,043	1,232	1,535	1,445	1,287	925	1,113

Source: Data from the General Statistics Office of Vietnam (2004, 2009) and Bac Ninh Statistical Office (2004, 2009).

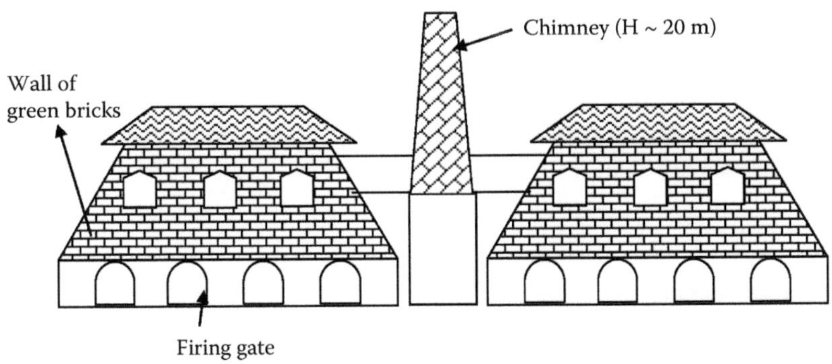

FIGURE 16.1
Sketch of a traditional improved brick kiln in Vietnam.

The brick production process in Vietnam still largely relies on simple technologies, especially in the nonstate sector, where the manufacturing process is still mainly manual. Overall, a wide range of brick kiln types exists in the country, including vertical, annular, and tunnel kilns (Dung and Son, 2004; RERIC, 2003). Simple traditional open-topped vertical kilns are still common. These kilns are small in size and low cost, so are more affordable for low-income owners. The second popular type is the traditional improved brick kiln with a fixed chimney. These kilns are normally constructed in twos, that is, two kilns are connected, sharing one chimney, and are operated on rotation (Figure 16.1). Advanced kilns such as tunnel kilns and vertical shaft brick kilns are gaining popularity. However, due to the high initial capital investment, their share is still small. These advanced kilns are normally made available through financial supports from international cooperation projects/organizations with the aim to stimulate the development of brick production technology in the country. Brick kilns in Vietnam normally are concentrated in separate clusters, so-called manufacturing villages, where both simple and improved kilns coexist. Various types of fuels are used in brick making in Vietnam. In Northern Vietnam, coal and partly wood fuel are used for brick firing.

16.1.2 Need for Integrated Air Quality Management

The large amount of air pollution released from brick kilns concentrated in a manufacturing village creates environmental and social problems that should be considered along with the economic benefits. In particular, effects on human health and crops may be substantial but not well studied in Vietnam or elsewhere in Asian developing countries. Good practices or "showcase" of air quality management applicable to a brick-manufacturing village should be developed, demonstrated, and multiplied in the country and the region.

The demonstration project Integrated Air Quality Management for Song Ho Brick Manufacturing Village was implemented by the Asian Regional Air Pollution Research Network (AIRPET)-Vietnam team (AIRPET-VN) in cooperation with the Asian Institute of Technology (AIT) during 2006–2007. Workable management measures were analyzed through implementation of emission and air quality monitoring, dispersion modeling, and emission control technology.

16.2 Methodology

16.2.1 The Study Area: Song Ho Brick-Manufacturing Village

The Song Ho village, the study area, is one of 17 villages of the Thuan Thanh District of Bac Ninh Province, which is located in the Red River Delta in Northern Vietnam. The village covers an area of about 50,000 m^2. The administration map of the study area is presented in Figure 16.2.

Air pollution emission from brick kilns in the Thuan Thanh District was believed to cause some visible effects on agricultural crops in the surrounding fields, which led to a conflict between the brick-manufacturing and paddy crop cultivation activities. The local authority, the provincial Department of Natural Resources and Environment (DONRE), initiated an experimental research project to develop limestone scrubbers for kiln stack emission control, mainly for SO_2 removal. However, the efficiency of the treatment system appeared to be insufficient to reduce the stack emission to the desired level. In addition, in many cases after installation the control devices were not operated as there was no rigorous enforcement or incentives for kiln owners to do so. Consequently, to protect rice plantations, the authority decided to ban brick-making activities in the Thuan Thanh District during the rice-growing period, from March to September each year (i.e., about 7 months)

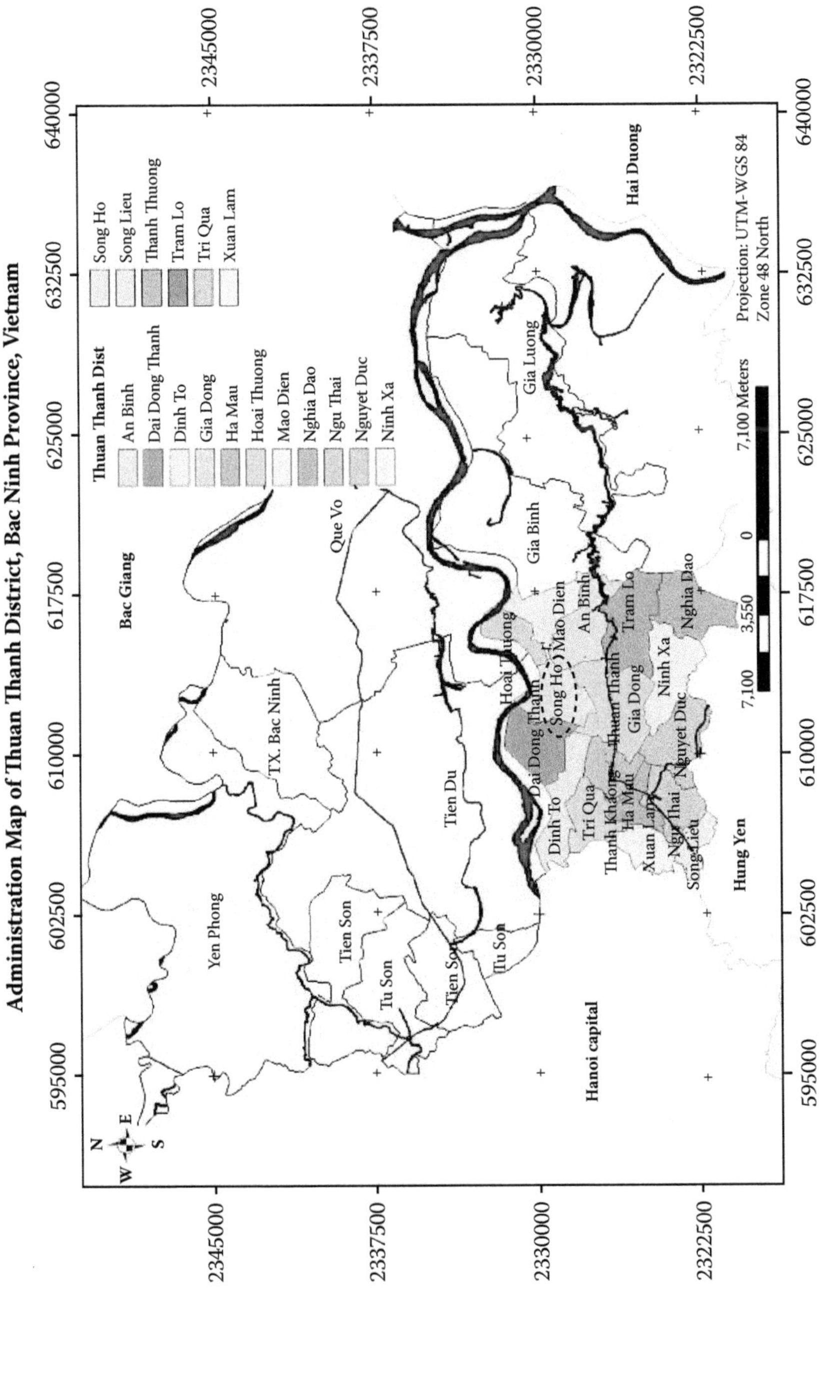

FIGURE 16.2
Map of the study area highlighting Song Ho village (circled) of Thuan Thanh district (shaded parts).

(AIRPET-VN, 2007). This certainly affects the income of brick owners and the overall income of the district.

Another measure that has been enforced in this district by the authority was to relocate polluting brick kilns further away from populated areas. However, the current relocations were only a few hundred meters, so they may not be sufficient to reduce adverse effects. Relocation over large distances (i.e., a few kilometers) may not be always practical due to the local availability of suitable clay soil for use as the raw material.

16.2.2 Integrated Air Quality Management Activities

The implemented activities of integrated air quality management for the Song Ho village consisted of the following steps:

1. Survey overall brick manufacturing in Bac Ninh with focus on Song Ho village (kiln types and kiln locations)
2. Monitor ambient air quality and meteorology
3. Monitor stack emission
4. Apply dispersion modeling tool to assess emission control scenarios
5. Install and operate limestone scrubbers for stack emission control
6. Analyze alternative measures for air quality improvement

The survey and monitoring activities (ambient air and stack emission) were conducted in two campaigns. Each monitoring campaign lasted 7 days. Campaign 1 was carried out in April 2006 when only one kiln was operated, and campaign 2 was in January 2007 when the brick production in the village was under normal operation (about 21 kilns were operating simultaneously).

16.3 Results

16.3.1 Survey Results: Current Status of Brick Industry in Song Ho, Bac Ninh

Bac Ninh Province has a huge clay source for brick making. This region also has a large demand for bricks for construction. The province has policies to encourage the development of the agricultural sector, industry, and craft villages, including brick making. The province has a population of 976,700 and an area of 804 km^2, of which 64.7% is for agricultural use. There are over 1,500 brick kilns in the province, which are distributed in six districts along the banks of the Duong and Cau Rivers. Normally, about 40–60 kilns form a brick-making village or commune (Le, 2007).

The brick production in Bac Ninh Province steadily increased in 2001–2005, decreased afterward until 2008, and increased again in 2009. The province brick production contributed from 5% to 9% of the total national brick production (Table 16.1). Similar to other parts of Vietnam, brick manufacturing in Bac Ninh brings in significant revenue, but it also creates many environmental and social problems. The most critical problem in the area was the conflict between brick manufacturing and agricultural production, as mentioned. Local people also have started realizing the negative health and crop effects from the brick making on their daily life.

Song Ho village had 45 operating kilns in January 2007, which was an increase by 10 kilns within 10 months since April 2006. The traditional improved kilns were the most popular, with a share of 90% as of 2007, while the remaining 10% were the simple traditional open-topped kilns. The production per batch (7 firing days) averaged at 450,000 brick pieces. The average coal consumption was about 100 tonnes per 1 million bricks (Le, 2007). There were no emission control measures applied in brick kilns. Coal briquettes were prepared from coal particles and different types of binding materials and placed between raw bricks for firing. Generally, coal of low quality and high sulfur content was used as there is no effective regulation for fuel quality. The flue gas was directly released to the atmosphere without passing any end-of-the-pipe emission control devices.

16.3.2 Ambient Air Quality in Song Ho Village

Ambient air monitoring was conducted at several sites in and surrounding the Song Ho village to capture the upwind, inside the village, and downwind levels of PM, CO, and SO_2 during the two campaigns (Le and Kim Oanh, 2010). The monitoring results at four ambient monitoring sites located about 500–700 m apart are summarized in Table 16.2. These included an upwind site, two sites located inside the village, and a downwind site.

During April 2006 when only one kiln was operating, the SO_2 and TSP (total suspended particulate matter) air quality in the village did not exceed the Vietnam National Ambient Air Quality Standards (NAAQS) at any monitoring sites. However, in January 2007 when the village had its normal production, serious air pollution was observed. Hourly SO_2 exceeded the NAAQS by 48% of the measurements, and hourly PM_{10} (particles with an aerodynamic diameter less than 10 μm) even exceeded the hourly NAAQS for TSP (300 $\mu g/m^3$). High levels of PM_{10}, which are inhalable, should be of concern due to their potential effects on human health. The drastic deterioration of the ambient air quality in the village in January 2007 suggested that operation of brick kilns was a main cause of the air pollution in this peri-urban area. Detailed measurement results for SO_2 are given in Figure 16.3, which shows that for 21 measurements, 10 times the levels exceeded the hourly NAAQS at any sites.

TABLE 16.2

Summary of Hourly Ambient Air Quality in Song Ho Village ($\mu g/m^3$, 25°C, and 760 mmHg)

Parameter	1-h CO	1-h SO_2	1-h TSP	1-h PM_{10}	1-h $PM_{2.5}$
April 2006	NA	45–112	150–280	NA	NA
January 2007	1,030–3,350	87–514 (48%)[a]	NA	35–388 (14%)[b]	11–68
NAAQS[c]	30,000	350	300	NA	NA

Source: From Co H.X., Dung N.T., Le H.A., An D.D., Chinh K.V., and Kim Oanh N.T. 2009. Integrated management strategies for brick kiln emission reduction, *International Journal of Environmental Studies*, 66(1): 113–124.

Note: NA, not available.

[a] Percentage exceeding the hourly Vietnam NAAQS.

[b] Percentage exceeding the hourly NAAQS for TSP.

[c] Ministry of Natural Resources and Environment (MONRE). 2005a. *TCVN (Vietnam Standards) 5937:2005—Ambient Air Quality Standards*, Ministry of Natural Resources and Environment, Hanoi, Vietnam.

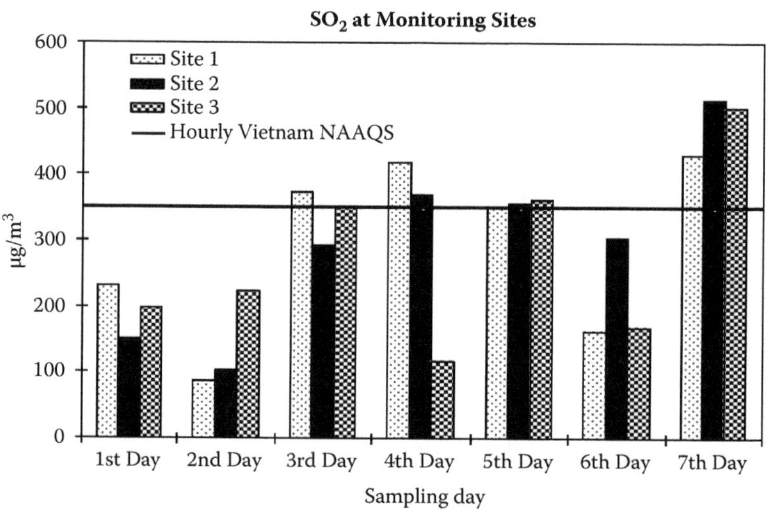

FIGURE 16.3
SO_2 in ambient air as compared to Vietnam NAAQS during 7 days of monitoring in January 2007 (site 1, upwind; site 2, inside village; and site 3, downwind).

16.3.3 Stack Emission

Stack gas emission was monitored for one kiln in each campaign over a firing batch (7 days). The monitoring parameters included PM, SO_2, and CO. The isokinetic stack sampling for suspended particulate matter (no size segregation) was conducted following the U.S. Environmental Protection Agency

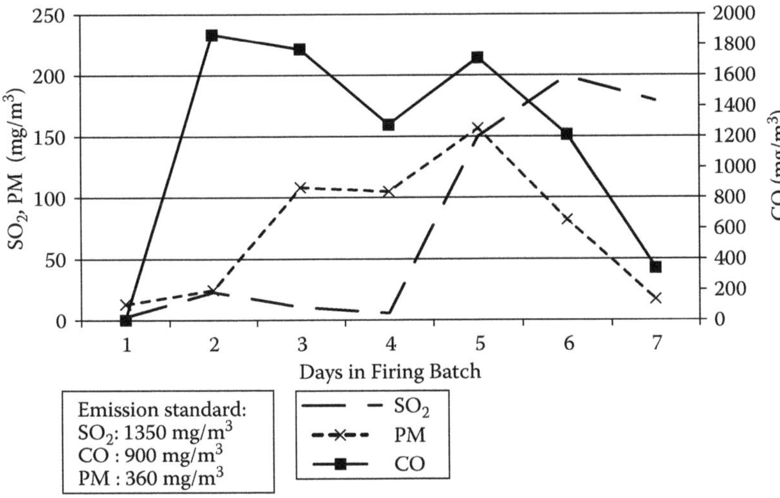

FIGURE 16.4
Variation of SO_2, CO, and PM concentration in stack gas during a firing batch in January 2007 campaign. *Note:* Emission standards given in the figure are values calculated for the emission flow rate and the source location region specified by MONRE (2005b).

(USEPA) guidelines (USEPA, 2000). Both monitored kilns were of the traditional improved type with a stack above 20 m high.

The concentrations of CO, SO_2, and PM in the flue gas varied significantly during a firing batch in January 2007 (Figure 16.4). CO was high and quite stable during the middle of the firing batch/cycle and starting from day 2 to day 6 was mostly above the emission standard (ES). PM was also high in the middle of the firing batch but did not exceed the ES set for this type of source (Ministry of Natural Resources and Environment [MONRE], 2005b). SO_2 did not exceed the ES during the firing period. Higher SO_2 concentrations observed toward the end of a firing batch may have been due to the initial sorption of SO_2 onto raw bricks in the top layers in the kiln, which were then released when the fire reached these top layers (Le and Kim Oanh, 2010).

The obtained average emission factors per 1,000 bricks (2.2 kg per piece) were 6.35–12.3 kg of CO, 0.52–5.9 kg of SO_2, and 0.64–1.4 kg of PM. Higher emission factors were obtained for a smaller kiln monitored during the April 2006 monitoring campaign. Large variations in SO_2 emission factors showed a possible high variation in the sulfur content of coal used in the two monitored kilns. The particle size distribution in the stack gas emission was monitored by a cascade impactor, which showed that most particles emitted were in the inhalable size range (PM_{10}), and of those about 30% belonged to the fine particle size range ($PM_{2.1}$), with further details reported by Le and Kim Oanh (2010). A high fraction of fine particles in the stack emission certainly would cause more health effects to the exposed population (Pope et al., 2009; Schwartz and Neas, 2000).

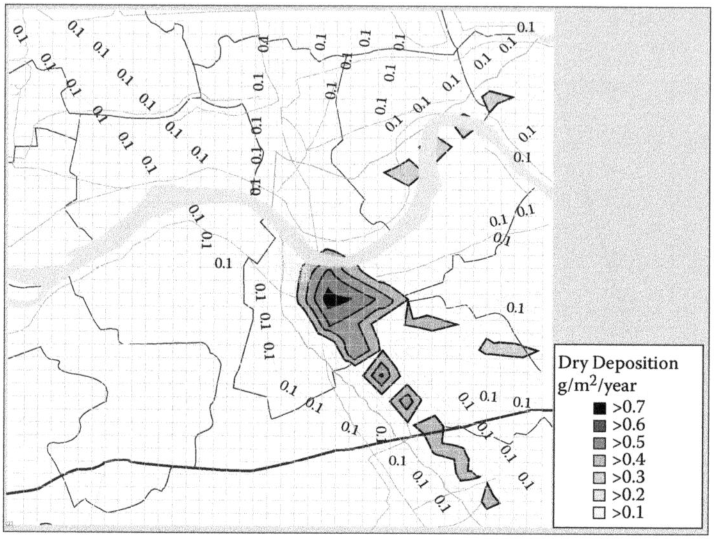

FIGURE 16.5
The highest annual dry deposition flux of PM in base case (g/m²/year). (Adapted from Le (2007).)

16.3.4 Stack Gas Dispersion Modeling

The Short-Term Industrial Source Complex (ISCST3) model (USEPA, 1995) was applied to estimate the ambient pollution concentration in the area surrounding the Song Ho brick-making village. The domain (Figure 16.5) was 100 km² (Cartesian 40 × 40 grids of 250 m) of flat terrain covering the village. The meteorology input file was prepared from the reanalyzed data for 2003 obtained from the National Oceanic and Atmospheric Administration (NOAA) for the Noi Bai Airport station, which is about 40 km away from Song Ho village. The data were processed by the PCRAMMET meteorological preprocessor (USEPA, 1999) to produce the meteorological input file for ISCST3. The emission input data were prepared based on the results of the survey (number of kilns operating per day, kiln locations, stack height, etc.) and the emission rates obtained from the stack-monitoring activities. The particle size mass distribution obtained in the flue gas monitoring was used in modeling the PM dry deposition.

The model output included the ambient concentrations of PM, CO, and SO_2 as well as PM deposition flux over the domain. Simultaneous measurements of wind speed and direction and of the stack emission and ambient air concentrations were used to evaluate model performance. Overall, the agreement between the measured and modeled values was reasonable, especially for SO_2, which was mainly contributed by the kiln stack emission (Le and Kim Oanh, 2010).

16.3.4.1 Base Case

The simulation results showed that for the normal operation in January 2007 (base case) the most serious pollutant in the study area was SO_2, which had the simulated maximum hourly concentration of 1006 μg/m^3 and 24-h concentration of 407 μg/m^3 (i.e., well above the respective NAAQS of 350 and 125 μg/m^3). The area with the maximum 1-h SO_2 exceeding the NAAQS was 63 km^2. A high concentration of SO_2 in the domain would be of concern not only for human health but also for vegetation/crops due to its phytotoxicity. The simulated ambient PM_{10} and CO levels were below the NAAQS (Le and Kim Oanh, 2010). This suggests that the high 1-h PM_{10} levels measured at the monitoring sites were possibly also caused by other sources (than kiln stack emission simulated in this study). It is interesting to note that although the kiln stack CO emission often exceeded the ES, the ambient air CO levels, based on both monitoring and modeling results, were in compilation with the NAAQS.

A significant deposition flux of PM on the surrounding area was simulated, with the annual highest flux of around 1.0 g/m^2/year (Figure 16.5). The PM deposition may cause adverse effects on the surrounding rice fields by both the physical blockage of plant leaves and the potential phytotoxic PM composition.

16.3.4.2 Emission Reduction Scenarios

Different emission reduction scenarios were simulated. The results showed that if only the kiln stack emission sources in the village under normal operation were considered, then to attain the Vietnam NAAQS over the domain the SO_2 emission should be reduced by roughly 65–70%. CO would be in compliance in all considered scenarios. However, PM_{10} would exceed the NAAQS in the most conservative scenario when all kilns in the village were operating simultaneously on the day of maximum PM emission of a firing batch. Note that if other emission sources, such as traffic, road dust, or residential cooking in the domain, are included, then further emission reductions would be required. Nevertheless, kilns appeared to be the major SO_2 source in the domain (Le and Kim Oanh, 2010), so kiln emission reduction would efficiently lower its ambient air levels. Thus, to achieve the SO_2 emission reduction to attain the Vietnam NAAQS in the domain, either SO_2 emission control measures should be applied or the number of operating kilns should be reduced by about 70%, say from the current number of 21 to 7 or fewer.

16.3.5 Emission Reduction Measures for SO_2

A regulation on the number of operating kilns in the villages is straightforward but would affect revenue. Alternatively, the SO_2 emission reduction can be achieved through stack emission control devices or use of low-sulfur fuels for brick firing. Once these measures are applied, more kilns can be operated even in the rice-growing season. Thus, the kiln operation ban during March–September every year could be partially relieved for those kilns that applied necessary emission reduction measures. This would create incentives for kiln owners to aim for emission reduction and is a win-win solution for both brick making and rice plantation in the area (Co et al., 2009).

16.3.5.1 Emission Control Device

A simple limestone scrubbing system was designed and applied as a showcase for two kilns in the Song Ho area. The SO_2 removal efficiency of the control device, however, was only 50% (AIRPET-VN, 2007), which suggests further improvement can be made to gain the common efficiency of such devices of 80–90% (Noll, 1999). Nevertheless, the system was simple, cheap, and easy to install and operate. If this simple scrubber (50% SO_2 removal) were applied, then the number of simultaneously operated kilns in the village could be increased two times (i.e., up to 14) while attaining the NAAQS for SO_2. The scrubbing system would also remove PM from the flue gas, but that was not quantified in this project. Further improvement of the device would need to concentrate on the optimum amount of limestone used, better contact between the limestone slurry and flue gas, sufficient retention time, and so on, which can increase SO_2 removal efficiency.

Brick kiln owners in general do not have sufficient knowledge of emission control technology. They also have limited financial resources to invest in the control systems if these are expensive. Thus, the simple, easy-to-operate, and low-cost limestone scrubbers developed in this project appear to be attractive. They are especially suitable for relatively permanent kilns that are built for long-term operation. Within this project, the brick kiln owners in Song Ho were trained on the design, the treatment mechanism, as well as the operation steps of the scrubbing system. After the training, more kiln owners expressed an interest in installation of such a treatment system. In addition, a workshop, Air Quality Monitoring Results of Brick Kiln Manufacturing in Song Ho Village, Bac Ninh Province, was organized in Bac Ninh in 2007 to disseminate the project findings to stakeholders.

16.3.5.2 Potential Application of Economical Tools for Emission Reduction

With the current weak enforcement of emission control measures, alternative approaches should be taken to provide economic incentives to kiln owners and hence to sustain the emission reduction efforts. The current ban on brick production during the period from March to September each year to protect the two major rice crops certainly seriously affects the income of brick owners and their employees. A mechanism should be created to allow them to operate the kilns even during this rice-growing period as long as emission is properly controlled.

Accordingly, if a kiln implements necessary emission reduction measures, then it should be allowed to operate even during the rice-growing period. This would give the kiln owners economic incentives, for example, to invest in control device installation and operation or to use better coal with a low sulfur content. Co et al. (2009) analyzed an emission fee scheme and showed that the total cost for the limestone scrubbing control device (operation and equipment depreciation) would be US$230–240 per production batch, which is only a small fraction (0.8%) of the gross income from bricks produced per batch. An emission fee can be proposed so that the total cost (the control device plus the fee charged for the remaining 50% SO_2) would be lower than the fee paid for 100% SO_2 emission (no emission reduction). Thus, application of the right fee level would provide incentives to kiln owners and certainly encourage them to use the appropriate emission reduction measures such as installation of emission control devices along with use of cleaner fuels with low sulfur content, and eventually application of advanced kiln technologies. It would provide even more incentives to implement emission reduction measures if the emission fee were charged progressively, that is, higher rates applied when emission exceeds certain limits. Similar measures could be applied for other pollutants emitted from the brick-manufacturing process.

16.4 Summary and Conclusion: Challenges and Opportunities

The brick manufacturing industry in Vietnam has rapidly developed throughout the country to meet the demand of fast-growing construction activities. Being mainly of small scale and virtually without any emission control measures, the manufacturing causes severe health and environmental impacts. This case study focused on the traditional brick-manufacturing village of Song Ho in the Thuan Thanh District, Bac Ninh Province, Vietnam. The conflict between the brick-manufacturing and agricultural cultivation factions in the area forced a ban on brick manufacturing for about 7 months of the rice-growing season every year. An integrated approach to air quality management was designed and implemented in this case study. It was comprised of monitoring and modeling activities to assess the current status of air pollution; design, installation, and operation of limestone scrubbing systems for stack emission control; and a comparative analysis of applicable emission measures.

Ambient air pollution (PM and SO_2) in the study area was serious when the brick-manufacturing activities in the village were under normal operation, for example, ambient hourly SO_2 levels exceeded the NAAQS by 48% of the measurements. Dispersion modeling (ISCST) showed that to attain the NAAQS for SO_2 the kiln stack emission should be reduced by 65–70%.

Various measures were analyzed to achieve the desired emission reduction, including a regulation on the number of operating kilns, use of a better quality of coal with low sulfur, application of limestone scrubbers for SO_2 removal from flue gas, and application of appropriate emission fee levels.

One of the most important achievements of the project has been the identification and analysis of various options to reduce emissions and improve air quality in the study area. The project has gained strong support from the DONRE of Bac Ninh, the head of the Song Ho village, and the kiln owners. With the existing interest created among the stakeholders, the demonstrated simple emission control technology has a good opportunity to multiply.

Vietnam already has a set of comprehensive ESs that are also applicable for brick kilns. The biggest challenge is to have the ES or emission fee strictly enforced for these small and scattered kiln stacks. Regular monitoring for emission and fuel quality is an expensive activity and needs to be externally funded by the environmental authority or other organizations as it is unlikely to be affordable to individual brick kiln owners. The emission fee should be correctly set, which should give substantial benefit to kiln owners if they apply emission control measures compared to paying the full fee. A mechanism should be created to use the collected emission fees for air quality management activities, including monitoring (ambient air and stack emission) and capacity building for installation and operation of emission control devices.

Once proper enforcement is in place, kiln owners can even pursue other, cleaner options, such as alternative fuels or advanced kiln technologies. Use of natural gas is certainly an advantage for emission reduction, but kilns would need to be redesigned. The advanced technologies would need external funding sources to be realized widely in the country, but they are certainly the most straightforward solution to the air pollution problems associated with the brick manufacturing.

There is a need to establish and multiply the "good model" of air quality management for brick-making villages. Technical and financial assistance should be made available to kiln owners for emission reduction, hence minimizing the adverse effects and maximizing the economic revenue.

References

Asian Regional Air Pollution Research Network-Vietnam Team (AIRPET-VN). 2007. *Regional Program on Environmental and Technology (ARRPET): (Annual Report 2006)*, Asian Institute of Technology, Pathumthani, Thailand.

Bac Ninh Statistical Office (BNSO). 2004. *Bac Ninh Statistical Yearbook 2004*, Statistical Publishing House, Hanoi, Vietnam.

Bac Ninh Statistical Office (BNSO). 2009. *Bac Ninh Statistical Yearbook 2009*, Statistical Publishing House, Hanoi, Vietnam.

Blackman A. 2000. Informal sector pollution control: What policy options do we have? *The Journal of World Development*, 28(12): 2067–2082.

Brick Industry Association (BIA). 2004. Technical notes on brick construction, technical notes 9—manufacturing, classification, and selection of brick, manufacturing, part 1, Brick Industry Association. http://www.bia.org/BIA/technotes/t9.htm.

Clean Air Initiative for Asia Cities (CAI-Asia). 2008. Clean brick making technology—success of VSBK in Kathmandu, Clean Air Initiative for Asia Cities. http://www.cleanairnet.org/caiasia/1412/article-70695.html.

Co H.X., Dung N.T., Le H.A., An D.D., Chinh K.V., and Kim Oanh N.T. 2009. Integrated management strategies for brick kiln emission reduction, *International Journal of Environmental Studies*, 66(1): 113–124.

Dung B.T., and Son T.T. 2004. *Low Cost Roads, Working Paper No. 16: Clay Brick Paving Investigations in Vietnam*, MoT, TEDI, Intech Associates. Hanoi, Vietnam.

Ferdausi S.A., Vaideeswaran S., and S. A. 2008. Greening brick making industries in Bangladesh, paper presented at the Better Air Quality Conference, Bangkok, Thailand, November 12–14.

Food and Agriculture Organization (FAO). 1993. *Status and Development Issues of the Brick Industry in Asia*, Regional Wood Energy Development Programme in Asia, Bangkok, Thailand.

General Statistics Office of Vietnam (GSO). 2004. *Statistical Yearbook of Vietnam 2004*, Statistical Publishing House, Hanoi, Vietnam.

General Statistics Office of Vietnam (GSO). 2009. *Statistical Yearbook of Vietnam 2009*, Statistical Publishing House, Hanoi, Vietnam.

Gomes E., and Hossain I. 2003. Transition from traditional brick manufacturing to more sustainable practices, *Energy for Sustainable Development*, 7(2): 66–76.

Le H.A. 2007. Integrated monitoring—modeling tool to develop air quality management strategies for a brick kiln manufacturing community in Vietnam, master's thesis, Asian Institute of Technology, Pathumthani, Thailand.

Le H.A., and Kim Oanh N.T. 2010. Integrated monitoring and modelling for air quality management of brick manufacturing village, *Journal of Environmental Monitoring and Assessment*, 171: 381–394.

Ministry of Natural Resources and Environment (MONRE). 2005a. *TCVN (Vietnam Standards) 5937:2005—Ambient Air Quality Standards*, Ministry of Natural Resources and Environment, Hanoi, Vietnam.

Ministry of Natural Resources and Environment (MONRE). 2005b. *TCVN (Vietnam Standards) 5939:2005—Industrial Emission Standards*, Ministry of Natural Resources and Environment, Hanoi, Vietnam.

Noll K. 1999. *Fundamentals of Air Quality Systems: Design of Air Pollution Control Devices*, American Academy of Environmental Engineers. Annapolis, MD.

Pope C.A., III, Ezzati M., and Dockery D.W. 2009. Fine particulate air pollution and life expectancy in the United States, *The New England Journal of Medicine*, 360: 376–386.

Regional Energy Resources Information Center (RERIC). 2003. *Small and Medium Scale Industries in Asia: Energy and Environment, Brick and Ceramic Sectors*, Regional Energy Resources Information Center, Asian Institute of Technology, Bangkok, Thailand.

Schwartz J., and Neas L. 2000. Fine particles are more strongly associated than coarse particles with acute respiratory health effects in schoolchildren, *Epidemiology*, 11: 6–10.

Singh A.L., and Asgher M.S. 2005. Impact of brick kilns on land use/landcover changes around Aligarh city, India, *The Journal of Habitat International*, 29: 591–602.

U.S. Environmental Protection Agency (USEPA). 1995. *User's Guide for the Industrial Source (ISC3) Dispersion Model*, Vol. 1, U.S. Environmental Protection Agency, Washington, DC.

U.S. Environmental Protection Agency (USEPA). 1999. *PCRAMMET User's Guider*, U.S. Environmental Protection Agency, Washington, DC.

U.S. Environmental Protection Agency (USEPA). 2000. *Current Knowledge of Particulate Matter (PM) Continuous Emission Monitoring (Final Report)*, Emission, Monitoring and Analysis Division, U.S. Environmental Protection Agency, Washington, DC.

Zhang Z. 1997. Energy efficiency and environmental pollution of brick making in China, *The Journal of Energy*, 2(1): 33–42.

Index

C

J

K

L

S

U

W

For Product Safety Concerns and Information please contact our EU representative GPSR@taylorandfrancis.com
Taylor & Francis Verlag GmbH, Kaufingerstraße 24, 80331 München, Germany

www.ingramcontent.com/pod-product-compliance
Ingram Content Group UK Ltd.
Pitfield, Milton Keynes, MK11 3LW, UK
UKHW022007180926
475701UK00003B/231
* 9 7 8 1 1 3 8 0 7 1 8 4 1 *